Comparative hydrology

An ecological approach to land and water resources

Edited by Malin Falkenmark
and Tom Chapman

Unesco

The designations employed and the presentation of material throughout
this publication do not imply the expression of any opinion whatsoever on the part
of Unesco concerning the legal status of any country, territory, city or area
or of its authorities, or concerning the delimitation of its frontiers or boundaries.

Published in 1989 by the United Nations Educational,
Scientific and Cultural Organization
7 place de Fontenoy, 75700 Paris

Printed by Imprimerie Floch, Mayenne

ISBN 92-3-102571-6

Preface

Although the total amount of water on earth is generally assumed to have remained virtually constant, the rapid growth of population, together with the extension of irrigated agriculture and industrial development, are stressing the quantity and quality aspects of the natural system. Because of the increasing problems, man has begun to realize that he can no longer follow a "use and discard" philosophy - either with water resources or with any other natural resources. As a result, the need for a consistent policy of rational management of water resources has become evident.

Rational water management, however, should be founded upon a thorough understanding of water availability and movement. Thus, as a contribution to the solution of the world's water problems, Unesco, in 1965, began the first world-wide programme of studies of the hydrological cycle - the International Hydrological Decade (IHD). The research programme was complemented by a major effort in the field of hydrological education and training. The activities undertaken during the Decade proved to be of great interest and value to Member States. By the end of that period, a majority of Unesco's Member States had formed IHD National Committees to carry out relevant national activities and to participate in regional and international co-operation within the IHD programme. The knowledge of the world's water resources had substantially improved. Hydrology became widely recognized as an independent professional option and facilities for the training of hydrologists had been developed.

Conscious of the need to expand upon the efforts initiated during the International Hydrological Decade and, following the recommendations of Member States, Unesco, in 1975, launched a new long-term intergovernmental programme, the International Hydrological Programme (IHP), to follow the Decade.

Although the IHP is basically a scientific and educational programme, Unesco has been aware from the beginning of a need to direct its activities toward the practical solutions of the world's very real water resources problems. Accordingly, and in line with the recommendations of the 1977 United Nations Water Conference, the objectives of the International Hydrological Programme have been gradually expanded in order to cover not only hydrological processes considered in interrelationship with the environment and human activities, but also the scientific aspects of multipurpose utilization and conservation of water resources to meet the needs of economic and social development. Thus, while maintaining IHP's scientific concept, the objectives have shifted perceptibly towards a multidisciplinary approach to the assessment, planning, and rational management of water resources.

As part of Unesco's contribution to the objectives of the IHP, two regular publication series are issued: "Studies and Reports in Hydrology" and "Technical Papers in Hydrology". Occasionally, documents such as the present one are produced by special arrangements with IHP National Committees or other national entities associated with the production of the basic report and are made available for distribution from the IHP headquarters at Unesco, Paris.

This textbook <u>Comparative Hydrology</u> owes much to the enthusiasm, devotion and hard work of the two editors Malin Falkenmark and Tom Chapman, and their collaboration is acknowledged with gratitude. Special thanks go to Professor Chapman who devoted very many hours to development of the text and diagrams in computer-compatible form and to preparation of the book as camera-ready copy.

Dedication

This book is dedicated to the memory of academician Dr. Gyorgy Kovacs, former Director-General of the Hungarian Water Resources Research Centre (VITUKI), Past-President of the International Association for Hydrological Sciences (IAHS), and Chairman of the IHP Council from 1981 to 1983, who died in Budapest on 21 April 1988.

George Kovacs enthusiastically supported the concept of a book on Comparative Hydrology and was a member of the Editorial Board which had the task of developing an unformed idea into a structured text. Through his broad knowledge and comprehensive understanding of the differences between hydrological processes and systems in different parts of the world, his contribution was critical in the selection of the general approach and regional classifications adopted in this book. He personally wrote two chapters and contributed to others, and the references to his work which have been cited in many chapters are a testimony to his contribution to what he felt is a new and exciting development in hydrology.

Editors' foreword

This book has been made possible only through the interest of the Intergovernmental Council for the International Hydrological Programme (IHP), the International Association of Hydrological Sciences (IAHS) and the World Meteorological Organization (WMO); and in particular, all the distinguished experts in hydrology with experience from different zones of the world who have participated as authors.

The Council transferred the task of implementing the book to an Editorial Board with the following members:

Professor Malin Falkenmark, Swedish Natural Science Research Council, Stockholm, Sweden
Professor Tom Chapman, University of New South Wales, Sydney, Australia
Ms. G.M. Chernogaeva, National IHP Committee, Moscow, USSR
Mr. J. Colombani, ORSTOM, Bondy, France
Dr. G. Kovacs, Research Centre for Water Resources Development, Budapest, Hungary
Professor Lekan Oyebande, University of Lagos, Abeokuta, Nigeria
Dr. Arthur J. Askew, WMO, Geneva, Switzerland
Mr. W.H. Gilbrich, Unesco, Paris, France.

After the outline had been first defined by the Editorial Board, Unesco invited selected experts to provide preliminary chapter texts to be discussed at an IHP Workshop on Comparative Hydrology, held in Budapest as part of the IAHS Assembly in July 1986. Many national IHP Committees contributed additional material which was included in the text. The final authors were invited to deliver their texts by February 1987. The Editorial Board, meeting in Paris in May 1987, assigned the final editing of the text to the two of us.

Thanks to a research grant from the Swedish Natural Science Research Council, an editiorial office was organized in Stockholm during the fall of 1987. Now that we are about to forward the final text to the IHP Secretariat for publication, we wish to make some explanatory comments.

We are fully aware of a number of deficiencies in the textbook which could not be avoided at the present time. In spite of - or maybe due to - the fact that so much more has been written about the temperate zone than about the various subzones of the warm and hot regions, considerable difficulties were found in locating authors willing to reduce the material into the required format of a single chapter. We acknowledge the assistance of the Institute of Hydrology in Wallingford in finding the final author and reviewers for the chapter on humid temperate sloping land, which is a contribution from IHP Project 6.1 Flow Regions from Experimental and Network Data.

Comparative hydrology

The problems were even greater in locating an author for the chapter on dry temperate sloping land. As the book would have been incomplete without that chapter, we decided to produce a text in the editorial office, considering it more important to produce a textbook of variable quality at this time, than to continue the search for interested authors.

Similar arguments attach to our acceptance of a number of regional biases in many of the chapters. The book has been written by many authors, and all would agree that their contribution is less comprehensive than they would like, due to inevitable limitations of their own experience. The editors have respected the texts of the authors, and have only added material which brings out links between chapters. Our understanding is that the texts probably fairly well reflect the present status of the ecological approach to hydrology.

The editors have however attempted to achieve a consistent terminology, and to avoid some words (like "losses" in relation to the difference between rainfall and runoff) which imply a preconception of the role of water. It should be remembered that the book represents no more than a first effort to draw attention to the field of comparative hydrology, and we sincerely hope that by doing so, further research in the field will be stimulated. In our understanding, comparative hydrology should develop into a basically analytical science. The heavy descriptive content in the late sections of this textbook should therefore be accepted as an infant disease, as few analytical studies stressing similarities and differences between hydrological zones are yet available.

We wish to convey our great appreciation to our colleague members of the Editorial Board for their continuing interest and support.

We acknowledge the invaluable assistance in the editorial work of Ms Irene Johansson including her checking of the references. The text was skilfully typed by Ms Britt Soderberg. All the text and figures have been prepared on a Macintosh Plus, generously put at our disposal by the Globe Tree Foundation in Stockholm.

The editors now commend the book to readers as a first attempt which we hope will increase interest in the comparative approach to hydrology and water resources management. We wish the readers as many challenges in reading the book as we have been experiencing ourselves in trying to merge, during the past few months, the hydrological paradigms from our two globally opposite home environments, Scandinavia and Australia.

Stockholm 16 December 1987

Tom Chapman
Australia

Malin Falkenmark
Sweden

Contents

Comparative hydrology

Symbols

The symbols used in this text have been selected to form a consistent unambiguous set that is as far as possible in conformity with other Unesco publications and international standards. Symbols used only once are defined in the text; those used more generally are listed below. Where more than one meaning is given, the correct interpretation should be clear from the context; no chapter has a symbol with more than one meaning.

Many symbols are modified by subscripts, to indicate a more particular meaning. Subscripts used more than once are also listed below.

General symbols

Symbol	Meaning
a,b	Parameters in Montanari equation
c	Concentration
d	Depth (of water-table)
D	Saturation vapor deficit
E	Evapotranspiration
E_o	Potential evaporation
G	Sensible heat flux into ground
h	Piezometric head, or water depth in a storage
H	Sensible heat flux from surface to atmosphere
i	Rainfall intensity
I	Infiltration depth
K	Parameter in flow routing eqyations, or hydraulic conductivity (permeability)
m	Index in nonlinear flow routing equation
n	Porosity
n_s	Specific storage
P	Precipitation depth
q	Flux
Q	Discharge
r_a	Wind-dependent resistance in combination equation
r_s	Stomatal resistance in combination equation
R	Groundwater recharge (as a depth)
R_n	Net radiation

S	Stemflow,
	or rate of change of saturation vapor pressure with temperature,
	or groundwater storage coefficient
t	Time
T	Duration,
	or throughfall depth,
	or transmissivity of an aquifer
T_r	Turnover time
V	Volume
α	Parameter in Priestley-Taylor equation
β	Bowen ratio
γ	Psychrometric constant
Δ	Change in value of folowing quantity
λ	Latent heat of vaporization of water
ρ	Density

General subscripts

g	Groundwater
i	Interception
p	Pan (evaporation)
s	Surface
u	Unsaturated zone
y	Year

Introduction

New perspectives

With satellites to observe phenomena on the global scale, on the one hand, and intensifying international trade, scientific knowledge, empirical experience and technical assistance, on the other, the world is becoming continuously smaller. This highlights not only similarities but also contrasts between different parts of the world. Comparisons and regional analyses are needed as a basis for improved understanding of vital differences, in order to make international exchange of information and techniques more efficient, and to give more consideration to the diversity of human behavior and aspirations.

At the same time, the ever-increasing scale of environmental degradation brings into focus the close link between the hydrological cycle, which provides water to the earth's surface, and the terrestrial ecosystems, which are fed by the water entering the plant root zone. Recent studies on ecological optimality (e.g. Eagleson 1982) are deepening our understanding of the water balances of some typical ecosystems. On the other hand, experience has demonstrated that man's interaction with the vegetation may produce unintended effects not only on the ecosystems as such, but also on the local and even regional regime of water in streams and aquifers.

Indeed, the presence of water distinguishes our world from the other planets (Dooge 1983; McIntyre 1977). The fundamental importance of water to plant and animal life makes differences in its occurrence in different environments crucial to human behaviour. It is a thought-provoking fact that most developing countries are located in the vast tropical and subtropical zone where at least part of the year is dry. In such arid, semi-arid and sub-humid environments, water management is particularly dependent on hydrologists for the provision of quantitative information on hydrological regimes and processes, necessary for the design and operation of on hydrological regimes and processes, necessary for the design and operation of water supply projects, irrigation, flood mitigation etc.

Decision-makers in the steadily increasing field of environmental protection are becoming aware of the role of the water cycle both in transporting soluble and insoluble pollutants, and in producing feedbacks to the water balance from intervention with the soil-vegetation system. Hydrologists are therefore required to develop efficient methods for prediction of these effects, in order to bring environmental costs into the decision-making at an early stage. Indeed, the continuous character of the water cycle implies that any meso- or macro-scale perturbation of the complex geosphere-biosphere system will influence a whole array of hydrological processes; any change in the precipitation, for instance, would be expected to produce consequential changes in soil water and therefore in the biological ecosystem, in groundwater recharge and therefore in the water-table and groundwater quality, in runoff production and therefore in the quantity and quality of stream flow, and water levels in rivers and lakes .

Consequently, not only engineers but also ecologists and land use planners have to rely on hydrologists to provide the information relevant to the planning of drainage schemes, water supply and irrigation projects, predicting the suitability of land for diverse purposes, and possibly designing countermeasures against a variety of water-related disturbances.

These water-related dependencies between different professional groups related to water add up into a need for improved understanding of the relation between hydrological processes and biological ecosystems. The need for this improved understanding of hydrology in relation to ecosytems is probably least in those areas of application generally referred to as operational hydrology, and greatest in relation to the long-term problems of the hydrologic consequences of changes in land use and management. Such changes typically alter the balance between the surface and subsurface regimes, and may change the character of both. However, because of the long turnover time (see Chapter 2) associated with most groundwater systems, the effects may not be evident for many years or even several decades, and may then be irreversible or at best ameliorated at great cost.

Hydrology and land management

Traditionally, hydrologists have been obliged to take a pragmatic approach in order to serve society with the background information needed to decide on various environmental enhancement measures and structures, which form the backbone of modern civilization. Hydrologists have therefore concentrated their interest on the runoff generation part of the water cycle, i.e. what can be called the horizontal or long branch of the terrestrial water cycle. Less attention has been paid to soil water, which provides the water flow consumed in the biomass production process of photosynthesis and returns to the atmosphere as evapotranspiration, in other words the vertical or short branch of the cycle.

Thus, hydrologists in the past have been mainly concerned with the runoff component of the water resource, since it is this component which can be managed (at least to some extent) to provide stable supplies for irrigation, industry and domestic consumption. Of these uses irrigation accounts overall for 70% of the total, and roughly a third of the world's harvest comes from the 17% of the world's cropland that is irrigated (Postel 1984). Irrigation is a very visible example of alteration of the water balance of an area, and its obvious benefits have been mitigated in many areas by mismanagement leading to waterlogging and salinization of land.

Often less immediately obvious are the effects on the hydrological regime of land clearing, over-grazing, and changes in agricultural practices. In the long term, these may lead to devastating erosion, loss of soil fertility, and increased salinity in the soil or water supplies. It has been suggested (Postel 1985) that deforestation may be diminishing the stable runoff in developing countries as rapidly as expensive new dams and reservoirs are augmenting it.

Paterson (1984), referring to the existence of a sustainable yield solution for the Australian environment, states:

> "The problem lies in the slow process of resource degradation which we will not, and more importantly, probably cannot, arrest. Creeping degradation, even at say 2 per cent per year, means a doubling period of about 35 years. Two per cent per year often escapes notice, yet in the lifetime of our children such processes can change the map. Salinity, soil loss, deforestation and other catchment degradation, destabilisation of riverine environments and aquatic ecosystems, and silting of storage, proceed gradually. These processes advance slowly, but slowly and surely modify the performance of natural systems that we depend upon and which our descendants have a right to inherit.

Some of the damaging processes are non-linear and their interaction effects are even more so. Long periods of superficial stability can precede what the mathematics of catastrophe theory describes as a "cusp"; a sudden shift from a "stable" equilibrium of an acceptable kind to a radically different one. The second equilibrium may be far less acceptable. Recent destruction of

3

European forests by acid rain, and perhaps our own die-back problems appear to be examples of a "cusp". Ignorance, as much as veniality, has been responsible for many environmental losses in the past, and unfortunately, on questions of long term stability of total systems, our ignorance remains substantial.

If this view is accepted, then it calls for major changes in the water management agenda. Explicit choices are required on matters which have been ignored or considered to be "state" or background variables. Trends which are of minor contemporary importance, but which have long run irreversible consequences, must be dealt with as high priority matters...."

In the light of these issues, hydrologists must become more ecologically oriented, since it is their task to predict, ahead of the event, the hydrological consequences of changes in land use and management. This requires a change in emphasis in engineering hydrology, from concentration on surface runoff processes to studying the interactions between surface and subsurface processes, and between humans, vegetation, land and water.

The comparative approach

Discussions on the need for a book with a comparative approach to hydrology started in Sweden in the 1970s, when the growing volume of technical assistance to low-latitude countries brought increasing attention to the considerable hydrological differences between the snow-controlled hydrology of the Scandinavian environment and the warm semi-arid environment of many recipient countries. The idea of developing a project on comparative hydrology within the framework of the International Hydrological Programme (IHP) slowly took form.

The term "comparative hydrology" was coined to describe the study of the character of hydrological processes as influenced by climate and the nature of the earth's surface and subsurface. Emphasis is placed on understanding the interactions between hydrology and the ecosystem, and determining to what extent hydrologic predictions may be transferred from one area to another.

At the local or regional level, a restricted form of comparative hydrology is almost as old as scientific hydrology itself. Due to the spatial sparsity of streamflow measurements, it has always been necessary to predict the flood and yield characteristics of ungaged catchments from data available on "similar" catchments. Similarly, predictions of the behavior of regional aquifers have been extrapolated from well logs and pumping tests at a few sites, using available knowledge of aquifer geology.

The distribution of the components of the water balance of an area depends usually on the size of the area. For example, runoff per unit area in a humid zone increases with area up to some limit, while in an arid zone it decreases as the area considered increases. The spatial scale is therefore important in comparing hydrological processes within a region, as well as between regions. The main emphasis in this book will be on processes at a catchment scale, though where possible these will be related to site parameters.

This textbook

This textbook therefore stresses the whole terrestrial part of the hydrological cycle, and gives as much attention to the return pathways of transpiration and groundwater flow as to stream flow. It is concerned also with the role of water as a solvent, and as a means of transport of sediment, nutrients and pollutants.

Of the precipitation which is the source of the world's water resources, about 65% overall is returned to the atmosphere by plant transpiration and evaporation from soil, while the remainder becomes surface or subsurface runoff (Postel 1984). The proportions vary from 100% evapotranspiration in most flatland areas to nearly 100% runoff on bare mountainous areas.

The textbook attempts to bring to earlier descriptions of world hydrology, such as L'vovich (1979), the understanding of the hydrodynamics of ecosystems developed, for example, in the text by Miller (1977). The editorial board and authors are well aware of the ambitious nature of the project, and their aim is as much to identify areas of lack of understanding and knowledge as to draw out generalizations which can be used in a prescriptive way in particular areas or situations. While not a guidebook in the sense of recommending specific techniques or models, the text should provide guidance on the hydrologic processes which are likely to be critical in a particular environment.

The need for a textbook of this kind is based on both scientific and pragmatic grounds. At a scientific level, the interpretation of catchment hydrology in relation to a broad range of environments can be expected to lead to clearer insights into the movement and behavior of water in ecosystems.

At a pragmatic level, hydrological data collection is expensive and time consuming, particularly when related to the understanding of hydrological processes. If results obtained in one part of the world can lead to more effective effort elsewhere, in the form of better designed experiments or data analysis, this will accelerate the rate of progress towards better informed hydrologic predictions.

The book has been written as a graduate text for hydrologists, ecologists, foresters, agricultural and irrigation engineers, soil conservationists, environmental scientists, and land use planners and managers. For the scientists it should aid the understanding of hydrological processes in different environments, and help to identify the most important gaps in knowledge. For the managers, it is hoped the text will indicate what questions should be put to the scientists and engineers to ensure consideration of the full range of options for land and water management, including the reduction of water consumption rather than the provision of increased supplies.

Structure of the book

This textbook on comparative hydrology is related to the different ecohydrological zones of the world and the different types of hydrological environments existing in those zones. The book is divided into three main sections, the first presenting the basis for the comparative approach, and the second and third describing the main hydrological environments.

In Section A, the theoretical and methodological base is provided. After an introductory chapter on the new concept of comparative hydrology follows a chapter analyzing the main hydrological systems and processes, and distinguishing between vertical and horizontal flows. Water cycle components and interactions are discussed together with transport and reaction processes. Chapter 3 discusses different approaches to regional classification, focussing on some problems of regionalization, and comparing climatic classification with land parameter classification. This chapter ends with a presentation of the hydrologically relevant regionalization, which has been selected for the purpose of this textbook.

Chapter 4 focusses on fundamental problems of quantification of hydrological processes under different conditions, discussing in particular analytical tools by which similarities and differences may be studied. Some pertinent scale problems are addressed. This chapter finally demonstrates and discusses some basic regional differences of the main water balance components.

Chapter 5 addresses man, and in particular his land-use systems, as hydrological factors, stressing the non-stationary character of interactions with hydrological processes. Three basic types of interventions are addressed: agriculture, forestry and the modifications typical of urban areas.

6

Chapter 6, the final chapter in Section A, outlines some techniques for inter-regional comparison which have been used in the past.

In Sections B and C, the focus changes from comparisons to descriptions of the fundamental characteristics of the main hydrological environments. These environments are seen as produced by the interaction between climate as the forcing function, geology as a basic constraint, and man as a major manipulating agent. These sections distinguish between the various environments on the basis of the parallel aspects of climate and topography. Section B discusses the hydrology of sloping land with catchment response, while Section C covers the hydrology of flatlands, defined as areas where catchments are poorly defined, at least on the small scale. Two specific environments, small islands and delta lands, are the subject of separate chapters .

It will be noted that the chapters on the cold regions (Chapters 7 and 13) cover both dry and humid zones, because of the relatively small extent of the dry zone in this region (with the exception of the Arctic and Antarctic, not covered by this book). The distinction between the two sections is also slighly different in this case, between mountainous areas in Chapter 7 and gently sloping land and plains in Chapter 13.

In general, each chapter of Sections B and C starts with the main occurrence of the particular environment on the global scale, and the current water balance components. Then follows a discussion of exogenous and endogenous interactions and the current patterns of water and land use, including the potential for short- and long-term effects of human activities. The chapters generally close with a discussion of the implications of the specific environmental characteristics for hydrological data collection and the management of land and water resources.

Closing comments

This book will be a disappointment to those who are seeking quick answers to water-related problems by the use of analogies or models developed elsewhere. As has been noted at several points, there is no substitute for comprehensive collection of hydrological data, and one of the most shortsighted current trends is the general reduction in hydrological networks, particularly in developing countries. This book can however provide guidance on the type of data which are most likely to be useful in the long term, and it stresses the importance of appreciating all the water pathways in the hydrological cycle, and their sensitivity to change.

The book will also disappoint those who are seeking a new global approach to hydrology, with an all-encompassing model that will provide quantitative solutions in any environment. While the book contains many quantitative examples of particular phenomena, the emphasis at this stage of our knowledge must be on a descriptive approach which seeks to understand differences and interrelationships.

In its interdisciplinary approach the book has something to offer to a wide audience, and we believe it will be useful in setting the context for the future work of both researchers and decision-makers in a broad range of water-related fields.

Section A

Theory and methodology

1. Comparative hydrology - a new concept

The need for transfer of information between regions

Current simplistic approaches to water and hydrology. Hydrology in its modern sense is a young science, focussing on various phenomena related to the hydrological cycle. The continuity of this cycle adds new perspectives to the study of issues related to environment and development. The view that development takes place with the natural environment as its main scene is now being widely accepted. The natural environment provides man with water, food, fuelwood, energy, minerals etc. Water is deeply involved in most of these resources in complex ways reflected in Fig.1.1. This broad importance is seldom reflected in the attention given to water problems in planning and decision-making. On the contrary, problems related to water are generally approached in a very simplistic way. Contributing to this situation is the sectored organization of most societies and administrations, which leads to focussing on one aspect of a problem at a time.

In the past, we have therefore seen broad biases in regard to water. In the field of environmental quality, for instance, water has been treated mainly as a victim, neglecting its role as a mobile and chemically active solute carrier. In terrestrial ecology, subsurface water phenomena have largely been hidden in various black-box approaches, which treat soil water as a climatic phenomenon. Ecologists interested in tropical soil degradation problems generally pay much less interest to soil moisture phenomena than to other problems, such as nutrient cycles and erosion, even though water acts as the main transport agent in both cases.

In the sector of technical assistance to developing countries, the view of water has been equally simplistic, i.e. mainly as an issue of supply of manageable water for irrigation, industry and domestic use. There has been a general tendency to underestimate the fundamental differences between water conditions in the subtropics and tropics as compared to the temperate zone, where water can often be taken for granted (Biswas 1984; Falkenmark 1981; Kamarck 1976). This tendency incurs a considerable risk for climatic bias in technical assistance and in the north-south dialogue in general.

Many of the present problems in developing countries are in fact water-related: the considerable poor health is mainly related to water-borne diseases; land degradation is closely related to soil water problems; droughts are an extreme form of lack of water; starvation is partly generated by water scarcity; and the hunger crescent in Africa is located on a hydrological margin, sensitive to changing relations between the short and the long branches of the water cycle (Falkenmark 1986 b).

In spite of the wide importance and involvement of hydrological processes in environmental and ecological problems, very few hydrologists have become involved in these broader issues. Un fact, hydrologists seem to have widely underestimated their responsibilities and roles. They have seen themselves mainly as data providers, and have largely concentrated on surface water data such as water-levels and stream flows. Water quality issues have been left to geochemists and limnologists, groundwater to geologists, terrestrial ecology to biologists, and technical assistance to engineers and agricultural scientists.

The fact that water availability is limited by the water cycle generates increasing water scarcity in dry regions with rapid increases in population. This makes hydrological studies urgent for the solution of water deficiency problems, and as a support for the inter-related problems of soil degradation and water quality.

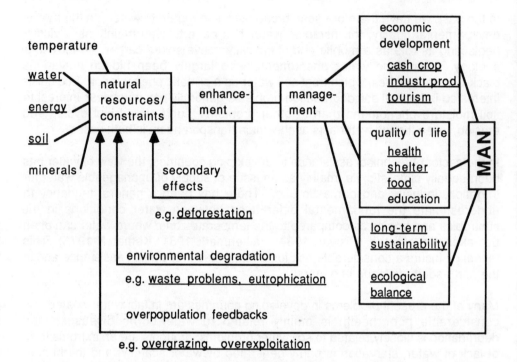

Fig.1.1 The complexity of water in relation to socio-economic development.

To facilitate and improve the productivity of the natural environment
it has to be developed,
i.e. enhanced to overcome basic constraints, and
    managed to operate the systems.

The ultimate aim is to support economic development and quality of life.
Long-term sustainability and ecological balance provide limiting criteria.

Underlined are all the places in this system where water is involved:
- it is part of the natural constraints
- it is involved in a multitude of societal activities
- it is active in the feedback processes

Problems of information transfer between regions. Water resource development and successful land and water management both depend on a general understanding of local hydrology. Classical hydrology was developed in the temperate regions of Europe and North America. The present understanding of hydrological conditions is therefore strongly influenced by the humid and sub-humid climate and sloping terrain in the temperate zone (Kovacs 1984 b). This context makes the drainage basin a fundamental concept.

Developing countries have tried to benefit from the technological skill and experience of developed nations, implemented in the form of technical assistance including education and training. Some local hydrological observations may have started during colonial times, but present data collection and research are generally poor, due to changed priorities and lack of manpower and resources. The present strategy of solving hydrological problems is therefore to benefit to the maximum possible degree from existing knowledge on hydrology and water resource systems, especially from generalizable theories (Ayebotele 1979). Local studies form an essential part of this strategy, aimed at making the general knowledge adaptable to local conditions.

Within the International Hydrological Programme (IHP) the basic ambition is to facilitate transfer of knowledge between regions. The present textbook on comparative hydrology is a contribution to this strategy. It will focus on basic similarities and differences between regions with different parameters in the hydrological cycle.

Methods used for practical applications of hydrology generally have a number of limitations inherent in the basic assumptions on which they are besed. It is therefore relevant to ask to what degree traditional forms of water balance calculation can be used under arid and semi-arid conditions, under conditions with deep groundwater or groundwater recharged by local rivers, or under conditions with poorly developed water divides. To what degree are models of runoff formation, developed for humid temperate conditions, applicable also under tropical or arid conditions? The concepts on which they are based may not be applicable in zones with very different climatic or physiographic conditions.

Indeed, Australian hydrologists, starting from the question "World hydrology: does Australia fit?" (McMahon 1982), have approached the issue of similarities and differences between world regions, focussing on rainfall-runoff relations and streamflow modeling (Finlayson et al 1986). Their global study, based on over 900 streamflow stations, showed that Australia and South Africa had a greater variability of both annual stream flow and annual peak discharge than all the other world regions. In Australia these parameters increased with an increase in catchment area, a trend opposite to that in the rest of the world.

Many water resource concepts are also inadequate for worldwide application. There is for instance a need to distinguish better between the amount of water hydrologically available in a country and the amount that could be made available in a particular locality, for purposes such as water supply. There is in fact today a considerable confusion on the meaning of such concepts as water availability (potential, dependable, readily available, regionally vs locally available), and water accessibility (what could be made available locally or what is actually made use of by the local population).

It is also pertinent to question how far graphs and equations describing interdependencies between meteorological, hydrological and landscape factors, based on temperate conditions, can be transferred to and applied in other regions. Rules of thumb, graphs and diagrams have to be checked by local studies of hydrological processes and by the operation of hydrological observation networks. It must also be stressed that the analytical techniques developed in advanced countries generally build on a much longer and more dependable data base than is usually available in developing countries. Moreover, such advanced technology often presumes ready availability of sophisticated computational and other aids.

Effects of neglect of climatic factors. It has been demonstrated by various authors (Kamarck 1976; Biswas 1984) that there has in the past been a sad neglect of the impacts of climate on the development process. This is even seen as a contributing reason for the poor agricultural performance of a majority of developing countries. It is argued that the numerous ruins of development projects in the tropics may be partially caused by a refusal to recognize the special problems of the tropics.

Furthermore, few development specialists seem to have realized the basic importance of climatic factors for the development of nations. The World Bank now clearly admits that economic development problems in the tropics are different from those in the temperate zone due to the effects of the climate. The rainfall rather than the heat determines the seasons so that, in large areas, periods of productivity must be sandwiched between droughts and floods (Kamarck 1976). Moreover, the humid tropics remain largely unfarmed, due to the inability to understand the dynamic nature of the soil. The impact on agriculture from diseases and pests, afflicting both animals and plants, is another climate-related factor which has hampered economic development in third world countries.

The general conclusion is that we need a more accurate diagnosis of the real

14

problems of development in order to avoid continuing disillusionment and despair over the prospects for developing countries. It is of course a fundamental aim to turn the tropical characteristics into assets by taking advantage of the heat, and to promote crop yields at least as high as in the temperate region (Kamarck 1976). This calls for research into the means of coping with the tropical obstacles, and for a recognition that solutions from the temperate region cannot be transferred automatically.

Need for understanding hydrological differences. In a large part of our society there seems to be a poor understanding of the fact that the amount of water made available through the hydrological cycle may indeed involve a distinct constraint to the development potential of a semi-arid or arid country. In such areas, the population capacity will be determined by the availability of water for plant production, whether rainfed or based on irrigation. The fact that water may be inadequate for the support of present populations on a decent level of life-quality and self-sufficiency in terms of food production, has given rise to the proposal (Chambers 1978; WCED 1987) that maximum productivity should be expressed in terms of one unit of water rather than one unit of land area as is the custom under humid conditions.

It is evident that hydrologists have an important role to play and a responsibility to contribute by clarifying the true implications of hydrological fundamentals in low latitude countries. As already stressed, the maximum benefit should be obtained from the hydrological experience and knowledge in temperate regions, and in those industrialized countries situated in arid and semi-arid climates. Many textbooks on hydrology tend to be descriptive and to focus mainly on conditions in Europe and North America. Furthermore, the role of water as a solute continuously moving through the landscape tends to be more or less neglected, possibly due to the fact that water quality is generally handled by people concerned with environmental pollution.

Understanding the similarities and differences between zones is also fundamental as a basis for selecting appropriate applications of engineering techniques. Climatic differences involving extremes of precipitation and evapotranspiration may raise problems for hydrometric methodology. Geological differences contribute to differences in soil wetting patterns, and to differences in groundwater recharge/discharge and flow patterns. Geology may also create local water supply constraints, so that the results from test boring for groundwater may only provide information about local water yield but say nothing about regional water availability.

Hydrological differences result in differences in water management approaches, in the importance to be given to salinity control, in the importance of water-related hazards to land use, in different aspects of water-related measures for soil improvement and agricultural development, in different interactions with vegetation, and in differences in terms of environmental resilience.

The great differences in water stress between high latitude and low latitude regions might indeed call for a shift in the basic approach to water. The technically biased approach, which is typical of large parts of the industrialized zone, may have to be transformed into a more ecological approach, recognizing the need for water for biological production as a major element in the national water balance (Balek 1983; Falkenmark 1984).

Hydrology and global change. The conclusion that a better understanding of hydrological differences is important for low-latitude development is but one strong argument for the new sector of hydrological science called comparative hydrology. Another very strong argument for this new perspective on hydrology is man's present problem of coping with the climatic change that is being forecast by the scientific community to cause significant effects in the next few decades (UNEP 1985; Eagleson 1986).

The key process that is generating major changes in the global climate is the greenhouse effect of the atmosphere. This change in the radiation balance is a result of the accumulation of carbon dioxide in the atmosphere in response to expanding use of fossil fuels. Pollutants added to the atmosphere strengthen this effect. The predicted result of the increasing content of these "greenhouse gases" is that the temperature will increase as less long-wave radiation leaves the earth.

It is evidently of major importance to be able to forecast what effects such a temperature increase will have on plant production and water resources in different parts of the world. When meteorologists can predict the primary changes in the wind systems and the aerial pattern of water vapor flux, hydrologists will have to estimate the responses of terrestrial hydrological processes to the changed precipitation and evaporation patterns, and their feedback to the atmosphere.

In this research it will be important to understand how regional hydrological conditions develop in response to climatic conditions as the forcing phenomena and geological conditions as constraints. In this context comparative hydrology will find another important area of application.

16

Some fundamental differences

Climatically induced water balance differences.   The interaction between climate, hydrology and vegetation is the basis for distinguishing various ecohydrological regions. L'vovich (1979) has analyzed the interrelations between the water balance elements in the main vegetation zones. Based on the features that each zone has in common on the different continents, and which have with some degree of generalization been extended to the entire globe, L'vovich has produced type values for the various ecohydrological zones (Table 1.1). Some of these data are shown in Fig.1.2, in order to illustrate some of the main differences between the zones.   The ecohydrological regionalization shown in Fig.1.3 presents the IHP macroclassification basis for representative and experimental basins, recommended for use in inter-basin comparisons (Hadley 1975).

Relation to aridity. Budyko (1986) discusses not only the water balance but also the biomes which dominate under different conditions of temperature and moisture. His starting point is a general geographical zoning, based on net radiation R_n and a radiation index of aridity (RIA), defined as the ratio of the net radiation to the heat (λP) necessary to evaporate the local precipitation. Combining this with general water and radiation conditions in the main ecosystems, he develops a classification of geobotanical zones (Fig. 1.4 a). He finally arives at a composite representation (Fig.1.4 b) of local runoff production in the main geobotanical zones. It should be noted that Budyko uses the term climatic runoff for the local runoff production, defined as the difference between mean precipitation and evaporation.

Resource base differences. The differences in hydrological conditions, intensity of human interference, and level of socio-economic development combine to produce a wide spectrum of methods, philosophies, tactics, and strategies to avoid tensions in human dependence on the interrelated land and water resource systems. The selection of adequate land and water development and management strategies in different parts of the world, including the best ways of utilizing water resources, needs strong support from the hydrological sciences to provide the fundamental understanding on which an intelligent choice should be based.Water is supplied to a country in one or two complementary ways: by the supply of endogenous water, provided by the precipitation on the country; and possibly by the inflow of exogenous water from upstream countries, the characteristics of which are subject to control by those countries.   Subject to international or inter-state agreements, the water is available for use during the time that it passes through the country.

Table 1.1

TYPICAL WATER BALANCES ($mm\ y^{-1}$) IN THE WORLD'S MAIN VEGETATION ZONES (After L'vovich 1979)

Climatic belt	Vegetation zone	Precipitation	Runoff			Evapo-transpiration	Potential evaporation	Annual biomass growth (t/ha)
			Total	Ground-water	Surface			
Subarctic	Tundra	370	110	40	70	260	400	1-2
Temperate	Taiga	700	300	140	160	400	500	10-15
	Mixed forests	750	250	100	150	500	700	10-15
	Wooded steppes, prairies	650	120	30	90	530	900	8-12
	Steppes	500	50	10	40	450	1300	4-8
Subtropical and tropical	Eastern broad-leaved wet forests near the ocean	1300	420	120	300	880	1000	10-15
	Desert savanna	300	20	2	18	280	1300	2-6
	Dry savanna	1000	130	30	100	870	1300	6-12
	Wet savanna	1850	600	240	360	1200	1300	10-20
	Wet monsoon forests near the ocean	1600	820	320	500	780	900	15-30
Equatorial	Perennially wet evergreen forests	2000	1200	600	600	800	800	30-50
Mountain	Wet monsoon forests in mountains near the ocean	2200	1700	700	1000	500	600	15-30

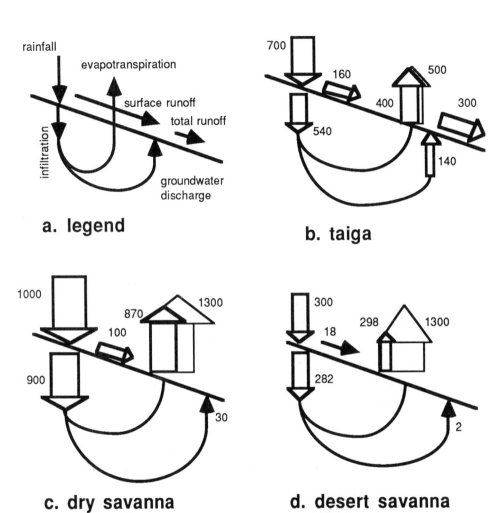

Fig.1.2 Schematic representation of flows in short and
long branches of the hydrological cycle.
(Based on data from L'vovich 1979)

19

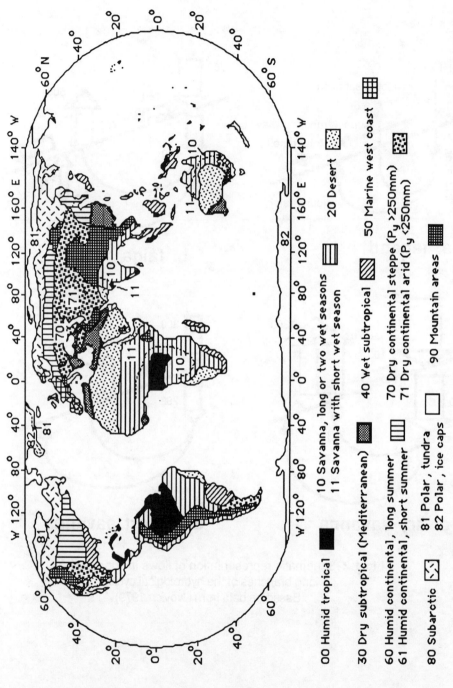

Fig.1.3 Regionalization of the world, based on climate and geography (from Hadley 1975).

■ 00 Humid tropical

10 Savanna, long or two wet seasons
11 Savanna with short wet season

▦ 20 Desert

▦ 30 Dry subtropical (Mediterranean)

▨ 40 Wet subtropical   50 Marine west coast

60 Humid continental, long summer   70 Dry continental steppe (P_y >250mm)
61 Humid continental, short summer   71 Dry continental arid (P_y <250mm)

80 Subarctic   81 Polar, tundra   90 Mountain areas
                 82 Polar, ice caps

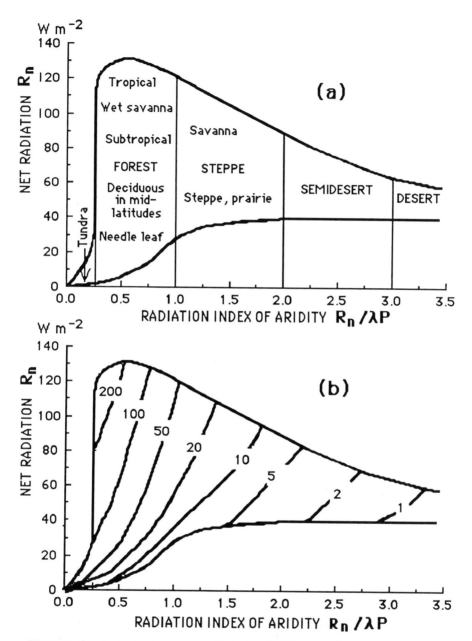

Fig.1.4 Geobotanical zones and local runoff (*mm* y^{-1}) as a function of $R_n/\lambda P$ and R_n (after Budyko 1986).

In water-scarce areas, the aim of self-sufficient food production may result in competition for water between plant production and man. The larger the amount needed for crop production, other things being equal, the less will be available to society in streams and aquifers. This makes it fundamental to take a more ecological approach to water in the future (Fig. 1.5).

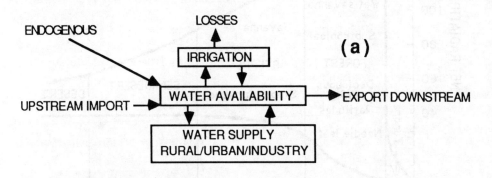

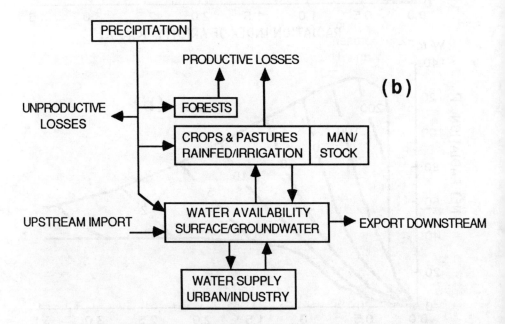

Fig.1.5 Renewable water supply and different categories of water
demands in a national macro-perspective (from Falkenmark 1984).
(a) Traditional approach (water as a technical factor)
(b) Ecological approach (water for plant production included)

22

Reductions in river flow, generated by intensively irrigated agriculture, are causing great problems in regions with dry climates, both in the USSR and the USA. In the USSR, the increased short-branch flow of water is referred to as"irrecoverable losses" (Unesco 1978 b), and great interest is being devoted to the prediction of such losses (Table 1.2). In the USA, the national water assessment compares the consumptive water demand with the renewable supply, defining the available water supply as the endogenous water renewal only (USGS 1984). This implies a tacit assumption that the exogenous inflow into a region may be cut off or fully consumed upstream. In the above discussion, national water availability has been seen from the aspect of the manageable use of water, by focussing only on the river runoff and its components. When considering also the water used in rainfed plant production, the total water supply has to include precipitation as well as the rain-generated runoff.

A general water balance approach to national water supply and use makes it possible to distinguish between vertical and horizontal components. Thus, the total renewable supply is composed of precipitation (vertical) and exogenous inflow (horizontal). The water leaving the country has two components: the evapotranspiration, composed of a productive part used in plant production and an unproductive part representing true losses, and the runoff in rivers flowing to other countries or the ocean. The vertical output represents water leaving through the short branch of the water cycle, whereas the horizontal output is the water leaving in

Table 1.2

IRRECOVERABLE LOSSES IN WATER USE (after Unesco 1978 b)

Continent	Water use ($km^3 y^{-1}$)			Irrecoverable losses ($km^3 y^{-1}$)		
	1900	1970	2000	1900	1970	2000
Europe	40	320	730	20	100	240
Asia	270	1500	3200	200	1130	2000
Africa	30	130	380	25	100	250
N.America	60	540	1300	20	160	280
S.America	5	70	300	3	50	130
Australia/ Oceania	1	23	60	0.6	12	30
Total	400	2600	6000	270	1600	3000

the long branch, which is available for managed use by society while it passes through the country.

Fig.1.6 shows regional resource balances for a number of countries. The diagrams, based on the 15 year old approximations of L'vovich, are shown here to illustrate the use of hydrological information as a basis for water supply comparisons between countries. It has to be pointed out that, although the more recent river flow data of Forkasiewicz and Margat (1980) are probably more precise, L'vovich's data had to be used as they include precipitation and evapotranspiration. Considerable differences (more than 10%) in the size of the flow leaving a country are frequent between the two data sets, and have been marked by an asterisk in Fig.1.6.

The water resource balances shown in Fig.1.6 include both developing and industrialized countries. It may be noted that some countries import a considerable fraction of their river flow resources: Federal Republic of Germany 15%, German Democratic Republic 50%, Hungary 75% and Egypt almost 100%.

The volumes shown in these resource balance diagrams have to be related to the population in order to allow a comparison of similarities and differences between countries. The relative availability (flow per head of population) is the conventional index of the potential stress on national water resources; it is indicated under each national column, based on Forkasiewicz & Margat's flow data. It can be compared with the average water demand in relation to the long-term perspective of socio-economic development in a country. For instance, most studies seem to arrive at close to 1000 $m^3y^{-1}p^{-1}$ as an index of long-term water needs in a modern society (Widstrand 1980). This is a gross demand, arrived at by just adding all demands without attention to any possibilities for reuse.

Water competition differences.   Parallel with the more or less unbroken population increase still characterizing many developing ocuntries, there is a widespread call for improved living conditions and increased food production. Most countries also aim at self-sufficiency in food production. Improved living conditions imply increased water demands on a per capita basis both for productive water "losses" (water consumed as evapotranspiration in the crop production process) and for managed use for domestic and industrial water supply. Population increases, however, reduce the relative water availability, i.e. the average share of water hydrologically available to each individual.

The stress on water resources can alternatively be discussed in terms of the water competition index, defined as the number of individuals depending on each flow

24

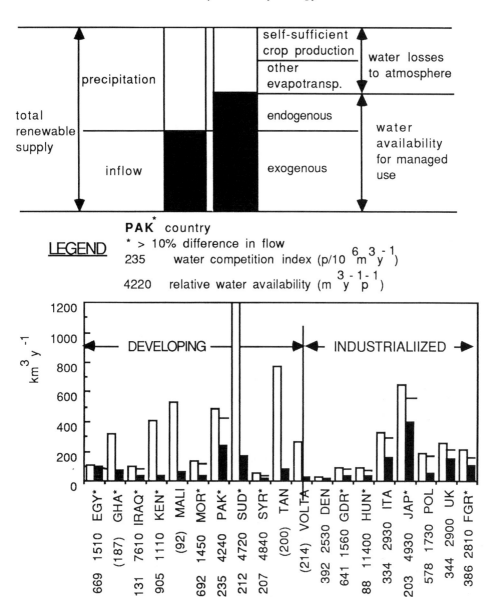

Fig.1.6 Regional water balances (data from L'vovich 1979), water competition index, and relative water availability (data from Forkasiewicz and Margat 1980; population of 1980)
* see text

unit of water available to the country (Falkenmark 1986 a). The present level of water competition in different parts of the world is illustrated in Fig.1.7, based on the data base compiled by Margat and Forkasiewicz (1980). Obviously a doubling of the population is equivalent to twice as many individuals competing for each unit of water flow ($10^6 m^3 y^{-1}$). The figure shows that even with the present population, the overall water competition index is comparable for developing countries like China and India on the one hand, and industrialized countries like France, Italy, the United Kingdom, Spain and East Germany on the other. In countries like Kenya, Egypt and Sri Lanka, the water competition index with the present population is of the same order as that in Poland (600-800*p/flow unit*).

The bottom part of Fig.1.7 shows conditions in the smallest and most exposed countries. Considerable water competition characterizes the Canary Islands, Israel and Tunisia with about 2000*p/flow unit*. Libya and Saudi Arabia have reached their present level of development in spite of a water competition index of 4000*p/flow unit*. This has been possible by relying on unconventional water sources, such as fossil groundwater (Margat and Saad 1984).

The dramatic increase in the water competition index which will be generated by population increase is illustrated in Fig.1.8 for a number of African countries. Evidently, water shortage is a rapidly expanding phenomenon in the world. Consequently, the importance of hydrology as a basis for water management is increasing dramatically. In striking contrast to these facts, hydrological networks of routine observation stations are falling apart and data are becoming less and less reliable. The services responsible for operational activities seem to get lower and lower budgets all over the world.

Development differences. The basic water-related problems of the present world are very different between high-latitude industrialized countries and low-latitude developing countries. In industrialized countries, the water resources are more or less developed and under control to support societal life and activities. The main problems are often related to water quality, due to an overestimation of the capacity of rivers to absorb introduced wastewater flows, and an underrating of the capacity of water to remove soluble waste products, surplus fertilizers and pesticides from the soil water zone. Groundwater pollution is developing on an ever increasing scale. In general, hydrological conditions are fairly well known as a result of observations and research studies.

In low-latitude developing countries, many of them situated in areas with a dry season of various length, the situation is vastly different. Crop production is severely constrained by considerable seasonal and interannual fluctuations in water availability. Due to overestimation of the direct transferability of a water

26

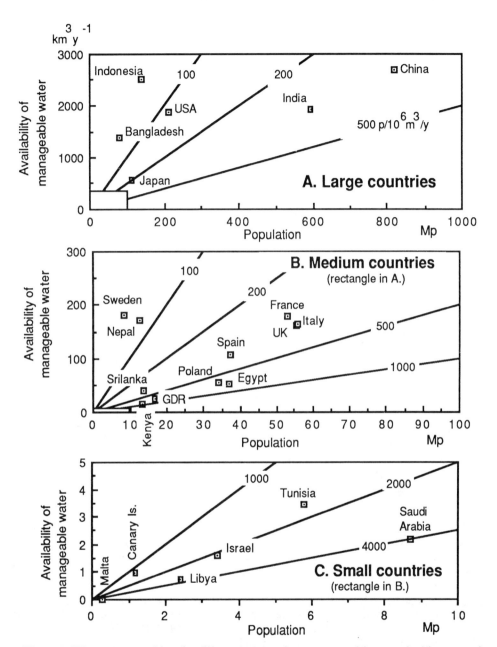

Fig.1.7 Water competition in different countries: renewable supply (rivers and aquifers) in relation to 1975 population. Sloping lines indicate water competition indices. (Data from Forkasiewicz and Margat 1980)

27

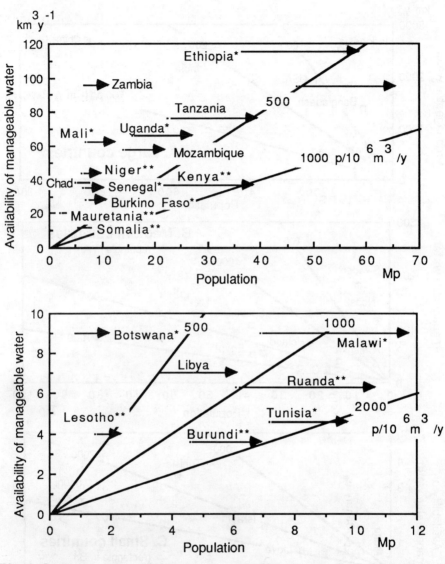

Fig.1.8 Water competition in some African countries, showing relation between total water availability and populations in 1975 and 2000. Water data from L'vovich (1979); population data from WRI/IIED (1986). Sloping lines show water competition indices.

* Critical population supporting capacity by 2000, with low level technology (FAO/UNEP/IIASA 1983). ** The same, with intermediate level technology.

technology developed in the temperate zone, these countries are often littered with unsuccessful irrigation projects. Due to the lack of solutions to various problems, soil erosion and land degradation is severe. Africa is now developing creeping disasters of continuously increasing dimension. Droughts and floods hit enormous populations. Water supply is not yet organized for more than parts of the urban areas, leaving urban fringes and rural areas to live on contaminated water carried home by women and children. Sanitation is generally very poor, causing massive poor health due to water-related diseases. In general, hydrological conditions are poorly known, and most water resources remain undeveloped.

Human interactions with the natural environment

Differences in environmental problems. As already mentioned, the natural environment provides man with water, food, fuelwood, energy and minerals. To support increasing populations, man is forced to manipulate the environment (Fig.1.9). Different types of natural constraints, which cause problems for societal development (water-logging, dry soils, obstructing vegetation etc) have to be overcome or compensated. To get water, man constructs water schemes and projects for storage, water supply and irrigation. To get energy man develops electric power from water, or fossil or nuclear fuels.

As all measures related to soil, vegetation and hydrology involve interventions with complex natural systems, feedbacks are bound to develop as responses. These secondary effects are often grouped together under the diffuse concept of "environmental effects". As they are (except possibly for outputs of polluting substances) impossible to avoid, the ultimate question in relation to environment and development reduces to the following: how does one define the ecologically acceptable balance between the benefits for societal development, made possible through the interventions, and the costs in terms of unintended changes of that environment, that these interventions cause, due to the way the natural systems operate?

From the hydrological perspective, current environmental terminology and concepts are very incomplete. A description has to be developed, that pays more adequate attention to the role played by water in land management, and man's interaction with that role.

Today's "no-change" voices among environmental activists in the temperate zone have difficulties in being understood by people from developing regions, where

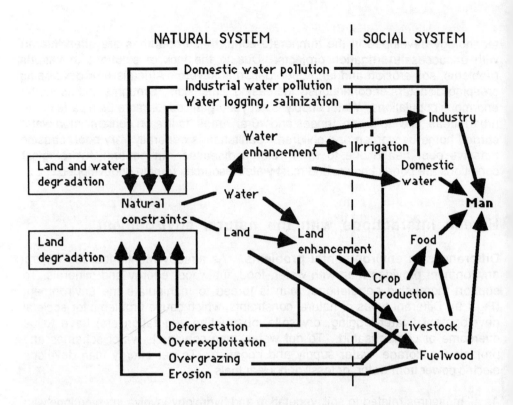

Fig.1.9 The complex interactions and feedbacks between the natural and the social systems (after Falkenmark 1986 b).

major modifications of the natural environment are necessary to approach self-reliance in fuelwood and food production for rapidly increasing populations. Laszlo David in UNEP, in discussing the relation between socio-economic development, ecological concerns and an environmentally sound development of the natural environment (David 1985), makes a distinction between seeing the natural environment in the untouched "paradise" form, and in its fully developed form. In socio-economic development, the attention given to socio-economic aspects must be related to ecological aspects.

Man as a hydrological factor. The interaction between rain, soil and vegetation is central to the main environmental and land degradation problems of the present world, which indeed imply a severe menace to man's future in many tropical and subtropical environments. Several hydrological processes are involved in this zone of interaction: interception, direct evaporation from moist surfaces including foliage, surface runoff formation, soil erosion, infiltration, soil water and nutrient flow, formation of soil surface crusts etc. Such processes are all fundamental for the understanding of soil productivity, land degradation risks, the effects of drought etc. Indeed, the successful future of low-latitude countries lies in successful development of control over this interaction system.

The soil zone with its cover of vegetation in fact constitutes a key zone of the hydrologic cycle. A large part of the precipitation passes through the short branch of the cycle with a turnover time (see Chapter 2) generally of the order of days, and then returns to the atmosphere at approximately the same location. The other part (the long branch) moves into the groundwater system and in the humid zone discharges into a river or the sea (Fig. 1.10a). In arid zones the discharge is to the sea or to the atmosphere through dry lake beds, usually with a turnover time of the order of centuries (Fig.1.10b). Land use changes strongly influence the short branch flow, which then results in long-term changes to the quantity and quality of the long branch flow. Through this connection, land use should be seen as an inherent part of the description of the hydrological cycle.

Degradation of life support systems. Current literature on the human environment indicates that the most pressing ecological problems of the present world are related to land degradation, caused by reduction or total exhaustion of the vegetation cover, which would otherwise provide shelter from erosion due to heavy rain, or by mismanagement of irrigation projects, causing waterlogging or salinization of previously productive soils.

Activities contributing to the first category of land degradation are deforestation for industrial purposes, clearing for agriculture or pastures, or simply fuelwood harvesting for local households. In lower rainfall areas, desertification is the final effect of the stripping of vegetation from the land surface. The second category is caused by an unawareness of the importance of achieving the necessary water and salt balances to avoid a destructive accumulation of groundwater or salt in the irrigated area.

All these problems are closely related to hydrological processes. It is therefore fundamental that land degradation should be approached from a hydrological viewpoint by studying and stressing the interaction man-vegetation-soil-water (Fig.1.11). Even though hydrologists are seldom consulted when major global environmental problems are discussed, the hydrological cycle is in fact the phenomenon that ties together the closed loop between vegetation, soils, water and man himself.

The hydrological cycle links vegetation-destroying human activities with reduced

31

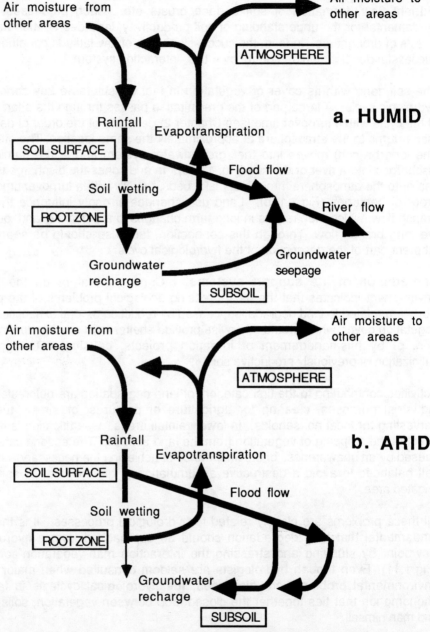

Fig.1.10 Continuity relations in the terrestrial part of the water cycle.

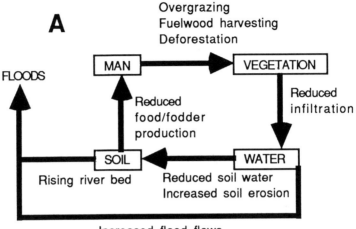

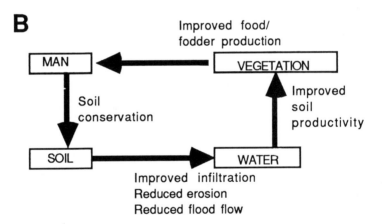

Fig.1.11 Interactions between man, vegetation, soil and water.
A. Deteriorating cycle with destructive interactions transforming human activities into various forms of environmental degradation.
B. Improving cycle with land and water conservation, and increased productivity.

land productivity and therefore reduced agricultural production on the one hand, and increased flood flows and inundation on the other. Increased flood flows are a consequence of reduced soil infiltration and hence increased surface runoff; increased inundations are a combined consequence of increasing flood flows and rising river beds produced by increased volumes of sediments which settle further downstream. If evapotranspiration remains the same, the increased surface runoff is reflected in reduced groundwater recharge and therefore increased risk for drought. However, reduction of the vegetation cover or its replacement by species with shallower roots will usually reduce evapotranspiration, and may therefore lead to increased recharge even when infiltration is reduced. The increased recharge may be beneficial, but often leads to waterlogging or, where the groundwater is saline, to salinization of stream flows or the plant root zone.

A basic countermeasure against continuing soil degradation is land conservation. By careful conservation measures, infiltration and soil-plant-water relations may be improved, groundwater recharge controlled, and erosion and floods reduced. Within limits, increased soil water in the root zone makes possible improved soil conditions and increased plant production. It can also lead to water conservation as an important secondary effect.

In other words, land conservation and water conservation should be integrated. Irrigation management is also a question of integrated land and water conservation, since neglect of the need to secure adequate leaching of salts as well as drainage of surplus water from the fields reduces crop productivity. At present, land conservation measures are carried out by ecologists, while hydrologists tend to be treated as experts on groundwater and surface water only. Due to the key role played by the soil-vegetation system in hydrology, it would however seem highly desirable to bring hydrologists into the field of land degradation by paying more attention to plant-water relations and the soil water balance.

Hydrology and human disasters. In the wake of increasingly frequent human disasters, particularly in Africa and India, great efforts have gone into more thorough investigations of the causes and effects of what are often referred to as "natural disasters". It has been shown that the type of disasters affecting the largest numbers of people are droughts and floods (Fig.1.12), both of which are hydrological events closely related to land use.

Drought is a natural hazard which is a recurrent phenomenon in all semi-arid and sub-humid climates, and is caused by the interannual variability of the rainfall (Fig.1.13). The drought disasters which developed during the 1970's and 80's

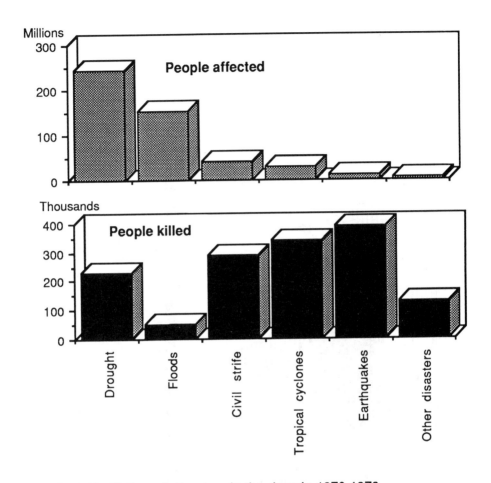

Fig.1.12 Victims of disasters in the decade 1970-1979
(from Swedish Red Cross 1984)

were the result of a chain process: the initial hazard (deficient rain) brought increasing populations to relief centres, producing local crises of water and food shortage. These developed into ecological crises by generating severe soil degradation due to the large population pressure in the surroundings of the relief centres. In other words, even if the basic hazard - drought - is quite natural and part of the characteristic climate, its consequences are to trigger land degradation in numerous ways, and reduce what may be called the "sponge effect" of the soil to the rain input. Floods are events at the opposite end of the rainfall scale. They,

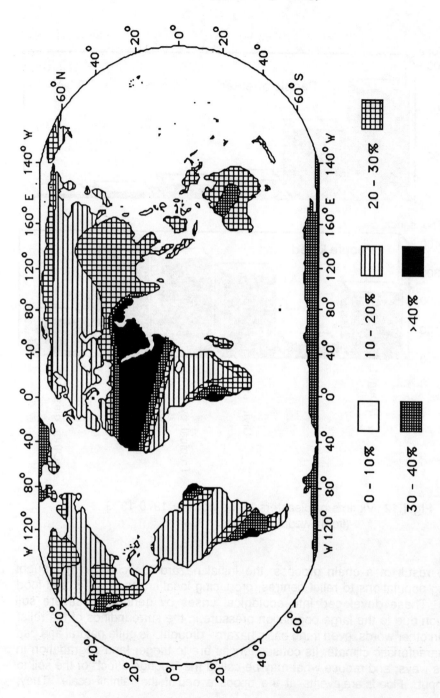

Fig.1.13 Global distribution of variability of annual rainfall, expressed as percentage departure from normal (from Landsberg 1975).

0 – 10%   10 – 20%   20 – 30%

30 – 40%   >40%

too, are natural features due to the fact that intense rains, with great interannual variability, are typical of the climate in tropical and subtropical regions. They also trigger land degradation by setting in motion ever increasing amounts of soil which settle downstream, raising the river bed so that the same flow reaches higher levels in the landscape.

Both these phenomena thus represent consequences of man's destructive interaction with soil and vegetation. It is therefore fundamental that these forms of human disasters should also be viewed from the hydrological aspect, including the role played by the hydrological events involved in the upstream interaction between rain, soil, vegetation and man. In modern disaster prevention programs, integrated soil and water conservation are seen as basic measures.

Differences in need for hydrological support

Hydrology and water engineering projects. Successful water management depends on hydrological data and the studies needed to identify the water-related problems and find solutions, whether by structural or non-structural measures. General information on the water systems involved is needed as a basis for the planning of water development and management. More detailed information is necessary for adequate design of various structures. Other types of water-related information may be needed in the construction phase. Finally, for the operation of a water resources structure, there may be a need for continuous information, including forecasting. The present tendency to degrade hydrological networks, referred to earlier, indicates a poor understanding of the fundamental importance of a reliable database on the water resources available in an area, for successful management of these resources.

Information support for water resources development and management projects is the everyday task of hydrological services. Operational hydrology involves a number of methodological questions: data transfer from measured to unmeasured sites, identification of past trends and man-made changes, forecasts of the hydrological characteristics at a given site, and (in the longer term) prediction of the future hydrological effects of human activities. Operational hydrology in other words focusses on the characteristics of both points and systems in the landscape, and on both past and future conditions. Since the hydrological characteristics at a site are the result of the operation of the system which brings the water to that point, the basic theme of a book on comparative hydrology should be the operation and characteristics of different types of hydrological systems seen both from a retrospective and a prospective viewpoint.

Information about local water conditions has to be based on local observations in hydrological networks. As it is impossible to take measurements at more than a limited amount of locations, techniques for generalization have to be developed in order to produce information on other locations where problems might develop or where water development measures might have to be taken. These techniques, which should allow transfer of local data both in time and in space, have to be based on research, which clarifies the general operation of the local hydrological systems. Forecasting techniques represent a special group of extrapolation tools, developed to assist the operation of water schemes.

Differences in water resources focus. Water resources have traditionally been viewed in relation to surface water. A main tool has been the water balance of a landscape or a river basin, i.e. determining how the precipitation reaching the ground is divided between evaporation "lost" and runoff retained. Rainfall-runoff models have become one of the most fundamental tools of modern hydrologists in the temperate zones. Their main effort has gone into serving the water resources development and management sector with data, predictions and forecasts.

Development in tropical countries has broadened this focus. Some countries in the semi-arid tropics have marginal hydrological conditions, in the sense that most of the precipitation returns to the atmosphere in the short brach of the water cycle. The water going into the long branch is particularly sensitive to interannual fluctuations in rainfall, which translate intermittent droughts into water supply problems and hardship for the local population.

The particular problems of arid and semi-arid regions call for much more attention to groundwater recharge and flow, due to the intermittent or episodic character of local water courses. This represents a rather recent branch of hydrology, not always well understood by people engaged in groundwater explorations. It still causes confusion, for instance, when pumping tests indicate good local "yield" in contradiction to hydrological commonsense, based on estimates of possible groundwater recharge from water balance considerations.

Hydrology and land conservation/management. Land management is generally water dependent in the sense that a certain type of activity may need water supply. Water may also pose constraints to a planned land use, e.g. high water-tables, poor aquifers, or the presence of disease vectors due to humidity. Agricultural yield depends heavily on an adequate amount of water being present in the root zone. Supplementary irrigation depends on water being locally accessible. Erosion and land degradation risks are also related to water. The risk of salinization on irrigated land is tied to the need to maintain the salt and water balance necessary

38

to avoid the accumulation of salt or water. Land use is often related to the production of various forms of waste, whether fluid or dry. The possibility of disposal of these wastes is also closely related to water: wastewater outlets in a river produce pollution which may call for treatment in order to maintain concentrations sufficiently low to allow the use of river water for drinking or to avoid toxic effects on local fish.

As a basis for land management there is a need for water-related information such as local soil water conditions, probabilities of soil water deficit and crop failure, drought risks etc. These risks constitute fundamental information in a country where overyear food storage has to be provided to cope with drought years when they arrive. Soil water conditions should also guide the general planning of the crop production pattern in water-short areas, where the water availability provides a fundamental constraint and the crop production policy should be to produce the maximum amount of biomass per unit of water.

Measures for soil protection, both from erosion and from salinization, have to be based on information about local hydrological conditions. Maintaining the salt and water balance in an irrigation project, for instance, needs continuous monitoring of the water-table and groundwater salinity to achieve efficient operation of the system. In dry regions, agricultural practices might be adapted to actual water availability by developing both a good weather code as a general base for local farmers, and a bad weather code to which they should turn during dry years (Biswas 1984).

Some fundamental process differences. Some important differences in the way hydrological processes operate have to be well understood.

Firstly, there are important differences in regard to the processes of rainfall genesis in different zones (tropical cyclones, monsoons, trade winds, convection, frontal movement etc.). Rainfall intensities may therefore be very different, and rainfall variability will change greatly between zones in response to the different genesis processes. Patchiness, or spatial variability, is another aspect, and differences in the seasonal pattern are fundamental to the whole organization of society.

Following on from the rainfall and the generating process behind it, the resulting infiltration will depend on the hydraulic conductivity of the soil which varies between soil types, and depends also on soil structure, density, and the pattern of macropores. The capacity to store soil water is also a function of the soil type, including again soil structure and organic content, and is often expressed as the difference between field capacity and wilting point. The depth from which the vegetation gets its water is also of great importance, as demonstrated by Balek

39

(1983) in his discussion of the different root types.

Due to difficulties in measurement, evaporation and transpiration are often quite poorly treated in textbooks on hydrology. In the present book a very clear distinction will be kept between potential and actual evapotranspiration. The difference between the two (see Chapter 2) is fundamental for understanding differences in crop production potential.

Comparative hydrology should facilitate an understanding of the influence of geology, rainfall characteristics and vegetation on the conditions for groundwater recharge. The genesis of groundwater varies greatly between zones (Fig. 1.14). In the humid zone, groundwater is recharged in upstream catchments, and discharges in local depressions or downstream through the beds and banks of rivers. In the arid zone, groundwater recharge occurs mainly from stream beds. For further discussion, see Chapter 2.

The aquifer types may also be quite different. The depth to the water-table is a vital factor for accessibility of water in wells, and may vary considerably between humid and arid conditions. In systems with a short turnover time (see Chapter 2), this depth may vary seasonally. The general pattern of groundwater movement may differ between different geological settings. The outflow area may have very different characteristics in different environments: temperate wetlands, chotts, playas, oases, coastal springs etc.

Runoff is generally referred to as an element in the water balance. It is however important to understand also runoff formation as a process. L'vovich (1979) distinguishes between two types of runoff emerging from the water balance of a humid zone catchment: a flood flow rapidly transferred along the river during wet periods, and a groundwater-produced less variable base flow which provides the dependable flow resource in a river. The base flow may be increased by contributions from water storages in lakes and reservoirs. In the arid zone there is no base flow, since the water-table is below the stream bed. As a result of the transmission loss from the stream to the groundwater, the stream flow decreases in the downstream direction and frequently does not reach a drainage terminal. Understanding interregional differences in the process of runoff formation is fundamental as a basis for water resource development and management. These aspects are covered in more detail in the next chapter.

Finally, river flow may also be classified according to its genesis (L'vovich 1979; Hadley 1975), depending on whether it comes from solid phase storages such as glaciers and snowpacks, from liquid phase storages such as lakes and groundwater, or from atmospheric water vapor as rainfall. The primary effect on river flow is that its

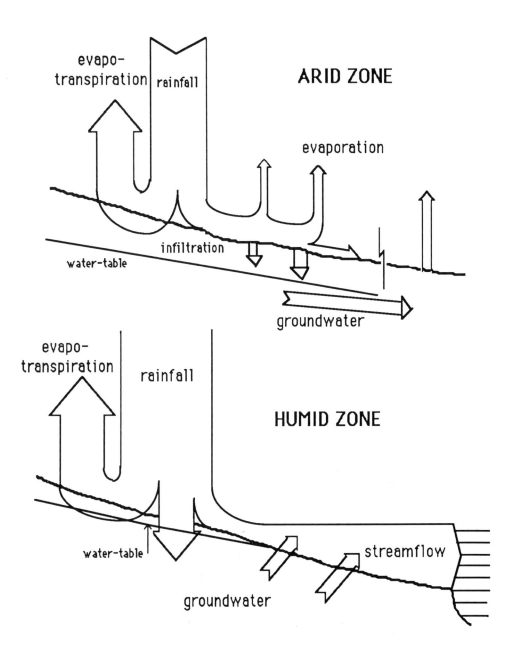

Fig.1.14 Groundwater formation and discharge under arid and humid conditions (after Erhard-Cassegrain and Margat 1979)

41

stability increases with an increase in the turnover time of the source storage. The secondary effect in the case of the solid storages is the melting due to seasonal temperature changes, causing spring-summer floods when all the snowpack is melted or a more regular variation in the case of permanent snow or glaciers.

Where snow and ice are not factors, there is a marked contrast between the perennial regimes of rivers in the humid zone which are fed by base flow from groundwater, and those in the arid zone which are the source of groundwater recharge, and exhibit flashy flows which may peter out before reaching their drainage terminals. A particular case of great historical and social significance is arid zone rivers which have their source in the humid zone, either in snowpacks (e.g.Indus and Colorado rivers) or in large lakes (e.g. the Nile).

Perspectives of comparative hydrology

In this chapter we have discussed a number of major hydrological differences between countries, which tend to complicate the transfer of hydrological understanding, methods and theories between zones. Basic differences in terms of climate and geology between the different regions of the world produce different soil water regimes, differences in potential evaporation, different genesis and behavior of groundwater, different regimes of river flow, and so on.

Differences in temperature, soil water and groundwater produce differences in vegetation biomes, in root development and root depth, in other words fundamental ecohydrological differences (Fig. 1.3).

Different population pressures and different intensities of human activities contribute to major differences in the environmental problems encountered, complicating international exchanges of experience and knowledge.

The aim of this book is to clarify some of the most basic differences in hydrological phenomena and processes in order to improve the understanding of people with various academic backgrounds, involved in interzonal transfer of hydrological knowledge, water engineering skills, definition and mitigation of environmental problems, land use planning etc.

2. Hydrological systems and processes

Hydrological systems

Systems of storages and flows.   The conceptual framework common to all
hydrological analysis, at whatever scale, is the systematic division of the water
environment into a number of storages, connected by water fluxes. The simplest
such description for global hydrology uses only three such storages (Fig.2.1),
while in some hydrological models, particularly for subsurface systems, there may
be hundreds of storages.

Fig.2.2 shows a framework consisting of twelve storages which is adequate for
description of the main hydrological processes occurring at catchment scale. For

43

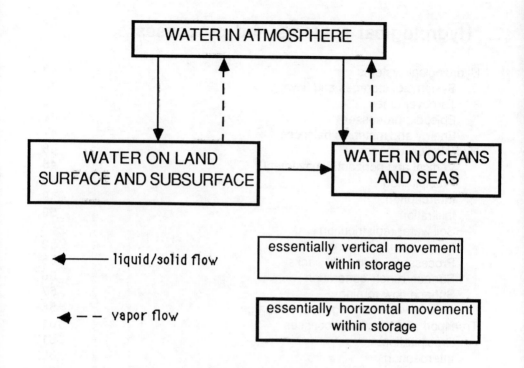

Fig.2.1 Generalized world hydrological system

each of these storages, the water balance (conservation of mass) equates the difference between the inflow $Q_i \Delta t$ and the outflow $Q_o \Delta t$, in a period Δt, to the change of liquid mass or volume ΔV as follows:

$$Q_i \Delta t - Q_o \Delta t = \Delta V \qquad (2.1)$$

Here V may be defined as mass per unit horizontal area ($kg\ m^{-2}$) or as the (numerically equivalent) average depth in *mm*.

The water balance equation may be calculated for different periods (day, month or year), and may be applied to a group of storages, so eliminating from consideration the fluxes between them. Techniques for computing a wide range of water balances have been described in an earlier Unesco publication (Sokolov and Chapman 1974). The advantage of using a long balance period, such as a year, is

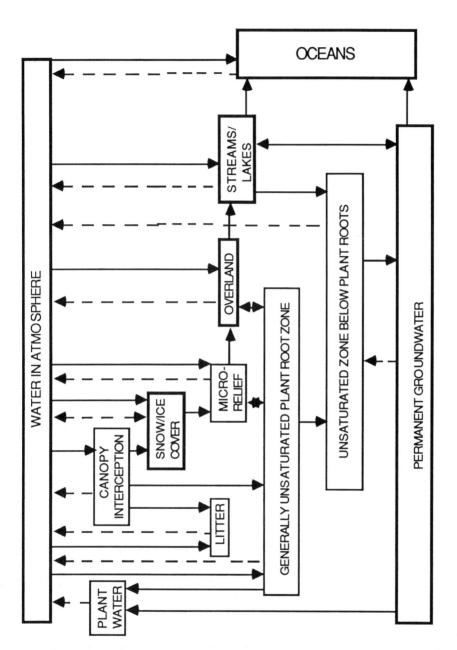

Fig.2.2 Hydrological system for a catchment
(Legend as for Fig.2.1)

45

that changes in storage are small relative to inputs and outputs, and may often be neglected as a first approximation. Long term water balances for the world have been computed and mapped by Baumgartner and Reichel (1975) and by Unesco (1978 a). The first shows the distribution of long-term precipitation, evaporation and runoff, while the second also shows runoff coefficients and the difference (positive or negative) between precipitation and potential evaporation.

Turnover times. The disadvantage of the long balance period is that most information about the hydrodynamics of the process is lost. To provide a simple measure of typical rates of change of hydrological storages, we can adopt a turnover time as used in ecology (Odum 1971), defined as the time required to replace a quantity of substance equal to the amount in the component. In the hydrological context the terms "conventional residence time" (Unesco 1971) and "rate of water exchange" (L'vovich 1979) have also been used. The turnover time T_r is defined by

$$T_r = \overline{V} / \overline{q} \qquad (2.2)$$

where V is the volume of water storage, q is the rate of output, and the bars indicate long-term averages.

Turnover times range over several orders of magnitude, but for each of the storages illustrated in Fig.2.2 there is a more restricted range, as shown in Table 2.1. These typical values are an indication of the dampening effect of the storage on inputs of water or energy, and show clearly why the effect of a change in input may not be detected for a long time in subsurface systems below the plant root zone. Klige (1981) has pointed out that the residence time of water in river channels has been substantially increased worldwide by the development of reservoirs. It should be noted that the typical turnover times listed in Table 2.1 are for precipitation occurring as rainfall, and may be much longer when the precipitation occurs as snow or the water in the storage is frozen.

Episodic processes. With the exception of storages with turnover times of the order of months or years, hydrological processes are essentially episodic, that is, there are periods of activity (flow between and within storages), usually called a hydrologic event, interspersed with usually longer periods of quiescence, during which many of the pathways between storages are inactive. A large part of the transport of water occurs during hydrologic events, and for each event there will therefore be a frequency distribution of the distances moved by water in the active storages. This frequency distribution will vary widely with the magnitude and

Table 2.1

TYPICAL PARAMETERS OF HYDROLOGICAL STORAGES IN SLOPING LAND

STORAGE	DEPTH (mm)	TURNOVER TIME	EVENT HORIZONTAL TRAVEL DISTANCE
ATMOSPHERE	25	8-10 DAYS	-
INTERCEPTION	0.04-5	<FEW HOURS	<FEW m
LITTER	2-7	1-4 WEEKS	<FEW m
PLANTS	5-50	HOURS - DAYS	<FEW m
DEPRESSIONS	0.2-40	MIN. - HOURS	<FEW m
OVERLAND FLOW	1-10	3-30 MIN.	0.5-50 m
STREAMS & RIVERS	3	WEEKS	1-100 km
PLANT ROOT ZONE	5-500	1-4 WEEKS	10-100 mm
VADOSE ZONE	$10-10^4$	FEW YEARS	-
GROUNDWATER	10^4-10^5	FEW DAYS - 10^6 YR.	-
LAKES & RESERVOIRS	-	MONTHS - YEARS	-
OCEANS	3.7×10^6	28 YEARS	-

intensity of the event, but an attempt has been made in Table 2.1 to assign characteristic values and ranges to some of the storages. Similar considerations apply to material, dissolved or otherwise, which is transported by the water in a hydrologic event.

Energy and materials balances. Each of the fluxes shown by broken lines in Figs.2.1 & 2.2 represents a flux of water vapor. For this flux to occur, sufficient energy must be provided to the source storage to cause the change in phase from liquid to vapor. The energy balance is therefore also critical in determining the partitioning of the water balance, particularly in relation to the transpiration and evaporation fluxes. While cloud cover and other factors impart some variability to incoming solar energy, there is an underlying diurnal and annual rhythm which affects the water balance, particularly on a seasonal basis.

To study interactions between land management and the hydrological system, the water balance equation must be supplemented by one or more materials balance equations, such as a sediment balance or a salt balance. However, these balance equations, which may be applied to each conceptual storage in the system, do not define the flows between the storages. For this purpose, another equation is required for each such flow, usually based on conservation of momentum or

47

energy, but taking a form which is apppropriate to the particular hydrological process. The more important processes are discussed in the next section.

Hydrological processes

Vertical and horizontal movement.   Comparative hydrology involves the study of the kind and extent of hydrological processes which act in one part of the earth, and differentiate it from another. Hydrological processses can be conveniently categorized into two groups, depending on whether the direction of water movement is mainly vertical or mainly horizontal (Chapman 1968; Kovacs 1984 a).

Vertical processes include interception, infiltration and water flow in the unsaturated zone, both within and below the plant root zone. The extent of horizontal water movement in these processes is typically of the order of a meter. While this movement may be critical to the water balance and nutrient status of vegetation, and so determine its survival and growth, it has no effect at catchment scale.

It is convenient to include in these vertical processes the output fluxes of evaporation and transpiration, since again the predominant water movement is vertical until it reaches the interface with the atmosphere.

The mainly horizontal processes are overland flow, channel flow and groundwater flow, including shallow saturated flow such as may occur above a relatively impermeable soil horizon.

The overall proportion of vertical and horizontal processes in an area depends on both the climate and the morphology of the land surface. With increasing aridity of climate, there is a decreasing opportunity for rainfall excess to occur at the ground surface, and a decreasing opportunity for water to move past the plant root zone to join the groundwater system. Again, with increasing flatness of land, overland flow and streamflow will move more slowly and the distance moved in all but extreme hydrologic events will decrease; because of low water-table gradients, groundwater transport will also be very slow. Thus predominance of vertical processes is associated with dry climates and/or flat land, while horizontal processes become more important as the climate becomes more humid and/or the land surface has increasing slopes.

These generalizations for the average climate in an area can be extended to extreme hydrologic events or non-events. For example, a sand plain or a dunefield in an arid zone may in an extreme flood display humid zone behavior, by becoming a shallow lake of water which may move considerable distances along the low regional gradients. Again, if there is a sufficiently long interval between hydrologic events, horizontal water movement may peter out even on sloping land in humid zones, and a regime typical of drylands will exist for a while. By analysing how hydrological processes depend on climate and morphology, some inferences can be made on the most significant parameters controlling the processes. This is the first essential step in ensuring that predictive techniques are relevant to the particular region under study.

The individual hydrological processes will now be discussed in more detail, starting with those in which vertical movement predominates.

Precipitation. Precipitation includes all forms of water falling from the atmosphere to the earth's surface. The main source of this water is evaporation from the oceans, other water bodies, and wet land surfaces, and transpiration by vegetation. The average amount of precipitation received at a particular location depends on such factors as its latitude, altitude, distance from upwind moisture sources, position within and size of the continental land mass, relation to other topographic features such as mountain ranges, and relative temperatures of land and oceans. Seasonal changes in prevailing winds and relative temperatures result in typical within-year distributions of precipitation.

The characteristics of the precipitation at a location in a given storm, in terms of intensity, duration and areal extent, are determined by the source of the water vapor in the atmosphere and the lifting mechanism which causes precipitation. In particular, convective rainfall typically has higher intensities for shorter durations, and affects a smaller area than cyclonic or frontal rainfall. Variations in these characteristics, and whether the precipitation falls as rain or snow, have a profound effect on the nature and extent of subsequent hydrological processes on and below the land surface. Variability of precipitation in space and time is discussed in Chapter 4, and numerous examples are given in the chapters on the main hydrological environments.

Interception. The process of interception by leaves was described by Horton (1919) in the following terms.

> "When rain begins, drops striking leaves are mostly retained, spreading over the leaf surfaces in a thin layer or collecting in drops or blotches at points, edges, or on ridges or depressions of the leaf surface. Only a meager spattered fall reaches the ground, until the leaf surfaces have retained a certain volume of water, dependent on the position of the leaf surface, whether horizontal or inclined, on the form of the leaf, and on the

49

surface, on the wind velocity, the intensity of the rainfall, and the size of the falling drops...."

The main hydrologic impact of interception is that it reduces the precipitation which reaches the ground, by the amount of water which is evaporated from the wetted parts of the intercepting surfaces during and after precipitation. The proportion of the precipitation so "lost" depends on the nature of the vegetation, categorized to a first approximation by its interception storage capacity, the precipitation regime and the potential evaporation. Other things being equal, the importance of interception loss increases with reduced intensity of rainfall and with increased temperatures and wind (input of advected energy). For some New Zealand forests, Blake (1975) quotes values of over 40% for the proportion of evaporation due to intercepted water.

Where plant litter occurs on the soil surface, it also intercepts both precipitation and throughfall from the leaf canopy. Although the hydraulic mechanisms of the interception process are similar in the canopy and the litter, there is typically a large difference in turnover time. Thus while turnover times for canopy water are of the order of minutes, turnover times for water in the litter are likely to be of the order of weeks. As noted earlier, these figures will be greatly increased if the precipitation falls as snow.

Infiltration. The process of infiltration (Fleming and Smiles 1975) has probably received more research effort than any other process in the hydrological cycle. It has been the common focus for surface water hydrologists, agricultural engineers and soil physicists. Infiltration data have been obtained from measurements in a wide range of infiltrometers, and experiments conducted in environments ranging from laboratory columns of uniform glass beads, through lysimeters and small plots to large plots and entire small catchments.

The descriptive model of Green and Ampt (1911), likening soil to a collection of capillary tubes, is still valid and the infiltration equation they derived is still in wide use. The rate of advance of water down a capillary tube is conditioned by the capillary tension at the air-water interface, which may be taken as constant, and the opposing force of the frictional resistance to the length of the advancing column of water. The rate of advance of the wetting front therefore falls exponentially with time. This characteristic is incorporated into the early empirical equation of Horton (Cook 1946).

The infiltration rate and the rate of advance of the wetting front are also a function of the initial soil water content. At higher water contents the rate of advance is increased and so the infiltration rate is reduced more rapidly with time and also with the amount infiltrated. Mein and Larson (1973) have shown how the Green and Ampt model provides excellent quantitative results and conceptual understanding of this interaction with initial water content.

Modern infiltration theory (Philip 1969 a) is based on the classical theory of one-dimensional flow in soil (Richards 1931). Philip (1957 a) introduced the term "sorptivity", which is a property of the soil related to the capillary potential concept of the Green and Ampt model, but which can accommodate a wide variety of (typically hysteretic) relationships between water content, soil water potential and hydraulic conductivity. Attempts have been made (e.g. Maller and Sharma 1984) to predict spatial variations in infiltration in terms of the measured spatial variability of the sorptivity and the saturated hydraulic conductivity.

These models assume that sufficient water is being supplied at the surface to keep it saturated, i.e. in a "ponded" condition. As long as the intensity of the rainfall reaching the ground is less than the rate at which it can enter the soil, there will be no ponding on the surface, and the infiltration rate will be exactly the same as the net rainfall rate. The typical situation is for the rainfall rate to be less than the ponded infiltration rate initially and possibly at later periods in the storm. However once the ponded rate falls below the rainfall rate, the surface saturates and ponding begins. The difference between the rainfall rate and ponded infiltration rate, called "rainfall excess", accumulates and moves to the local lowest points (microdepressions) in the landscape. Since depression storage capacity decreases with increasing surface slope, there may be a tendency for systematic variation along a catena, of the proportions of ponding and non-ponding infiltration. This does not appear to have been studied to date.

Increasingly complex mathematical models have been proposed to take into account spatial heterogeneity, expressed in terms of scale-heterogeneous media (Philip 1967; Watson and Whisler 1972; Whisler and Watson 1972), vertical and horizontal variations in soil properties (Philip 1969 a; Beven 1984), swelling and cracking soils (Philip 1969 b; Giraldez and Sposito 1985), surface crusts (Smiles et al 1982), and entrapped air and multiphase flow (Morel-Seytoux and Billica 1985).

An additional complexity is the existence of macropores (soil cracks, root channels, earthworm holes etc) in sufficient numbers to have a material or even dominant influence on the infiltration process (Beven and Germann 1982; Beven and Clarke 1986). In such situations, there may be a relatively rapid movement of water down the macropores, followed by a redistribution from the filled macropores to the soil matrix (Davidson 1985). For a given input, the effect of macropores is to deepen the wetted zone and to distribute the water more uniformly with depth, so decreasing the depth of saturated soil near the surface. Germann and Beven

(1985) have attempted to combine the sorptivity approach of Philip in the matrix with a kinematic wave approximation to bulk water flow in the cracks and macropores.

The phenomenon of inhibition of infiltration by the formation of a surface seal or crust is associayed with the absence of macropores and a smoothing of the surface microtopography, which occurs in fine-textured bare soils. Lack of vegetation both reduces its protection of the surface from the full kinetic energy of the rainfall, and removes the main instrument of macropore development (Raudkivi 1979).

Soil water redistribution. The term soil water redistribution is usually applied to the processes which occur between infiltration events and which may affect the whole soil profile down to the water-table. Redistribution in the deeper layers may continue while infiltration occurs near the surface.

These processes are profoundly affected by the presence or otherwise of active plant roots, to the extent that for hydrological purposes the unsaturated zone must be divided into two storages on this basis. The presence of plant roots provides a sink for removal of water, which results in a relatively rapid turnover time, of the order of a few weeks. In the underlying vadose zone, turnover times are more likely to be of the order of years, and values of the order of 50 000 years have been established for a semi-arid zone environment (Allison et al 1985).

It follows that the vadose zone can be conceived as a zone of one-dimensional unsaturated flow, with episodic inputs at the interface with the plant root zone, and quasi-steady flow towards the water-table. The important parameters are therefore the soil hydraulic characteristics and their variation with depth.

Most analytical studies of soil water redistribution have either ignored the presence and effects of plant roots, or have treated them in a way which fails to account for the interactions between roots and soil. Except for a few crop plants which have been studied in some detail, there is no predictive model of these processes which properly accounts for the physics and the chemistry, let alone the microbiology. At our present state of knowledge of soil water redistribution, we can only say that the process depends on both plant and soil characteristics, and that the single most important parameter appears to be the depth (or possibly the water-holding capacity) of the plant root zone.

Evaporation and transpiration. Evaporation has a primary meaning of the change in phase from liquid to vapor, which requires the input of energy. Transpiration implies the emission of water vapor from living plants and animals. For most natural surfaces it is difficult to apportion vapor fluxes between different

52

sources, and the term "evapotranspiration" or "total evaporation" (Monteith 1985) is used for the aggregate of all vapor sources in an area.

In the late 1940`s Budyko, Ferguson and Penman independently developed an expression for total evaporation from natural surfaces, which combines the surface energy balance with the transfer equations for the exchange of sensible heat and water vapor with the atmosphere (see Webb 1975; Brutsaert 1982; Monteith 1985). Many variations of this "combination equation" now exist, common ones being the Penman formula and the Penman-Monteith equation which particularly applies to vegetation (Monteith 1985).

For a shallow free water surface, the general expression can be written

$$E = \frac{S}{S+\gamma} \frac{R_n}{\lambda} + \frac{\gamma}{S+\gamma} \frac{\rho D}{r_a} \qquad (2.3)$$

where E is the evaporation flux, R_n is the net radiant energy absorbed and D is the saturation vapor deficit of the air, S is the slope of the curve of saturation vapor pressure against temperature, λ is the latent heat of vaporization of water, γ is the psychrometric constant (ratio of specific heat at constant pressure to the latent heat of evaporation), ρ is the density of moist air, and r_a is a wind-dependent resistance. The terms $S/(S+\gamma)$ and $\gamma/(S+\gamma)$ may be seen as temperature-dependent weighting functions of the two components of the evaporation flux.

In humid and sub-humid regions there is a high correlation between the terms in (2.3), so that a valid approximation (Priestley and Taylor 1972) is

$$E = \alpha \frac{S}{S+\gamma} \frac{R_n}{\lambda} \qquad (2.4)$$

with the parameter α taken as ≈ 1.3.

A fundamental assumption in (2.3) is equality of exchange surfaces and processes for sensible and latent heat. Land surfaces, with a mixture of soil and vegetation which may be differentially wet by precipitation (or may be partialy covered with snow or ice) have more complex and variable exchange processes. However in humid regions of uniform vegetation, an effective modification to (2.3) is by changing the psychrometric constant to

$$\gamma' = \gamma\,(1 + r_s / r_a)\qquad\qquad(2.5)$$

where r_s is another series resistance which accounts for the vapor path through the leaf stomata. This resistance depends not only on vegetation species, but also on carbon dioxide supply, light, vapor pressure deficit, leaf water status and leaf area index (defined as the ratio of single-sided leaf area to horizontal area of ground surface).

For short well-watered grass, r_a dominates over r_s , so that we can define a potential total evaporation E_0 (also called potential evapotranspiration or reference crop potential evapotranspiration). Values of potential total evaporation appropriate to other surfaces can be obtained by the use of "crop coefficients" (Doorenbos and Pruitt 1977).

There is also a need to calculate actual total evaporation, where evaporation is limited by available soil water. Just as actual infiltration rates are the smaller of the net rainfall rate and the ability of the surface to admit water, so are actual evaporation rates the smaller of the potential evaporation and the ability of the surface to supply water. If the actual evaporation is reduced, extra absorbed radiant energy goes into heating the air, so increasing D in (2.3) and therefore increasing the potential evaporation. Morton (1983) has developed the concept of "complementary ratio actual evaporation", which offers the possibility of estimating actual evaporation from meteorological data by making allowance for the raised potential evaporation. Soil water accounting (Fleming 1964) can also be used to estimate the supply rate to evaporating surfaces.

Current research (Hatfield et al 1984) suggests that remote sensing of land surface temperatures can offer direct estimates of sensible heat fluxes, and hence of the evaporation flux through use of the energy balance.

The land surface characteristics which determine total evaporation rates are of two kinds, the above-ground characteristics which determine surface roughness and the area of evaporating surfaces, and the below-ground characteristics which determine the rate at which water can be supplied to the evaporating surfaces.

For regularly spaced vegetation with a uniform height (row crops, orchards, planted forests), the effect on the wind velocity profile in the lower atmosphere can be estimated (Webb 1975), and estimates made of the effect on transpiration rates.

The situation with irregularly spaced vegetation (natural forests, woodlands and shrublands) appears too complex for such analysis.

The area of evaporating surfaces is conveniently categorized by the leaf area index.

The below-ground characteristics affecting total evaporation rates are the plant root characteristics (depth and morphology) and the soil characteristics which determine flow rates under hydraulic gradients caused by the suction heads in the roots, together with nutrient and biological factors which determine rates of root growth and death. Although the processes are qualitatively well understood, the three-dimensional and heterogeneous nature of the flow field have to date proved too complex for quantitative modelling without unrealistic simplifications.

The particular case of evaporation from bare soils has been successfuly modelled (Philip 1957 b; Van Bavel and Hillel 1976; Prevot et al 1984), and shows a rapid flux while the soil surface is wet, decreasing sharply when the surface is sufficiently dry that flow occurs in the vapor phase; applications are however very restricted.

Evaporation from open water surfaces is expressed by (2.3). Due to heat storage effects, the surface water temperature is a vital parameter; for example, heat storage in the Great Lakes of North America so dominates energy supply that maximum evaporation rates occur in midwinter. Local calibration is also required in arid regions, since the very dry air masses passing over open water significantly enhance evaporation, but are themselves modified to an extent depending on the size of the water body.

Processes within snowpacks. As noted by Eagleson (1970), snowpacks typically exhibit an increase of density with depth, with fresh snow having a relative density around 0.10 (ranging from 0.06 to 0.34 depending on wind conditions (Beskow 1947)), ranging to about 0.40 at the bottom of a deep seasonal pack and ultimately towards 1.0 in a glacier. This variation in density is associated with corresponding changes in the water and heat transfer and storage properties, so causing considerable complexity in modelling the energy and water balances of snowpacks.

If a snowpack is regarded as a porous medium composed of ice particles (Motovilov 1986), infiltration of water through it can occur as a result of rainfall or melting of the surface layers. This infiltration process may be arrested (with the creation of impermeable layers of solid ice) if heat exchange causes the temperature of the infiltrating water to fall below freezing point. During a snowmelt season, there may be diurnal initiation and cessation of infiltration, accompanied by a decrease in depth of the snowpack. This infiltration process will continue into underlying soil which is not frozen or saturated. Marsh and Woo (1985) describe the spatial variability of flow within a snowpack, and relate it to the physical properties of the pack and to the melt rate.

Evaporation from a snowpack is determined by the energy balance, and Kuusisto (1986) has shown that the net radiation and turbulent exchange processes play a major role, while the contributions from precipitation and heat exchange at the ground surface are relatively small. Variations of albedo are particularly important in the radiation balance, so that with the same incoming radiation, the melt rate at a snowfree glacier surface (albedo 0.3) may be two to three times that of a glacier covered with fresh snow (albedo 0.8) (see Chapter 7). It should be noted that the reverse process of condensation of water vapor from the atmosphere may add to the water content of a snowpack, and may be numerically much larger than the corresponding process of dew formation in warmer climates.

We turn now to those hydrological processes in which horizontal movement predominates.

Surface runoff generation. The classical view of the surface runoff process, developed by Horton (1933), is that it results from rainfall excess which occurs when the rainfall intensity exceeds the infiltration rate. While this concept is valid and useful at a site, its extension to catchment scale can be quite misleading. As indicated in the discussion on infiltration, rainfall excess accumulates first in microdepressions. When these are surrounded by areas of non-ponding infiltration, runoff occurs from the ponded areas across non-ponded areas, so that the catchment consists of a patchwork of runoff and runon areas. As surface slope is critical in determining both depression storage capacity and the rate at which very shallow water will move downslope, runoff into streams may occur from only favourable parts of the catchment, called source areas. These source areas will vary in extent during the progress of a rainfall event.

An alternative mechanism of surface runoff generation occurs in shallow texture-contrast soils, when a highly permeable surface soil overlies a less permeable subsoil. Again, this mechanism leads to saturated source areas of surface runoff generation, which can be predicted from information on surface topography, supplemented by data on soils and vegetation where these are not spatially uniform (O'Loughlin 1981, 1986).

These relatively recent developments in understanding the complexity of the runoff generation process at catchment scale cause fundamental difficulties in transferring hydrological relationships between small and large drainage basins and between regions (Pilgrim 1983). They also highlight the problems in trying to identify those properties of the catchment surface and subsurface which are most significant in affecting the surface runoff process.

The processes of surface runoff generation from a melting snowpack are qualitatively the same as those from direct rainfall, but have different spatial and temporal characteristics. Source areas for such runoff typically cover a large part of the catchment, and the flow continues over an extended period, often with a diurnal variation. Major floods may be caused by warm rain falling on a melting snowpack.

Surface flow routing. In contrast with the difficulties in predicting the space-time characteristics of surface runoff generation, there has been considerable success in modelling the runoff hydrograph as it moves down the catchment. At this stage, the problem is essentially one of open channel hydraulics with given inputs. However, although the full hydraulic equations have been used for routing flow on main rivers, the complexity of the channel pattern in upland catchments usually dictates a simpler approach with smaller data requirements. The main techniques used have been systems methods, storage routing, and the kinematic wave assumption.

The best known systems approach is the unit hydrograph, which assumes the system is linear. For a gaged catchment, the unit hydrograph can be determined from analysis of distinct flood hydrographs and the associated rainfall excesss, estimated by assumptions about the time distribution of infiltration and other "losses". Unit hydrographs may be calculated for a given duration of unit rainfall excess, or for a unit pulse of vanishingly short duration; this latter form, which should depend solely on catchment characteristics, is called the instantaneous unit hydrograph (IUH).

The parameters of such unit hydrographs may then be related to catchment characteristics (area, length and slope of main channel etc) by statistical techniques, in order to attempt a generalised prediction of unit hydrographs for ungaged catchments in a "region" defined by climate and/or land form characteristics. The errors of prediction of such methods are relatively high, due to the arbitrary choice of catchment parameters and the unknown nature of their interactions with the hydrograph parameters. Recent developments (Rodriguez-Iturbe and Valdes 1979; Troutman and Karlinger 1985) attempt to make better use of established morphometric properties (Smart 1978) of the stream channel network.

Recognition that the linearity assumptions of unit hydrographs are at best approximations, led to the extension of the systems approach to nonlinear techniques, such as the Volterra series approach of Amorocho and Orlob (1961), recently further developed by Napierkowski and O'Kane (1984). Although these methods may achieve better fitting of hydrographs on gaged catchments, they have not been applied in regional generalisations.

The storage routing approach attempts to integrate channel flow parameters into a simple mathematical relation between the water stored in a channel reach (or in the overland flow paths and network of channels in a subcatchment) and the output from the channel or subcatchment. Sometimes, as in the well-known Muskingum technique (Singh and McCann 1980), the input is also included in the equation. If linearity is assumed, there is one catchment parameter K (the ratio of storage to output, or to a weighted average of input and output) for each reach or subcatchment. These parameters may be estimated by fitting predicted output to observed flood hydrographs, and/or from measurements of channel geometry. Frequently the ratio of the K-values for subcatchments is estimated from catchment data, and the single value required to establish their magnitude is obtained by fitting.

The extension of the storage routing approach to nonlinearity is relatively simple, and usually involves a power relationship (Laurenson 1964) between storage S and discharge Q of the form

$$S = K Q^m \qquad (2.6)$$

where m is an exponent which is usually less than 1 for catchments and more than 1 for river reaches, and is normally taken to have the same value for all reaches or subcatchments. As the parameters K and m are not independent, relating them to general catchment characteristics is difficult. Recent work on the flood behaviour of river channels in Australia has caused Bates and Pilgrim (1983) to question the ability of power functions to describe catchment response over a wide range of discharge.

The hydraulic approaches to flow routing have had most utility in long reaches with low gradients, and particularly in estuaries. Because of the site-specific nature of such applications, they do not lend themselves to regional generalisation.

Groundwater recharge and discharge. There are two main mechanisms of groundwater recharge, general downward percolation of soil water which moves past the plant root zone, and percolation from the beds of streams, rivers and alluvial fans. With some qualification, the first is typical of humid areas and the second of arid areas.

In the first situation, the parameters determining the groundwater recharge will be the climate parameters of rainfall and potential evaporation, and the catchment parameters of the plant-soil complex and general land slope. Except in some flat catchments with uniform soils and vegetation, there is likely to be considerable spatial variation of groundwater recharge, with much of it having its source in local runon areas. There have been few attempts to quantify such spatial variations and relate them to the characteristics of the land surface or subsurface.

Although water balance considerations suggest that rainfall in arid zones will seldom move beyond the depth (typically two meters) above which it may be transpired or evaporated back into the atmosphere, general percolation to groundwater may occur in extreme events in which the land surface is generally flooded. These very rare recharge pulses, which may then take decades or centuries to reach the water-table, may be the only source of groundwater recharge in flat areas with no stream channels, and in dunefields.

The second recharge mechanism, percolation to the water-table from stream beds, takes two forms, depending on whether there is a saturated connection between the stream and the water-table (Dillon and Liggett 1983). Where no connection exists, water moves downward from the stream bed to the water-table, forming a linear groundwater mound which then dissipates laterally away from the stream. As long as the mound is recharged by unsaturated flow, there is no hydraulic connection between the groundwater and the stream flow, in the sense that the recharge rate is unaffected by the groundwater levels. The parameters determining the recharge process are then the width, depth and duration of stream flow, and the hydraulic characteristics of the local material in and below the stream bed.

Recharge without hydraulic connection is typical of arid zones, where water-tables are typically deep (tens of meters). In less arid areas, water-table levels tend to rise closer to the stream bed. In these situations, a hydraulic connection will usually exist between the stream and the groundwater, and the recharge rate will decrease as the water-table rises. The recharge process will be dominated by horizontal rather than vertical flow, and will have a much shorter turnover time than the disconnected flow process. In these less arid environments, there is likely to be also some recharge from general catchment percolation, and the mix between the two mechanisms may be hard to predict. It should also be noted that recharge may occur from the low order channels in quite humid areas.

There are also two main mechanisms for groundwater discharge, largely depending on whether the environment is humid or arid. The first consists of approximtely horizontal flow to springs, streams, lakes or the sea, while the second consists of vertical movement by transpiration from phreatophytes of evaporation from shallow water-tables.

Groundwater discharge into streams is their source of flow between hydrologic events, and constitutes 30% of runoff from the world's land surfaces (L'vovich 1979), the proportion increasing as annual precipitation increases. Information about the storage and the turnover time of the groundwater system may be readily obtained from the stream base flow recession hydrograph by expressing it in the form:

$$Q_t = Q_o \, e^{-t/T_r} \qquad (2.7)$$

where Q_o, Q_t are the stream discharges at times 0 and t.

Groundwater discharge by phreatophytes in low rainfall areas is usually concentrated along the banks of stream channels or in local floodout areas. As these locations are also the source of groundwater recharge, there may be little or no net effect on the regional water-table a few ten of meters away from the recharge-discharge zones.

In more humid areas, transpiration from vegetation in marshes and swamps, and on flatlands with shallow water-tables, is a major mechanism for groundwater discharge, operating over the whole region. Much of the groundwater is probably returned to the atmosphere not far from where it reached the water-table. This contrasts with evaporation from playas and salt lakes in arid zones, which often are the terminal point of an extensive groundwater system, with a travel time typically of hundreds or thousands of years between points of recharge and discharge.

The discussion so far has assumed that the groundwater system is contained within a uniform climatic zone. Many important systems have their recharge sources in zones of high precipitation, and then move through drier areas. As a result, stream base flow may occur in areas where the local water balance would predict a transmission loss situation. Due to the size of the groundwater system, turnover times are very long, and the groundwater flow is essentially steady. However, it is frequently also saline, if the lithology is such that the groundwater has had the opportunity to dissolve salts in the aquifer material, particularly if that material is of marine origin.

At the other end of the space scale, mention should be made of groundwater systems which occur on small islands, and frequently form the main or only water resource. Such systems usually have a zone of fresh groundwater (the Ghyben-Herzberg lens) overlying seawater, with a transition zone which may vary from less than a meter to over a hundred meters in thickness, depending on the recharge rate, the porosity and the lateral dispersion coefficient. Water in the fresh zone moves vertically into the transition zone and laterally towards the coast, sometimes emerging as springs on the beaches.

Two general types of small island hydrology can be distinguished, relating to flat coral or sand islands on the one hand, and steep islands of volcanic origin on the other. In the first case, there is no surface runoff, and the annual groundwater recharge is equal to the difference between rainfall and evapotranspiration. For the few islands for which data have been published, there is a close relationship between recharge and mean annual rainfall (Chapman 1985). The steep slopes typical of volcanic islands result in some surface runoff discharging to the sea, so that the groundwater recharge is less than would be the case in a flat island under the same precipitation and potential evaporation.

Transport and reaction processes

As this book is concerned with an understanding of hydrology in relation to ecosystems, it is necessary to consider the role of water as a solvent, and as a means of transport of sediment, nutrients and pollutants (Chapman et al 1982). In its role as a solvent, it is often possible to study a particular chemical or biological reaction from the viewpoint of either equilibrium or kinetics, depending on the rate of reaction relative to the turnover time of the system (Pankow and Morgan 1981). For example, many chemical and ion exchange equilibria are reached in periods of less than a second to minutes compared with hydrologic turnover times which generally range from hours to years, while other reactions are so slow as to warrant neglect except over periods of geological time. However, there is a wide range of reactions (including most biological processes) where the reaction rates are comparable with hydrologic turnover times, and these must be considered on the basis of reaction kinetics (Dutt et al 1972).

It is convenient to describe these processes on the same basis as hydrologic processes, starting with those in which vertical transport predominates.

Precipitation. Atmospheric constituents which may be transported to the earth's surface by precipitation occur as trace gases and aerosols, which consist of solid or liquid matter in the form of dispersed particles ranging in size from a few molecules to about $20\mu m$ radius. The incorporation of constituents into cloud droplets, and their removal by precipitation, is called "rainout". The term "washout" (Junge 1963) is used to distinguish the processes of removal by rain below the clouds.

Far from land, the inorganic aerosols consist mainly of sea-salt particles (radius $0.1\mu m$). By contrast, remote continental rain contains a mixture of vegetation and mineral dust components, and a small but variable sea-salt component.

In areas where emissions of particulates and gases cause atmospheric pollution, significant changes in precipitation chemistry are usually observed, with large increases in the concentrations of sulfate and nitrate, as compared with areas unaffected by air pollution. As a result, precipitation is typically acid, and may have an adverse impact on vegetation and bodies of surface water. However, except where tha area of the water body is a large part of the contributing catchment, most of the chemical and biological loading occurs indirectly as a component of catchment runoff, rather than directly from precipitation.

Interception. Significant changes in water quality occur in the interception process. Leaf surfaces may absorb some nutrients contained in the rainfall, particularly NH_3 and SO_4 when they occur in high concentrations, as in urban areas (Tamm 1958), but the most important processses are the washing off and leaching of materials from the leaves (Ovington 1968). As a result of these processes, the concentrations of most constituents are higher in throughfall than in the source rainfall, and are highest in the early stages of an event. Stem flows are enriched even more than leaf drip, partly due to the different nature of the surface and partly to the longer turnover time. The quality of stemflow water has a considerable impact on the soil around trees, with high concentrations near their base and decreasing progressively outwards.

While washing and leaching are the main processes for water quality transformations in canopy interception, the main process in litter interception is biodegradation, with subsequent leaching. These processes usually result in a return to the soil water of a large part of the nutrients that have been extracted earlier by the vegetation (Ovington 1968). These nutrient cycling processes become more complex in hilly country, where there may be significant downslope tranport of nutrient-rich water through the litter layer (Chorley 1978).

Water in plants. The flow of water through plants is a major pathway involving water quality transformations. The volume of water stored in plants ranges from an equivalent depth of about 5 *mm* for grasslands to 50 *mm* in forests (Miller 1977). Turnover times range from a few hours for grasses and crops with ample supplies of soil water, to several days for tropical forests and probably much longer times for arid zone vegetation.

The water in plants is the means of internal transport of minerals and photosynthetic assimilates. While the main pathways are an upward movement of water and absorbed mineral elements in the xylem and a downward movement of photosynthetic assimilates in the phloem, both are subject to recirculation and redistribution, and excess minerals may be transported back down to the roots (Richardson 1968).

Plants also provide a medium for exchange of nitrogen, mainly as NH_3, with the atmosphere. Measurements over a corn crop (Denmead et al 1978) showed a continual interchange of NH_3 between crop and atmosphere, with ammonia absorption by the plant-soil system predominating when the soil surface was dry, but sustained losses to the atmosphere occurring when it was moist. Atmospheric transfer is therefore a significant process in redistributing nitrogen over the earth's surface.

Water in the plant root zone. Water in soil may undergo significant quality changes by a great variety of physical, chemical and biological processes. These processes are related in a very complex way to the climatic and hydrological characteristics of the area, the geological history of the underlying rock formations, and the hydraulic characteristics of the soil rock system, and are further affected by land use and management.

Because plant root extraction excludes the majority of salts (other than nutrients necessary to the plant) and evaporation of water from bare soil excludes all non-volatile constituents, the effect of transpiration and evaporation is to increase the total salt content in the soil water and hence in any drainage water which may percolate downwards.

Superimposed on physical concentration changes are a variety of other physico-chemical changes brought about because the constituent composition of the input water is seldom in equilibrium with the mineral composition of the soil. Thus many reactions between the water and soil minerals may occur in order to restore equilibrium, with the result that constituent ion concentration tends to increase and the ion species distribution may change - processes which contribute in the longer term to weathering and soil formation (Van Breemen and Brinkman 1976).

Of major significance is the presence of carbon dioxide at considerably higher concentration than in the atmosphere. Since this results mainly from root respiration and the activity of soil micro-organisms on organic matter, it is evident that biological processes contribute significantly to the overall equilibrium in the soil.

63

Soil water below the plant root zone. Below the plant root zone, the absence of significant biological activity and the usually minor role of adsorption-desorption processes result in effective predictions of water quality by relatively simple models which assume no interaction between salinity and soil. Such models are also applicable to upward transport of salts by evaporation from a shallow water-table, provided the flow is in the liquid phase; if a transition from liquid to vapor flow occurs in the soil, there will be a slow continuing removal of water and hence a gradual increase in salt concentration in the liquid flow zone.

Again we now turn to those hydrological storages in which horizontal flow predominates.

Water in depressions. The role of depression storage in relation to water quality is mainly related to entrainment and deposition of sediment, and therefore is most significant where the land surface contains appreciable proportions of bare ground. In the absence of vegetative cover, the kinetic energy released by raindrops on impact results in the entrainment of soil particles into the surface water flow; other particles may be entrained in areas of local high velocity. As depressions provide areas of low velocity which can become preferred sites for deposition of these particles, there will be a tendency towards reduction of depression storage, and a sorting of particle sizes according to the microtopography. The reduction with time of artificially formed depression storage, such as caused by various forms of land treatment, has been documented in a number of studies (e.g. Bruce et al 1968), but for a given combination of slope, surface soil and rainfall regime there appears to be a minimum depression storage which is self-sustaining.

In vegetated areas, depressions form preferred sites for lodgement of litter, whether air-blown or transported by surface flow, so that the effects of litter storage on water quality described earlier are accentuated at depressions in the microtopography.

Overland flow. The typical temporal pattern of water quality in overland flow is for high initial concentrations, particularly of organic material, in an event. This has been associated with splash erosion by raindrop impact before a protective layer of surface detention has developed (Smith and Wischmeier 1972), but at least equal importance must be attached to between-storm processes for providing a supply of readily transportable material, such as the breakdown of vegetable litter, surface traffic, soil wetting and drying, and weathering. In some situations, high sediment concentrations may be associated with rain falling on soil which has been wetted in an earlier storm.
The entrainment of soil and other materials in overland flow also has an effect on its

64

chemical quality, but as soluble rock material requires at least days, rather than minutes, to come into equilibrium with water, the amount dissolved in the overland flow process is usually minor. However, sediment transported to a stream in this way will continue to dissolve and increase the solute concentration as it moves downstream.

The water quality processes involved in sheet flow over artificial surfaces such as roofs, roads, runways and parking lots are essentially a washing-off of surface material, and its transport and possible solution in the surface runoff. On roofs the main source of material is atmospheric contaminants which have accumulated by sedimentation or impaction, but on areas subject to traffic there is a comparatively high rate of accumulation of degraded material from the surface itself and from motor vehicles (leakages of fuels, lubricants and hydraulic fluids) and windblown litter. The rapid transport of this material leads to high constituent levels at the start of runoff, with generally much lower levels later in the event.

Stream flow. According to purpose, emphasis in describing stream water quality is often placed on one of three groups of constituents, dissolved salts, nutrients and sediment. These groups are usually inter-dependent to some extent, but their relative contribution to the total constituent load may vary over a wide range, depending typically on the magnitude of the flow relative to its long-term mean.

The main sources of dissolved salts in streams are solutes in inflow from groundwater, constituents in rainfall and material dissolved in the surface runoff process before it reaches the stream, and partial solution of sediment as it is transported dcwnstream. One-dimensional approaches (Chapman 1982) have been used to model nonconservative reactants subject to such processes as precipitation and sedimentation and adsorption on to stationary reactive surfaces.

The process of sediment transport introduces complexities due to interactions between the flow and the mobile stream boundaries, particularly the stream bed. In addition, there is a continuous exchange of particles between the two modes of transport, bed load and suspended load.

Of the nutrients, phosphorus transport in streams has been most extensively studied, because of its worldwide implication in the process of eutrophication (Barabas 1981).

Groundwater. The range in turnover time of groundwater systems with different horizontal extent is associated with a corresponding range in water quality, as a result of the opportunities for solution of minerals in the soil-rock material. Another source of groundwater salinity is in aquifers which have formed in the sea bed.

Rising water-tables, due to irrigation or other changes in land use, may bring these saline waters into the plant root zone or the surface drainage system.

Much of the attention now given to groundwater quality relates to the load of nutrients from intensive farming and chemical wastes from industrial processes and landfill sites, which may have been moving towards or entering the groundwater body for some years, but often have not created an immediate problem because of the long turnover time involved. Water quality in unconfined aquifers, in particular, is strongly influenced by land use and geohydrological factors (Burden 1982).

While biological activity is generally at a low level in groundwater, it may locally be at levels or take forms which are of immediate concern. A large proportion of waterborne disease outbreaks are due to contaminated groundwater.

3. Classification of regions

Problems of hydrological regionalization

The discussion of hydrological processes has brought out the complexity of the environmental factors which determine the hydrological regime in a particular catchment. To establish a framework for a comparative hydrology of the world, we must now attempt to derive a broad scheme of regionalization which will take into account at least some of these factors. It is clearly impractical to take account of more than a few factors of the climate and land surface-subsurface characteristics, and the discussion has shown that the important factors change with location and time. What can be achieved will therefore be little more than a general indication of regional hydrological characteristics, with dramatic exceptions to be expected in particular areas.

A second requirement of any scheme for regionalization is that the necessary data must be available on a uniform basis for the whole world.

The ideal scheme for hydrological regionalization would be based on an ecological classification, since it is the natural ecology of an area which most effectively integrates the different features of the hydrological regime. Such a classification has been developed as a basis for study of the hydrology of smooth flatlands in Australia (Australian Water Resources Council 1972; Chapman 1984). However, the integrated land system mapping (Christian 1958) on which it was based is not generally available elsewhere, and the approach would not be feasible in the large part of the world where the natural ecology has been altered by human activities.

Eagleson (1982) has developed a hypothesis for short- and long-term ecological optimality in water-limited natural soil-vegetation systems, in which the peak vegetation canopy density and plant water use coefficient are related to the climate characteristics and the effective porosity of the soil. Based as it is on a statistical-dynamic model of the average annual water balance, this approach may lead to the type of hydrological regionalization which is being sought here. However, more verification and further development, particularly for arid climates, is required before such an application could be envisaged.

Another recently developed approach (Laut et al 1984) uses numerical taxonomic classification techniques to classify subcatchments from readily available land attributes, of hydrological significance, defined on a *4km* grid. The degree of similarity between catchments can then be determined by their relative location on the axes of a principal coordinate analysis. In its present form, the technique is too data intensive for application on a regional scale, but the tool of numerical taxonomy may well prove useful in developing new appraches to broad-scale hydrological regionalization.

In view of the difficulties in applying the above approaches, it appears that any currently feasible regionalization must rely on overlaying two classifications (Kovacs 1984 b), one based on climate and the other on the nature of the land surface and subsurface. Such overlaying techniques are of course notorious for defining many regions of relatively small extent, the occurrence or existence of which is highly sensitive to changes in the category limits of each classification.

Climatic classifications

Climatologists and ecologists have developed bioclimatic classifications based on readily available monthly rainfall and temperature data. Such schemes include Koeppen (1936), Thornthwaite & Mather (1957), and many others recently summarized by Mueller-Dombois (1979).

In contrast, schemes by hydrologists have used long-term average values of components of the water balance and/or energy balance (L'vovich 1979; Budyko 1971,1977). According to purpose, the main determinant of the classification is usually the precipitation or the runoff.

For our present purposes, with our strong emphasis on the relations between ecology and hydrology, it has been decided to adopt a primary classification based on annual potential evaporation, divided into three categories as follows:-

< 500*mm*	(cold)
500 - 1000*mm*	(temperate)
> 1000*mm*	(warm and hot)

The choice of these categories is quite arbitrary, and based on subjective judgement as a result of examining continental maps of potential evaporation (Unesco 1978 a). Potential evaporation in these maps has been calculated by the

Budyko method (Budyko 1956). These estimates have the advantage of being readily available in map form. Alternative measures of potential evaporation, such as Penman (1956) or Ferguson (1952), or of wet environment areal evapotranspiration (Priestley & Taylor 1972; Morton 1983) would require extensive data compilation and computation.

It should be noted that the "warm and hot" category defined above encompasses a greater range of potential evaporation (from 1000 to well over 2000 *mm*) than the other two categories, and therefore shows as the predominating area on a world map (Fig.3.1). While this provides a dramatic illustration of the potential problems of transferring temperate zone hydrology to warmer areas, a sub-classification of this category is provided by the 1500 and 2000 *mm* isolines of potential evaporation also shown on the map at the front of this book.

The discussion of hydrological processes has highlighted the distinction between dominating processes in humid and dry areas, as generally indicated by the elevation of the water-table relative to the stream bed. As data are unavailable for this hydrological definition of "dry" against "humid", a surrogate based on climatic data has been adopted. This uses the ratio of annual precipitation to annual potential evaporation as an index of hydrologic aridity.

Two world maps of this index have already been published. The first, in Unesco (1978 a), uses the ratio of average precipitation to potential evaporation calculated by the Budyko method, with the classification determined by an index value of one. The second (Unesco 1979) uses the ratio of average precipitation to potential evaporation calculated by the Penman method, with zones defined by the following index values:

Zone	Range of P/E_o
Hyper-arid	< 0.03
Arid	0.03 - 0.20
Semi-arid	0.20 - 0.50
Sub-humid	0.50 - 0.75

The upper limit of the sub-humid category was judged subjectively to be the best indicator of the distinction between hydrologically dry and humid regions (Fig.3.2), and has been used in preparing the world map for this publication.

The main shortcoming of this scheme of climatic regionalization is that it takes no account of seasonal variations of precipitation or potential evaporation, nor of variability from year to year. Both these features are critical in determining the

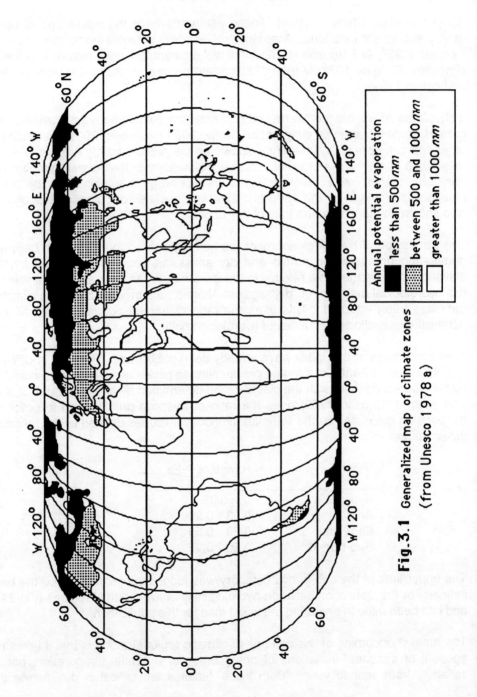

Fig. 3.1 Generalized map of climate zones (from Unesco 1978 a)

Annual potential evaporation
- less than 500 *mm*
- between 500 and 1000 *mm*
- greater than 1000 *mm*

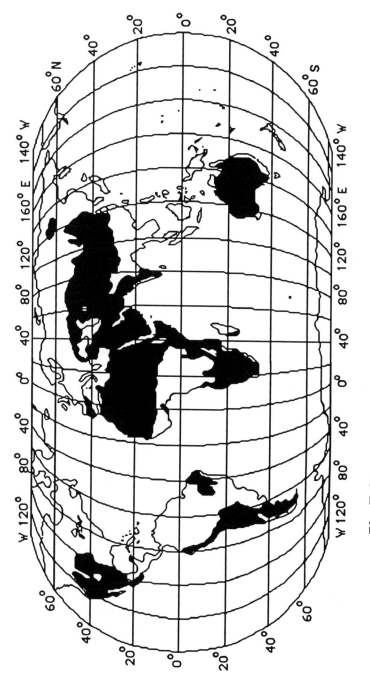

Fig. 3.2 Generalized map of dry regions (from Unesco 1979)

hydrological regime. As the classification already developed has six categories (three of potential evaporation by two of dry/humid), it would not be feasible to introduce additional variables of seasonality or variability into the system. However, in order to indicate some typical climatic regimes, small graphs are displayed with the foldout map, for some twenty representative locations selected by WMO. These graphs show monthly and annual values of the average potential evaporation and the median and upper and lower deciles of the precipitation. Similar graphs for selected locations will be found in some of the chapters on the main hydrological environments.

As noted earlier, the form of precipitation (as rain or snow) has a quantum effect on the nature and rate of hydrologic and materials transport processes. In about half the world's area, the precipitation occurs in one form only. In the remaining area, there is a range from general snowfall with rain only in spring and summer, to general rainfall with an occasional snowfall in winter. A generalized map of snow cover, such as that displayed by Hall and Martinec (1985), is therefore a useful adjunct to the primary climatic classification described above and used as the basis for the organization of this book. This map shows four categories of snow cover:

1. Permanent cover of snow and ice
2. Stable snow cover of varying duration every year
3. Snow cover forms almost every year but is not stable
4. No snow cover

Land parameter classifications

The discussion on hydrological processes has shown that hydrologically significant land parameters include measures of slope, vegetation canopy density, and root depth, and soil hydraulic characteristics. Techniques aiming at an integrated classification of such characteristics were described by Christian (1958), and a scheme for hydrological purposes was developed for the Australian Representative Basins Program (Australian Water Resources Council 1969). In this scheme, six relief categories were mapped for the whole continent, and overlaid with five broad categories of rock types, differentiated on the basis of assumed hydrological significance.

Data are not available on a uniform basis for extension of such schemes to a world map. After consideration of a number of possibilities for developing a classification on the basis of readily available topographic data, it was finally decided to use the nature of the surface runoff system as an expression of the hydrologic regime.

The two categories in this classification are "areas with catchment response", broadly defined as areas with an organized natural drainage network, and "flatlands", defined as areas with a less organized natural drainage network (except for that which may be determined by exogenous factors). These categories are broadly related to land slope, but the sub-division is based on the hydrology rather than surface morphology. In this way a dunefield is properly classified as a flatland from the hydrological viewpoint.

A world map by De Martonne (1927) shows areas of uncoordinated drainage, but does not represent the flatland concept envisaged here. There is a current IHP project for production of a world map of hydrological flatlands. However, in many parts of the world there is a mosaic pattern of flatlands and areas of catchment response, and Colombani (see Chapter 17) suggests that a scale of 1:200 000 would be required to identify significant areas. No attempt has therefore been made in this book to show a map of flatlands, but the larger regions are of course identified in the appropriate chapters on the hydrological environments.

Endogenous/exogenous classifications

One of the recurring themes in the chapters on the hydrological regions is the importance of distinguishing between areas where the hydrology is influenced only by inputs within the area (endogenous conditions), and those where the regime depends on external inputs (exogenous conditions). It should be noted that this definition depends on the size of the area under consideration, as is illustrated by the differences in usage between Chapter 15 on dry temperate flatlands and Chapter 17 on dry warm flatlands.

In Chapter 15, endogenous conditions are seen as occurring within a repetitive landscape with microdepressions at a scale of tens of meters. Exogenous inputs are surface runoff from adjacent sloping land within the same climatic zone.

In Chapter 17 the scale is larger. Endogenous conditions are seen as occurring within a mosaic of adjacent areas of flatland and sloping land, at a scale of a few kilometers. Exogenous inputs to this mosaic are river flows from more humid areas which may be hundreds of kilometers away.

Yet another definition of the area under consideration is used in Chapter 1, where it is taken as the national boundary of a country, which may receive exogenous inputs by way of river (and/or groundwater) flows from adjacent countries.

In spite of these differences, the concept of distinguishing between endogenous and exogenous conditions is critical to understanding the spatial interrelationships between the hydrology and available water resources of different areas.

Map projection

In order to give an unbiased picture of the relative extent of the different hydrological regions, an equal area projection has been used for the world maps shown at the front of this book and as text figures. The projection selected for economy of page size is the Eckert IV projection. The central meridian has been selected as 40°E, which has been used in previous Unesco publications (Unesco 1978 a,b) and results in relatively small land areas at the margins.

4. Measurement and estimation of hydrological processes

Introduction

One of the main tasks of hydrologists has traditionally been to provide information on hydrological conditions for the planning, design, construction and operation of hydraulic structures and water resource systems. The users require this information in an integrated form, aggregated into relatively few numbers instead of long series of observations distributed areally over the catchment. The type of information depends on the purpose for which the data are to be used. Planning and design usually need a description of the hydrological conditions which may occur within the area independently of the time of occurrence (design data like mean and standard deviation of hydrological variables, extreme values expected with a given probability, etc). For construction and operation, the present state of the hydrological systems and the changes which may develop in the near future should be known (hydrological forecasting).

Numerous hydrological models have been developed to use the available data to provide the integrated information required by the users. The easiest task is perhaps the derivation of design data from long records. Statistical models are used to calculate mean and standard deviation, trend and periodicity as well as to fit probability distribution curves to the empirical distribution constructed from recorded point values. More versatile methods have to be applied when design data have to be transferred from the observation station to other locations. Such transfer is needed when areal averages are computed from point values or when information is required at ungaged locations. In the first case empirical relations may be applied to increase the accuracy of interpolation, while models based either on hydrological similarity or on relationships between recorded hydro- meteorological variables and hydrological data may assist the generation of hydrological data series at a location on a river. Finally, the most sophisticated models are used to forecast the hydrological conditions at a given time in the future. In contrast to the models used to determine design data, the forecast does not necessarily require long recorded data series, as the methods are based usually on the analysis of responses of the catchment to various previously observed events.

There are several factors influencing not only the quantification of model parameters but also the structures of the models themselves. This variability of models should be considered even when the investigation is limited to the determination of design data and the more versatile methods of hydrological forecasting are not included. Perhaps the five most important and closely interrelated factors having a decisive role in the selection of suitable models are the following:

- the type of design data to be determined;
- the appropriate time and space scales;
- human activities modifying the natural hydrological cycle and influencing the hydrological variables;
- socio-economic constraints influenced by hydrological conditions and having feedbacks on the analysis of hydrological processes;
- the ratio of the water balance elements within the region.

The first two of these factors are discussed in this chapter to give guidelines for the quantification of hydrological variables.

Types of hydrological variables

In the characterization of hydrological processes, the hydrological cycle is usually simulated with a series of interconnected storages (see Fig.2.2,p.45). On this basis the hydrological variables may characterize

- changes in the volume of water in the storages
- the transport of water inside one of the storages
- water exchange between neighboring storages

Hence the purpose of hydrological observations is either to estimate the amount of water held in the various storages at a given point in time or to measure the mass transported through a selected section or surface during the investigated time interval. Although only water transport was mentioned in the previous grouping, the processes have to be analyzed in a broader aspect, since hydrology investigates not only the continuously renewing movement of water along the hydrological cycle but also its interactions with the environment through which this movement takes place. The hydrological variables have to describe not only the transport of water in different phases (water, vapor or ice), but also the mass of solid materials carried by the water either as solid particles (suspended and bed load) or in dissolved form (dissolved solids and their various organic or inorganic components).

Characterization of storages in the hydrological cycle

In water balance studies efforts are usually made to select a long period, because the storage terms of the balance equations may become negligible if the difference between the amount of water stored in the system at the beginning and the end of the period is small relative to the integrated values of the other watar balance components. There are, however, several situations in which short period balances are required. For such studies the amount of water stored in the various reservoirs and its change in time must be known. Perhaps the most characteristic examples are the calculation of precipitation reaching the ground surface, which requires knowledge of the amount of water stored in interception, and the design of reservoirs, where the transformation of flood waves depends basically on the change in storage.

There are situations where the measurement of storage is considerably easier than the direct measurement of the movement of water between storages. In such cases the change in the stored amount of water is determined in order to quantify the transport processes. Thus, the storage in either artificial (pans and lysimeters) or natural (soil water and groundwater) storages is measured instead of the vapor flux to estimate evaporation. The inflow to lakes either from local surface runoff or groundwater cannot be measured, but it is estimated usually from the change of the amount of water stored in the lake.

The methods suitable for the determination of the amount of water in various storages depend mostly on the character of the systems investigated. A distinction should be made according to whether the water completely saturates the storage or the system is unsaturated.

The most simple examples for the first group are rivers and lakes, where observation of the water level provides sufficient information to calculate the storage provided the geometry of the bed is known. In saturated aquifers the conditions are similar, but it should also be taken into account that a large part of the volume is occupied by the solid matrix and only the pores are available for storage, so that the change in storage ΔV is characterised by a product. For an unconfined aquifer,

$$\Delta V = n_s \, \Delta h$$

$$= (n - W_o) \, \Delta h \qquad (4.1)$$

where Δh is the change in the height of the water-table, n_s is the specific yield, n is the porosity, and W_o is the water retention capacity, which depends mainly on the grain size of the material (Fig. 4.1). For a confined aquifer,

$$\Delta V = S \, \Delta h \qquad (4.2)$$

where the storage coefficient S expresses the compressibility of both the water and the solid matrix, and is determined either by pumping tests or from the relationship between the void ratio and effective stress in the material.

Characterization of the storage in unsaturated systems is more complicated. In the soil moisture zone the degree of saturation is expressed usually as the ratio of the volumetric water content to the porosity. Measurement of the profile of water content between the surface and the water-table is required for the determination of the stored volume of water. The changes in storage can be calculated from the difference between the water content profiles determined on two occasions (Fig.4.2).

The unsaturated condition of the atmosphere is usually characterized by either the relative humidity (vapor content relative to that which would saturate the air at the some temperature) or the vapor pressure deficit (differences between saturated and actual vapor pressure). Simultaneous measurement of dry and wet bulb temperatures is used to determine the actual and saturated vapor pressure.

The water stored in the solid phase, specially in the form of snow, is also an unsaturated system. The thickness of the snow cover is multiplied by the density of the snow to get the water equivalent stored in the layer. The calculation is made more complicated by the fact that the density varies in space and time. Fig 4.3 shows an empirical frequency distribution of snow density constructed from the

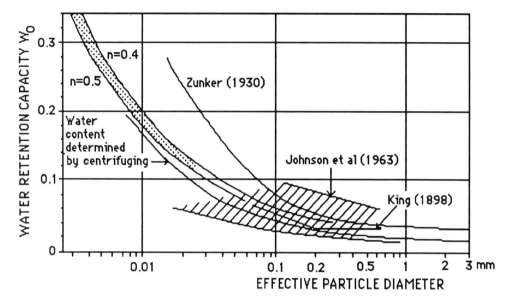

Fig.4.1 Some relations between water retention capacity
and effective particle diameter of a soil.

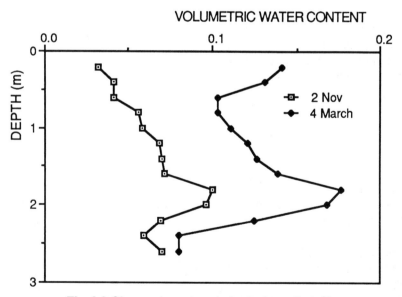

Fig.4.2 Change in water content of a soil profile
at the Poona research station, India.

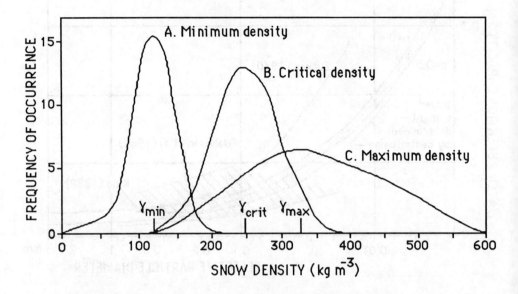

Fig.4.3 Frequency distributions of snow density in Hungary.

10-year-long records of the Hungarian snow-observation network of 74 stations (Kovacs and Molnar 1973). Curve A represents fresh snow, the density of which depends on meteorological conditions (temperature, vapor content, wind velocity) prevailing at the time of snowfall. In spite of these influences, the coefficient of variation (CV) of the density ρ_{min} is only 0.237 (ρ_{min}= 0.118 ± 0.028 *kg m⁻3*).

The density increases gradually due to the compaction of snow (and partly because melting water is also stored in the pores in this period). A critical condition (curve B) develops just before the beginning of the outflow of melted water from the snow cover. The density is doubled (it also depends on the number of layers within the snow cover) but the increase of variability is negligible (ρ_{crit} = 0.245 ± 0.060 ; CV = 0.255). The relationship between ρ_{crit} and the number of layers N can be estimated by ρ_{crit} = 0.153 + 0.050N ± 0.025 with a coefficient of correlation r = 0.906. The density increases further after the outflow of melting water, but the layer thickness decreases more rapidly, resulting in the gradual decrease of the volume of stored water. The maximum density observed in this period (curve C) shows considerable variability (ρ_{max} = 0.331 ± 0.137; CV = 0.414).

Horizontal and vertical water transport

As mentioned in Chapter 2, hydrological transport processes can be divided into two groups acording to the main direction of flow. As the most important storages (the atmosphere, land surface, soil water, and groundwater) are located above each other and their horizontal extension is much larger than their vertical size, the transport processes within such storages may be usually characterized as horizontal flow, while the water exchange between the storages is vertical flow.

This distinction between horizontal and vertical transport has an important practical aspect: horizontal flows are quantified generally by calculating the discharge (i.e the volume of water per unit time, crossing a section perpendicular to the main flow direction; its dimension is L^3T^{-1}), while in most cases the vertical transport is described by the average flux (i.e. the volume crossing unit area of the interface between two storages in unit time: the dimension is LT^{-1}). In some cases the cross-section of the horizontal flow is well defined (e.g. in rivers) but in other cases it can only be estimated, either from the few available data (e.g. geological sections of aquifers) or on the basis of theoretical assumptions (e.g. the section of the atmosphere through which vapor transport develops). The cross-section of the vertical transport is the interface between neighboring storages. These sections are usually large and local conditions may vary considerably, so that the realistic approach for quantifying vertical water transport is to calculate the average flux through the interface instead of the discharge.

There is also a basic difference between the interpretation of the discharge data of horizontal flow and the flux of vertical transport.

In horizontal flow the discharge value at a given section represents the integrated actions influencing the development of the hydrological process in question within the entire flow domain lying upstream of the section. This flow domain may be a well defined catchment (as in the case of overland or stream flow, independently of whether the material transported is water, solid particles or dissolved solids) or an open system without sharp boundaries, in which case the contributing area can be delineated only by assuming a decrease of the influence of perturbations with increasing distance from the section (as in a groundwater system). There are even cases where the boundaries of such open systems may change in time (e.g. the extent of the domain conveying water vapor to a given section depends on the circulation prevailing in the atmosphere).

Regardless of the character of the upstream domain, it can be stated that the discharge data of horizontal flow do not represent the conditions at a section, but provide average information on the catchment transporting water (or solid particles or dissolved material) to that section. It follows that the representation of such data (or usually their specific values, e.g. discharge data divided by the area of the cachment) on maps and the interpolation of isolines between the points is a misleading concept which is inconsistent with the nature of the data. It also follows that changes in horizontal transport should not be investigated areally, but analyzed along the path of the movement (e.g. the construction of longitudinal sections of rivers showing changes in discharge under different conditions).

In vertical flow, the water exchange between neighboring storages depends mostly on the conditions prevailing along and in the close vicinity of their interface. The flux describing the intensity of the vertical transport is therefore an areally distributed variable. The point values measured at observation stations provide samples of these areally changing variables, and both the mean and the standard deviation of the vertical flux within the study area, and their change in time, should be determined from the samples.

The method of calculation depends basically on whether interpolation can be applied to estimate the local change of the variable between the measuring points. This requires the fulfillment of two conditions:

- the variable investigated should change gradually between the
 measuring points; and
- the observation network should be dense enough to minimize the
 probability of local extreme values occurring between the stations.

The fields characterizing the areal distribution of climatic variables (precipitation, temperature, vapor pressure) usually meet the first condition. The required density of network depends on the character of the process analyzed and on the length of the period used as a time unit in the investigation (as will be discussed in the next subsection). Fig.4.4 is an example of the areal distribution of the field of 12h precipitation, obtained from a relatively dense network of stations.

In the case of vertical transport influenced by conditions changing sharply and step-wise along the interface (e.g. evapotranspiration determined by the vegetation covering the surface) it is impracticable to construct a sufficiently dense network to ensure gradual changes between the observation points. For the characterization of such fields, interpolation cannot be applied and a mosaic-like pattern of areal distribution should be determined instead of the construction of isolines.

To summarize this analysis of the processes conveying water and waterborne

82

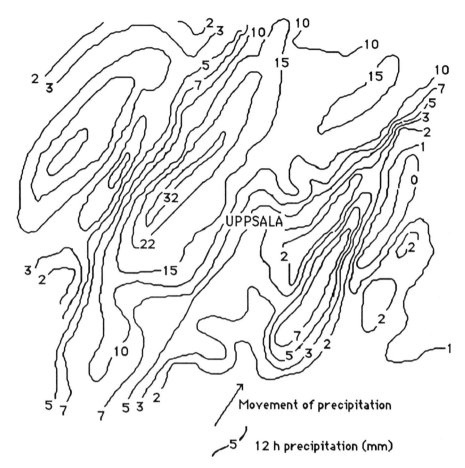

Fig.4.4 Isolines of 12 hour precipitation obtained from a dense observation network near Uppsala, Sweden (from Dahlstrom 1978).

materials, the hydrological variables describing the transport can be divided into two groups each containing two further subgroups (Kovacs 1980):

1. Discharge data give the volume of water (or other materials) conveyed per unit time by horizontal transport processes, through a cross-section perpendicular to the main flow direction. They fall into subgroups according to the nature of the contributing catchment:

(a) Catchments with well defined boundaries (e.g. river catchments);

83

(b) Catchments with unknown boundaries, which may be estimated (as in groundwater systems) or which may change in time (atmospheric vapor transport).

2. Flux data characterize the water exchange between neighboring water horizons, quantifying the vertical water transport per unit time through a unit area of their interface. They fall into subgroups according to the nature of their spatial variability:

(a) Gradual changes between the observed point values which may be represented by isolines (e.g. rainfall, temperature or vapor pressure);

(b) A mosaic-like pattern of areal distribution if the process is influenced by conditions changing sharply and step-wise along the interface (like vegetation and soil type in the case of evapotranspiration).

Variability of transport processes

Since all the forces causing the movement of water in the hydrological cycle are continuously changing in time, from a hydraulic viewpoint the transport processes can be characterized as unsteady flows. The type and magnitude of the variation may however be very different in the various processes. Some of the transport processes are episodic, with long quiescent periods occurring between hydrological "events" (this expression indicates periods of active movement). Precipitation is the best example with rainy periods usually interspersed by relatively long dry spells. In other processes, the movement does not stop between hydrological events, but the rate of transport is lowered considerably. An example is the floods developing in perennial rivers, where the base flow (low water discharge) maintains some limited water transport between the flood waves. Finally, there are systems where the transport is continuous and the time variation of the discharge is almost negligible, as for example in large groundwater systems. The variation of transport processes is influenced by both deterministic and random events. Since the main source of energy in the hydrological cycle is the sun's radiation, diurnal and seasonal fluctuations are the most characteristic deterministic components. Some investigations have indicated a periodicity of some hydrological variables with a longer wave length (11-14 years), but the reason has not yet been explained. A difficulty in recognizing periodicity longer than one year is that any such deterministic effects are not strong and are therefore usually masked by random events. The example in Fig.4.5 indicates that even in this case (which is characterized by an exceptionally high autocorrelation), about 70% of the variability of the yearly average runoff is explained by random events and only 30% can be

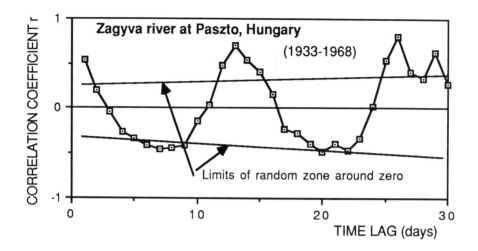

Fig.4.5 Autocorrelation against time lag for annual average discharges of the Zagyva river (from Rudas 1973).

explained by the existence of a 13-14 year period (Rudas 1973).

The year-to-year changes of hydrological variables are caused by the randomness of hydrological events. The calculation of the mean m and standard deviation s of the annual data may give sufficient information on the magnitude of the variability. To separate the deterministic seasonal fluctuation from the random changes it is advisable to carry out the calculation with monthly data as shown in Fig.4.6.

Apart from the regional differences, the figure clearly indicates how the character of hydrological processes influences the variability of data. Precipitation is a naturally episodic transport process, the input of which originates from the water stored in the atmosphere. In the case of pan evaporaation, water is always available for vaporization, and the variation of the process is caused only by the deterministic and random changes in energy available in the system and in vapor transport in the receiving air mass. As a rough approximation it can be assumed that variability is inversely proportional to the turnover time of the storage supplying the transport process in question.

The statistical analysis of hydrological data also depends on whether the purpose of the investigation is the characterization of general hydrological conditions (mean and variability for a long period) or the probability of extreme events (e.g. expected

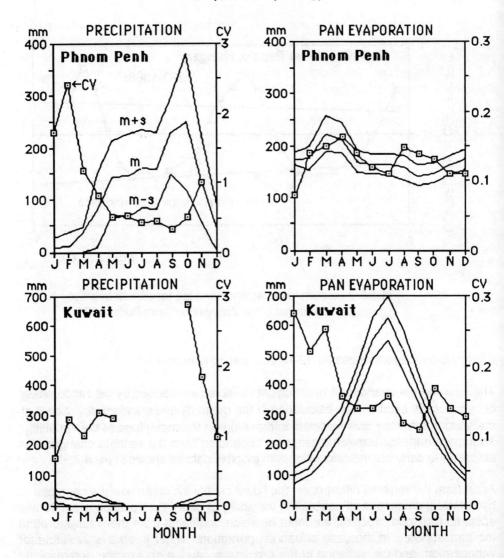

Fig.4.6 Monthly fluctuation of mean and variability of precipitation and pan evaporation in a humid and an arid climate.

flood discharge with a given recurrence interval, probable number of flood peaks within a period of the year, etc.). In the first case, all the recorded data are used. For the investigation of extreme events some representative data are selected from the records and only this limited set of data is analyzed further.

The determination of the probability of floods exceeding a given level (or discharge) is the best example for demonstrating the selection of representative data. Each flood can be sufficiently described by the maximum level developed during the flood (in other cases, the maximum discharge or the total volume of flood can be selected as characteristic data, and sometimes supplementary parameters are used like the time base of the flood or the time needed to reach the peak). The data set composed of the maximum levels of all floods is used to determine the empirical distribution function which may be fitted to the data.

In practice, the number of data selected is further limited to make the computation easier. Either only the yearly maxima are included in the set (annual series), or all the peaks exceeding a threshold level (partial duration series). Although the first method is more widely used, the second gives more representative samples. In the example presented in Fig. 4.7, the first graph shows the frequency distribution constructed from all peaks, and the gamma distribution function fitted to the data. A threshold level of 10.24 m was selected and the truncated empirical distribution was approximated by an exponential function. The probability of high floods calculated by using the two different sets was practically the same, proving the applicability of the more simple approach.

Time and space scales

Hydrological processes are continuous both in time and space. In practice, however, they can be characterized only by discrete samples. Although continuous recording in time can be achieved, such records have to be discretized for data processing and storage. Reliable hydrological information can be provided only if data are collected with sufficient density in both time and space, taking into account also the interrelation between the two scales.

The required sampling density depends on the practical purpose of the observations and on the type of hydrological processes to be characterized. The method of data utilization can be considered, however, only in the case of special purpose networks.

In country-wide networks the user's needs cannot all be determined, because providing hydrological data is an infrastructural service, which should give information for any type of water management activity (planning, design, construction and operation). The basic principle for determining the sampling density in such networks is that the samples should be dense enough to facilitate the reconstruction of the observed process with sufficient accuracy (Kovacs 1986a).

87

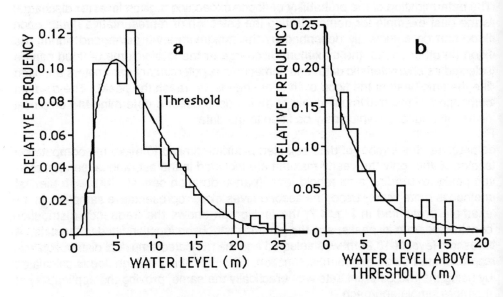

Fig.4.7 Peak water level data for the Narmada river at Garndeshwar, India.
  a. Frequency distribution for all peaks, with fitted gamma distribution;
  b. Frequency distribution for peaks above 10.24*m*, with fitted
    exponential distribution.

To demonstrate the interrelationship between the scales used for discretizing the continuous hydrological processes in time and space, four cases have been selected, corresponding to the four main types of transport processes.

Location and observation frequency of river gages.   The discharge data at a river section provide integrated information on the hydrological events developing in the catchment.   The investigation of their variability poses a two-dimensional problem (changes occur in time and along the course of the river). An ideal spacing would require a gaging station at each section where step changes in quantity or quality may occur (i.e. at tributary junctions and effluent release points).   To follow this principle would result in an uneconomical and unmanageable number of stations.   It is usual therefore to consider locating gaging stations only at those sections where the change of the hydrological variables (discharge, suspended or dissolved load) exceeds a threshold value (e.g. 10%).

Even this restricted principle can be followed only in the case of large rivers (catchments > 5000-6000km^2).   Below this limit the number of significant confluences increases rapidly. In the range of medium rivers, efforts are made to regionalize the hydrological variables. At least 10-15 gaging stations with long records are needed in each medium-size basin to facilitate the analysis of both the dependence of the variables on the size of the catchment and the areal variability of the hydrological conditions. This concept indicates not only a lower limit for medium rivers (for which the regionalization of variables is an acceptable approximation), but draws attention also to the interrelated character of time and space scales.

Fig.4.8 shows a relationship between catchment area and relative specific runoff, defined as the ratio of the specific mean annual discharge (Q/A) of a subcatchment to that of the whole basin. The data clearly demonstrate the increase of areal variability with decreasing area (Kovacs 1984 c). In Table 4.1 this variability is quantified for different ranges of catchment area.

Taking 10% uncertainty as a limit for hydrological variables, it can be concluded that the regionalization of mean annual discharge can be applied within the investigated region only for catchments larger than 400-500 km^2 .

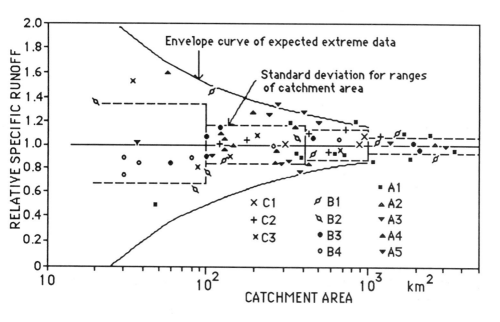

Fig.4.8 Relative specific runoff for 74 catchments in 12 basins A1 to C3 in Transdanubia, Hungary (from Kovacs 1984 c).

Table 4.1

COEFFICIENT OF VARIATION OF RELATIVE SPECIFIC RUNOFF IN HUNGARY
FOR DIFFERENT RANGES OF CATCHMENT AREA (after Kovacs 1984 c)

Area (km^2)	CV
> 1000	0.07
400 - 1000	0.11
100 - 400	0.16
20 - 100	0.33

It is obvious that even greater areal variability should be expected when the hydrological variable characterizes extreme events instead of the multiannual average which was analyzed in this example, because the range of scattering increases with decreases in the time interval used to describe the process in question. There are further problems when hydrological variables with an episodic character (e.g. flood peaks) have to be regionalized.

After selecting the location of river gages, three different methods can be applied to calculate the sampling density in time required for the reconstruction of the actual hydrograph:

(i) Nyquist's critical sampling interval is determined by replacing the actual hydrograph with a series of harmonic functions (Nyquist 1924). The Nyquist sampling interval Δt_N required for the accurate reconstruction of the hydrograph is given by

$$\Delta t_N = 1 / 2 f_o \qquad (4.3)$$

where f_o is the frequency of the harmonic function with the shortest wavelength in the series.

(ii) Another method is the investigation of the loss of information due to decreasing sampling density. Applying this method to calculation of the mean from discharge data, Szollosi-Nagy (1976) found that the loss is relatively small and almost constant below a limiting value of the sampling interval, but increases gradually when this interval exceeds seven times the Nyquist value (Fig. 4.9).

(iii) The third approach analyzes the differences between the recorded peaks of a hydrograph and the values calculated for the time of the peak from the polygon interconnecting points on the hydrograph at fixed sampling intervals. The results of this investigation have proved that the error caused by approximating the hydrograph by its chords (assuming a given sampling interval) decreases with increasing catchment area (Simonffy 1979) up to a certain limit, above which the decrease is negligible and the error has only random fluctuations (Fig. 4.10). When this limiting value is plotted against sampling interval Δt, a relationship can be derived (Fig. 4.11) for a reasonable sampling interval as a function of catchment area.

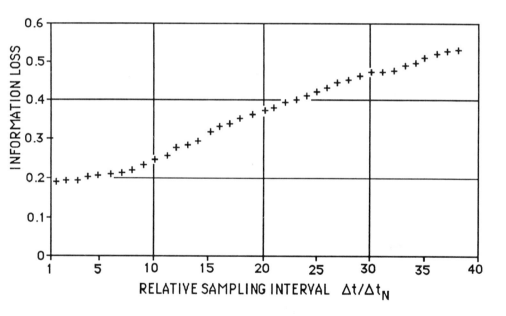

Fig.4.9 Relation between sampling interval and information loss in calculating the mean discharge (from Szollosi-Nagy 1976).

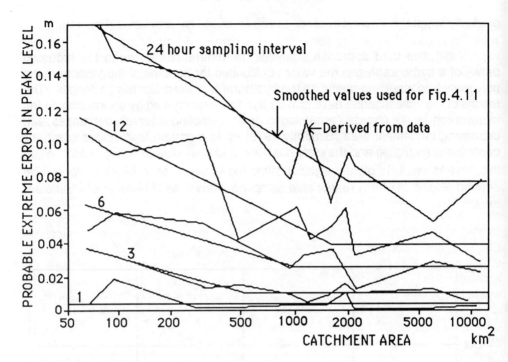

Fig.4.10 Relation between catchment area and error in estimating flood peak levels of rivers in Hungary, with different sampling intervals (from Simonffy 1979).

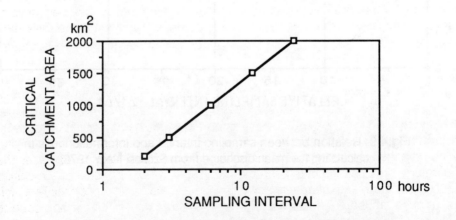

Fig.4.11 Recommended sampling intervals for catchments in Hungary, derived from Fig.4.10.

Areal evaporation estimation. The second large group of hydrological variables is composed of data giving the discharge of horizontal transport processes with a recharge area which is not well defined or changes in time. One of the representative processes belonging to this group is atmospheric vapor transport, which is regularly calculated in meteorology by applying circulation models. The wind speed and vapor content data needed for the models are determined usually by remote sensing, and the results are sometimes used for quantitative forecasts of precipitation.

There are hydrological problems for the solution of which the vapor discharge is used (e.g. the combination of the atmospheric and surface water balances to calculate the actual areal evaporation from large areas). Another example, also related to the determination of actual evaporation, is selected here to demonstrate the uncertainty caused by the undetermined character of the area recharging the vapor transport in the near-surface layer of the atmosphere. The example is the complementary appproach to the interrelation between potential and actual evaporation (Bouchet 1963; Morton 1965, 1983).

It is well known that the potential evaporation (the vapor receiving capacity of the air overlying the evaporating surface) is determined by the energy available for vaporizing the water and by the condition of the interface between the vapor producing and vapor receiving systems. The energy is provided by radiation and by horizontal advection. The importance of the second component was demonstrated by Mukammal and Neumann (1977), who analyzed the multiplying constant α in the Priestley-Taylor equation (Eqn.2.4, p.51). The results (Fig. 4.12) show that the evaporation from a class A type pan, which can be regarded as representative of potential evaporation from a well-defined surface and from which the daily data were used to recalculate the α factor, depends on wind speed and on the soil water content in the field surrounding the pan.

As shown in Chapter 2, the combination equation used to estimate potential evaporation has two terms, the first expressing the influence of direct energy (the radiation term) and the second depending on the wind speed and vapor pressure deficit of the air (the advection term). The actual evaporation from the area lying upwind from the measuring station increases the vapor content and decreases the temperature. The change in temperature and vapor content considerably modifies the advection term, and the temperature change also has an effect on the radiation term. These feedback effects are the basis of the complementary approach to estimation of actual areal evaporation. This approach states that actual evaporation is not proportional to the potential value, as usually assumed in traditional methods, but it influences the potential value calculated from the climatic variables measured

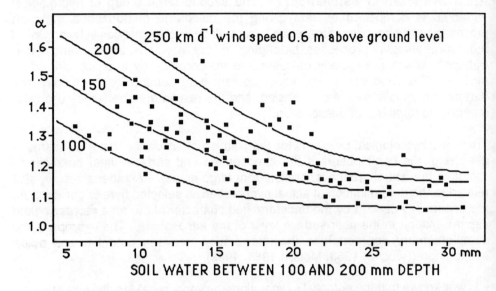

Fig.4.12 The Priestley-Taylor parameter α as a function of wind speed and the soil water content in the surrounding area, for a Class A evaporation pan (from Mukammel and Neumann 1977).

at a meteorological station. Among the several comparative studies demonstrating the new approach, Fig. 4.13 shows the average values of E and E_0 as a function of annual precipitation, which is regarded as a variable characterizing indirectly the availability of water in vapor producing systems, determined for several river basins in Malawi and Puerto Rico (Solomon 1967; Giusti 1978).

The model derived on the basis of the complementary approach expresses actual areal evaporation as a function of the potential value and the wet environment areal evaporation (in some times estimated by the radiation term). For practical application it is necessary to consider the different interpretation of the space scales used for the calculation of the radiation and the advection terms. The radiation term is an average variable (depending also on surface conditions) describing the actual amount of energy available within the whole area, while the advection term gives the integrated effects of actual evaporation from the varying surface of the area at the downwind edge of which the wind speed and vapor content are measured. Hence, the advection term represents the physically averaged impacts from the different parts of the area providing vapor for atmospheric

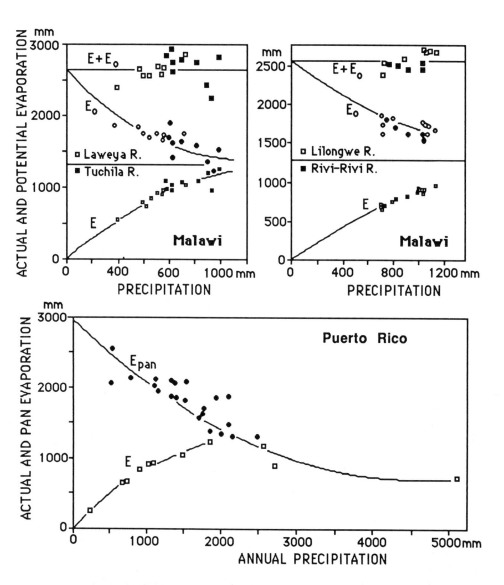

Fig.4.13 Actual annual evaporation (calculated from rainfall and runoff) and potential or pan evaporation, as functions of annual precipitation, for catchments in Malawi (from Solomon 1967) and Puerto Rico (from Giusti 1978).

transport near the surface. One of the most important aspects is, therefore, to delineate the borders of the influencing surface in order to determine the size of the area characterized by the actual evaporation calculated in this way (Kovacs 1987).

Variability of precipitation in time and space. The samples taken in observing vertical transport processes cover only a very small fraction of the continuous field characterizing the hydrological process in question (e.g. the measuring area of a rain gage is 0.02-$0.03 m^2$ and the depth of precipitation measured in this way is often taken to represent several tens of km^2). The samples can be regarded therefore as point values giving only very limited information on the vertical transport. The integrated effect of the areally distributed process (i.e. the whole mass of precipitation on a catchment) cannot be measured directly, although these values are required in practice. Methods are therefore needed to characterize the variability of the field and to estimate the change of the flux between the point values measured at the observation stations. Structural functions have been derived to describe statistically the internal structure of the field, and efforts have been made to improve the interpolation between the point values (Czelnay 1969, 1971; Gandin 1960; Clarke 1977). For recent developments in this field, see Rodriguez-Iturbe (1986).

When the conditions for the applicability of interpolation were discussed, a sufficient density of the observation network was mentioned as one of the requirements. It is necessary, however, to analyze the relationship of time and space scales for continuously distributed vertical transport processes, to give a more precise definition to the term "sufficient density". Precipitation data are used to demonstrate the scale effects on interpolation since the rain is the most important element of the water balance among those which can be characteried by isolines. The first aspect to be considered when the area-depth-duration relationship of precipitation is analyzed is a well-known thesis of meteorology: the lifetime of meteorological events producing precipitation is proportional to their horizontal extent. Hence, the areal distribution of the depth of rain originating from convective storms is more uneven than that from a cyclone or monsoon activity. The internal structure of a field of precipitation of short duration is therefore very uneven, as shown in Fig. 4.4. Reliable estimation of the areal distribution of the depth of rain produced by such events requires dense spacing of rain gages or a combination of field measurements and remote sensing (radar or satellite images).

It is necessary to consider also that the variability of the field of precipitation has a random character and, therefore, the differences between the point values are smoothed out when more than one strongly uneven field, covering the same area

96

but representing consecutive time intervals, are superimposed. Such an integration of point values in time is performed in practice when the depth of precipitation accumulated in longer periods is analyzed. In an analysis of the areally equalizing effect of longer time units in a 20 km^2 experimental area in Transdanubia, Hungary, Starosolszky (1977) concluded that the average precipitation on the area can be approximated by the data of only one rain gage when monthly or longer precipitation is required. It is necessary to recognize that this density of 1 station per 20 km^2 is about 5-10 times higher than the usual spacing of national observation networks. Reliable areal averages cannot be calculated therefore from standard meteorological data for shorter time intervals than one month; perhaps one year is a more acceptable lower limit.

Sampling density for evaporation measurements. The most representative hydrological variable in the last group of transport processes (i.e. vertical transport with a mosaic pattern of areal distribution) is the flux of actual evaporation. Since the amount of water vaporized per unit time from unit area strongly depends on the conditions prevailing on and near the evaporationg surface, the horizontal scale is *a priori* determined by the horizontal distribution of the influencing factors, like the type and development of vegetation, the texture and structure of soils, the availability of water determined by the soil moisture content, and the depth of the water-table. The point values measured at a given place represent only those areas where the conditions are similar, and models have to be developed to transfer the data to other areas. In these models the factors influencing the flux of actual evaporation should be considered as independent variables. Finally, the areal average can be calculated by summing the amount of water removed by evaporation from different areas. It is evident, therefore, that the problem of the temporal sampling density should be analyzed independently from the horizontal units used to calculate the areal sum or average of evapotranspiration.

In selecting the time unit for the determination of evapotranspiration data at a given location, two conflicting interests should be considered:

(i) Correct reconstruction of the variation of evapotranspiration in time would require dense sampling because of the high daily fluctuation of the process.

(ii) The various methods applied to measure the amount of vapor transferred into the atmosphere may provide data with a high random error, when short time intervals are used.

By surveying the complete system of evapotranspiration (which is composed of the

97

vapor-producing and the vapor-receiving subsystems as well as their interface) it becomes obvious that the most decisive factor influencing the probable upper limit of evapotranspiration (i.e. the potential value) is the amount of energy available to transform water into vapor. Further influencing factors are the transport processes in the vapor-receiving air mass and the conditions of the evaporating surface (e.g. roughness, leaf area index, plant resistance).

The amount of water actually removed by evapotranspiration may be limited by the availability of water in the soil-plant systems. Its variation in time is also determined by the daily fluctuations of energy. This is demonstrated in Fig. 4.14, which shows comparisons, for grass under different soil water conditions, of the diurnal variation of the four terms of the energy balance equation

$$R_n - G = H + \lambda E \tag{4.4}$$

where R_n is the net radiation received at the earth's surface, G is the soil heat flux into the ground, H is the sensible heat flux from the surface to the air, E is the evaporation from the surface to the air, and λ is the latent heat of vaporization of water.

It is obvious that at least hourly data are necessary for accurate reconstruction of the variability in time of the evapotranspiration. On the other hand, such frequent sampling would require the measurement of the height or weight of very small water columns. Actual evapotranspiration rarely exceeds $10mm\ d^{-1}$ even in humid tropical regions. The hourly data are, therefore, in the range of $0-0.5mm$. Hence except with precise weighing lysimeters, the random measurement errors may have the same order of magnitude as the value to be measured. This uncertainty can be eliminated by selecting longer time units (i.e. measuring only daily evapotranpiration) or by summing the data from short periods so that the random errors compensate each other within the integrated value.

In some cases the method applied to measure evapotranspiration determines the measurement time unit *a priori*. For example, in applying the heat balance by determining the Bowen ratio

$$\beta = H / \lambda E \tag{4.5}$$

from the gradients of temperature and vapor pressure, a precondition is that changes in the two gradients should be negligible within the measuring period.

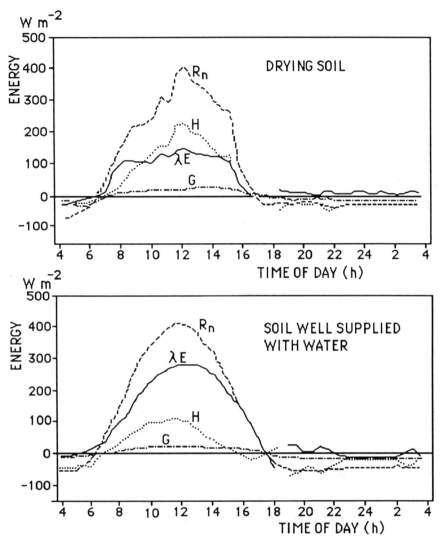

Fig.4.14 Elements of the energy balance over a grass cover in the Netherlands (from De Bruin and Kohsiek 1977).

Therefore, a time unit longer than one hour is not acceptable when this method is used. The measurement of pan evaporation represents an opposing example. Recent research has shown that the different thermal expansion factors of the water and the container may cause a regular error in the measurements when a time unit shorter than one day is selected (Gallo 1985). Fig. 4.15 shows the relationship between the change in temperature and the rise of water level in an INEP pan, obtained from laboratory measurements with the water surface sealed by oil to exclude evaporation. The results indicate that a change of 10º C in temperature causes an error of about 1.0*mm*, which is of the same order of magnitude as the value to be measured if the pan were used for 12-hour measurements. Although the shortest time interval for which reliable evaporation data can be determined by standard equipment is one day, daily values provide sufficient information for most practical purposes. More frequent sampling is required only for special research activities, when more accurate measuring devices have to be applied. Since random events (precipitation, wind, cloud cover, etc.) influence the process of evaporation apart from the availability of energy and water, it should be expected that even daily values will have high uncertainity. It is advisable, therefore, to use even longer sampling periods (5 days, 10 days or one month) whenever the practical application makes this possible.

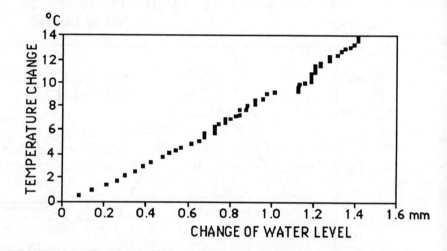

Fig.4.15 Relation between temperature and water level in
an INEP evaporation pan (from Gallo 1985).

Hydrological and socio-economic interrelations

One of the most important tasks of hydrology is to provide information for the rational use and development of water and other natural resources. The evolution of the science and practice of hydrology, including data collection and processing as well as the analysis of water-related processes, has therefore tended to follow the demands of society within a region. It should be emphasized however that hydrological conditions also influence socio-economic development, and therefore there are several feedback links characterizing the interrelations between hydrological conditions and socio-economic conditions.

The development of rice-producing areas in the alluvial plains of warm humid regions can be mentioned as an example of these interactions. The climate and the hydrological conditions are favorable for paddy cultivation, and the high yield provides a strong population-supporting power. That is why these areas are always densely populated. The society which has developed in these regions requires the maintenance and possibly the extension of paddy farming, which causes a demand for the solution of special water management problems (protection of densely populated flood plains against flash floods caused by tropical cyclones; providing water for industry and the community in areas where most of the river flow is reserved for irrigation; maintaining controlled flooding over large areas in accordance with the consecutive phases of paddy cultivation, etc.), and their solution needs special hydrological information. In this way the society, the development of which is influenced by the hydrological conditions of the region, determines the development of hydrological activities, starting from data collection through the analysis of hydrological processes and toward the development of forecasting services.

The transfer of technology has been seen as a tool for decreasing the gap between the socio-economic development in industrialized and developing countries. In the field of hydrology and water management, one important form of technology transfer is the application of hydrological models developed in other parts of the world. There are two serious obstacles to this kind of transfer. One of these is the incompatibility of hydrological conditions, which is being stressed throughout this book, and the other is the lack of adequate data for model calibration and application. In the past, the collection of hydrological data has been responsive to the needs of society, as is demonstrated in the following brief account of the history of hydrological observations.

Historical development of hydrological observations. Regular water-level recording dates back to about 2200 BC when gages were carved into rocks in the Nile valley. The purpose of these gages was the assessment of taxation, which depended on the flood level of the river. Discharge measurement is more complicated because it requires also the determination of velocity. Although orifices of different diameters (so-called callix) were used in the Roman Empire to estimate the amount of water provided for the users, the measurement of velocity in rivers did not start until the 15th century when Leonardo da Vinci used

101

floats. Current meters were developed at the beginning of the 18th century, first to serve the demands of navigation, but the technique provided the basis for the development of discharge records in the 19th century. This change in the character of hydrological records was initiated by socio-economic developments. Previously the dominating branches of water management, navigation and flood control, required only the knowledge of water levels, but discharge data were needed for the assessment of available water resources as tensions developed between resources and demands.

The transformation of water levels into discharge data was perhaps the first hydrological model the application of which required modifications according to the local morphological conditions. It became evident that the normal rating curve relating water level and discharge provides ambiguous results in the case of rivers crossing large plains, because the effect of changes in the slope of the water surface are not negligible below values of about $0.1 \, m \, km^{-1}$.

The next phase of the development of hydrological networks was the construction of observation wells for the assessment of groundwater resources. In this case the differences in the observation facilities and in the density of the network caused by the geological structure of the aquifers (sedimentary basins, karst etc.)are quite obvious. Regular observations of groundwater levels started at the beginning of this century, the extent of the development depending on the importance of groundwater resources to a region's water supply. In the 1920's there were countries where a comprehensive network of observation wells was already in operation. Information collected by the observation wells was presented in the form of hydrological maps. The large amount of data now available and the better knowledge of the regime facilitate the development of regional groundwater models which not only describe the present energy and flow conditions in the aquifers but can also be used to predict changes due to various human activities.

When the pollution of natural water bodies increased rapidly in the middle of this century, and the maintenance of the self-purification ability of rivers became essential to ensure the multiple reuse of water, a new branch developed in the hydrological information system, the collection and evaluation of water quality data. This change caused a basic modification in the structure of data management. Until then only two variables were involved in the system: water level and discharge. In contrast, the characterization of water quality requires the determination of a large number of chemical and biological parameters. The oxygen balance, dissolved chemicals, suspended solids, biological, toxicological and bacteriological characteristics are the main groups into which these parameters can be distributed and each group is composed of several variables. The evaluation of water quality depends also on the requirements of the users, as shown in Table 4.2.

102

Table 4.2

RELATIONS BETWEEN WATER QUALITY AND USE
(after Gras 1985)

Uses	Physico-chemistry				Biology			Hydraulic parameters		
	T°C	O$_2$	N P/Si	Toxics	Bacteri-ology	BOD	Eutrophic-ation	V	H	Q
Navigation	*	*		*		*	*	*	#*	
Irrigation										*
Fishing	#	#	#	#		#		#	#	
Swimming	#	#		#	#*	#				
Canoe, kayak					#			#	#	#
Drinking water	#			#	#					
Pollution carrier	*	*	*	*	*	*				#
Cooling	#*	*								#*
Chemical proc.	#*	*	#*	#*	#*	*				*
Aesthetics						#	#	#	#	#
Change in drainage		*	*	*	*	*	*			*

\# This use needs these qualities
* This use alters these qualities

The present situation. Unfortunately, rapid developments of the information system have widened the gap between hydrologists and policy-making authorities. In ancient Egypt where the basis of taxation was the annual maximum water level in the Nile, the decision makers were able to judge correctly the value of the hydrological information. When a special observation system is constructed to assist a given operation (e.g. to measure the levels in the reservoirs of a water supply scheme for optimization of its management) the benefit-cost ratio can be precisely measured. In the case of a national network however the collection and processing of hydrological data is an infrastructural service intended to provide information for any type of water management activity. The advantages provided by the system cannot be estimated directly in monetary terms, although the lack of information must cause serious deficits in the long term. When economic difficulties occur, the policy makers usually reduce the investment and operational cost of the non-productive infrastructure, because the economic losses due to such a reduction are generally not observable in a short period. This negative socio-economic impact on the operation of hydrological information systems is now observable all over the world. It is specially serious in developing countries where rapid increases of population, poverty, and the instability of policies almost completely exclude the consideration of long-term objectives in the formulation of policy related to infrastructure (Oyebande 1986).

The conditions of different basins cannot be compared without sufficient hydrological information. The maintenance of national networks is therefore a precondition of technological transfer. Falkenmark (1983) made a timely call for a renewed attempt by hydrologists to "develop large efforts in making themselves understood by planners by influencing the way they see water". Due to the interdisciplinary character and the rapidly increasing volume and complexity of hydrological data, the decision makers cannot be expected to understand the details of processes represented by the data, but hydrologists should learn how best to present their results in the form of summarized relevant information for planners and policy-making authorities. The preparation of decision support systems is advisable, therefore, to provide such integrated information on present conditions and the expected changes in the hydrological regime, for the different hierarchical levels of decision-making organisations (Kovacs 1986 b).

5. Human interventions in the terrestrial water cycle

Introduction

The natural environment encompasses the atmosphere, hydrosphere, lithosphere and biosphere. Fig.5.1 visualizes the relationships between these spheres and demonstrates the role of the hydrosphere in transporting the impacts of human activities between environmental media. The biosphere includes man and his society (the socio-economic sphere). The figure clearly shows how the hydrosphere, the large natural hydrological circle which includes the smaller one indicating the movement of water in the socio-economic sphere, weaves through and interconnects the other spheres.

Three phases of development. Although a general distinction may be made between the natural and influenced forms of the water regime, it should be recognized that natural conditions hardly exist on the continents, because any utilization of the land surface by humans influences the elements of the water balance. It is obvious that the size of such changes depends on the level of development in the region in question (Fig.5.2). In the first phase of water resources development, when the available resources are far greater than the demands, small structures (e.g. single wells, slight modifications of river beds, small barrages) causing only local impacts are sufficient to meet the demands of society.

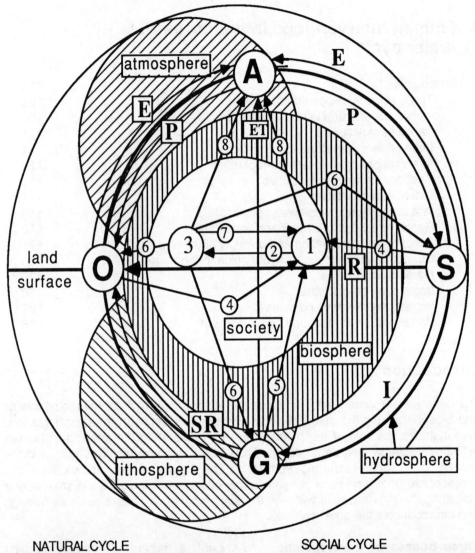

NATURAL CYCLE
A Atmospheric water S Surface water
G Groundwater   O Oceans and seas
P Precipitation   E Evaporation
ET Evapotranspiration
I Infiltration   R Surface runoff

SOCIAL CYCLE
1 Water users 2 Wastewater disposal
3 Wastewater collection and treatment
4 Water intake 5 Groundwater use
6 Effluents   7 Recycling
8 Evaporation

Fig.5.1 Interrelationships between the natural and social hydrological cycles.

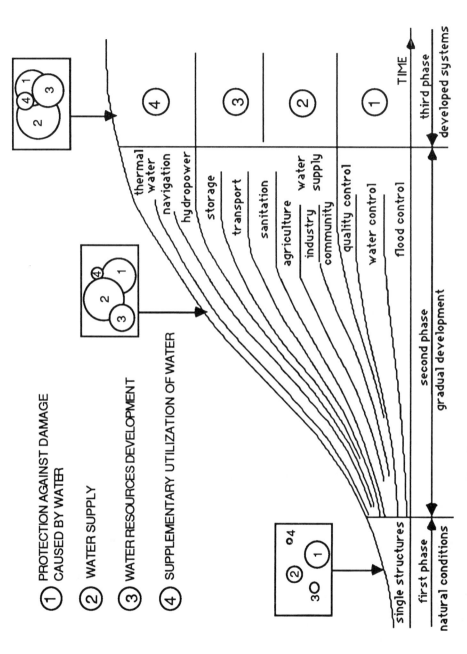

Fig.5.2 The development of water management

The second phase is characterized by the evolution of the various branches of water management (flood control, drainage and irrigation, water supply and sanitary engineering, river training, navigation and utilization of hydropower, water quality control and protection of the environment) when the construction of large systems is required to satisfy the increasing demands. The integration of the systems and their multipurpose utilization become necessary in the third phase, to eliminate undesirable interactions between the systems and to minimize the investment and operational costs (David 1981).

In studying the influence of human activities on the quantification of hydrological variables, the most serious problems to be solved are those related to the modification of land use, because in these cases the hydrological changes are not directly measurable and their development usually requires a long transition period, and therefore the hydrological records may lead to erroneous conclusions if the data relate to such changing conditions. These aspects are specially important in developing countries where both water and land management are in the rapidly changing second phase of development. Before any long records are analyzed it is essential therefore to check the stationarity of the processes and the homogeneity of the sets of data.

Interventions from rural land use activities

Agricultural activities. Intensive agriculture requires the maintenance of optimum conditions in the system composed of soil, water and plants. The soil should have a well defined structure which can store a sufficient amount of water and at the same time have an efficient transport capacity to carry away the excess water from the pores. The nutrients needed for the development of plants should be present in the soil, but the latter must not contain materials toxic for the plants or elements decreasing the stability of the soil structure. Water, which passes through the transpiration pathway soil-roots-stems-leaves, is the main carrier of dissolved nutrients. Lack of water causes the wilting of vegetation, but serious damage may also be caused by waterlogging (except in the case of aquatic plants like rice), because air is also required in the root zone for root respiration and other biological processes. Intensive agriculture requires also the selection of high yielding varieties and the protection of plants against pests, insects and weeds.

Many of these requirements are met by applying modern methods of agronomy and agro-chemistry. Water management has also a very versatile role in intensive agriculture. Its main task is to maintain a sufficient water/air ratio in the pores of the root zone through the control of soil moisture, requiring in most cases the simultaneous application of irrigation and drainage. A further requirement is the

108

protection of soils against erosion, waterlogging and salt accumulation. The flooding of cultivated areas should also be prevented (or in special cases like paddy fields, a continuous but controlled flooding should be maintained).

The soil water zone can be regarded as the heart of the land phase of the hydrological cycle, which distributes the water arriving from the atmosphere between the three main pathways, i.e. surface runoff, groundwater recharge and evapotranspiration. All activities involved in the intensification of agriculture, whether they have an agronomic or a water-managing character, modify the hydrological processes developing in the soil water zone. Hence, changes in the natural vegetation to cultivated land, and later the intensification of agriculture, have a basic influence on both the quantity and quality of water in all pathways of the hydrological cycle. The impacts strongly depend on the structure of soils and on the character of the transport processes between the surface and the water-table.

In a geological formation composed of silt and clay, the fine (mostly colloid) particles are evenly distributed around very small pores. The air entry pressure in these pores is extremely high and therefore the porous medium remains saturated even under the highest suction which can be produced by the roots of plants. This condition would exclude the development of agriculture in heavy soils, because of the lack of aeration in the root zone.

Fortunately, man is assisted by nature. Several bio-physico-chemical effects act near the surface, causing the aggregation of colloid particles. Inside the aggregates formed by the small particles, the primary porosity is the same as that of the original homogeneous material. The aggregates build the layer into larger grains, and the secondary porosity of the pore space between them is greater by some orders of magnitude than the primary porosity. This structure, which is usually characterized by a decreasing number of aggregates with increasing depth, ensures a high storage capacity in the primary pores and a high hydraulic conductivity through the secondary pores. Some physical characteristics of a chernozem soil are shown in Fig. 5.3.

There are areas where the natural aggregation is not sufficiently effective or other processes have destroyed the secondary structure. Here physical techniques (e.g. deep ploughing) and/or chemical treatment (e.g. applying lime to change the Na/Ca ratio) can be applied to increase aggregation. Independent of whether the soil structure is developed artificially or by natural processes, considerable efforts are required in most cases to maintain the large pores in the upper part of the soil profile. Since the two most serious factors causing the disintegration of aggregates are extended saturation of the profile and the accumulation of Na near the surface,

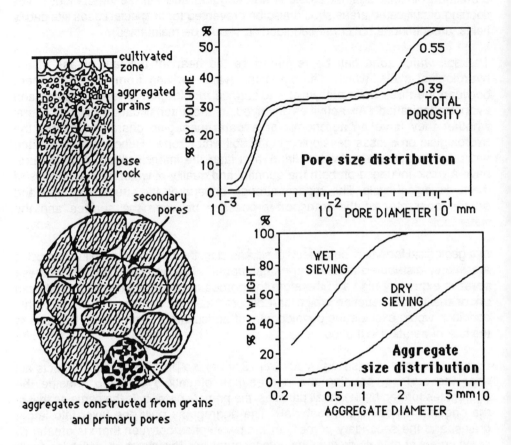

Fig.5.3 Aggregation of fine particles in a chernozem soil developed over a loess plateau near Titel, Yugoslavia (from Vucic 1966).

the drainage of excess water from the soil is the most effective action to protect the desirable structure and maintain sufficient aeration of the root zone.

Both components of the vertical water transport below the plant root zone (i.e. drainage to the water-table and the flux raised from the groundwater by capillarity to replenish evaporated or transpired soil water) decrease with increasing depth of the water-table. The difference between the two components is described by a characteristic curve of the groundwater balance (see Fig.6.3) as a function of the average depth of the water-table (Kovacs 1977). The conclusion of the analysis summarized in the characteristic curve is that the resultant of the vertical movement is directed downwards and the groundwater is recharged above a deep water-table. The excess water is transported laterally by groundwater flow to lower-lying areas, where the water-table is closer to the surface and therefore evaporation exceeds drainage.

An important consequence of this hydrological process is the migration and accumulation of salts in the soil. Precipitation infiltrating through the soil dissolves salt. As a result of the different mobility of various positive ions, Na is leached out first followed by K, Mg and Ca. Leached calcareous brown soils therefore develop first above a deep water-table, and the next stage (if the leaching is intensive) is characterized by the fact that all exchangeable cations are washed down and H+ ions are complexed by the free negative charges of soil particles, resulting in the development of acidic leached soils. Calcium can be found only near the water-table and sodium accumulates in the groundwater, which transports it away.

When evaporation from the water-table exceeds drainage to it, groundwater rich in salts occurs near the surface, and is drained partly by rivers and partly by capillary rise and evaporation. This second process raises both water and dissolved salts from the groundwater, but the evaporated water is free of salts, so that an excess of salts precipitates in the profile. Sodium, being the most soluble, reaches the surface, while calcium is deposited near the water-table. This process of salt accumulation explains the development of alkaline soils.

Interventions generated by irrigation. Although irrigation is only one sector of agricultural water management, its importance in food production and its versatile environmental impacts require the separate analysis of problems occurring in connection with this type of land use.

It is inevitable that agricultural production must be increased considerably to ensure the food supply for the prospective world population. This objective can be achieved by increasing the productivity on land already cultivated and by expanding the area of arable land. Irrigation has a paramount role in both activities. It has been estimated (Worthington 1977) that about 13% of the arable lands of the world are irrigated, using about $1400km^3$ of water per year. In the developing countries the yearly increase of productivity on irrigated land is about 2.9%, while it is only 0.7% in non-irrigated areas. Apart from this very positive primary impact, it is necessary to recognize, however, that the secondary effects of irrigation may be undesired, harmful and irreversible.

To determine the impact of irrigation on the hydrological regime, both the primary and secondary effects involved in modifying the water regime should be

111

considered. These changes can be divided into three groups according to the water horizons where they develop:

(i) changes in the interactions between the atmosphere and land and water surfaces;

(ii) quantitative and qualitative modifications of surface runoff;

(iii) changes in the regimes of soil water and groundwater.

The flux of actual evapotranspiration is raised considerably by irrigation, because its primary purpose is to maintain a relatively high water content in the root zone and prevent subjecting plants to water stress. Hence there is always water available for vaporization from the surface and the actual evapotranspiration approaches the potential value in each season of the year.

Due to the higher evapotranspiration, the vapor content stored in the atmosphere increases. The modification is not considerable, however, compared to the amount of vapor raised continuously from the oceans. This was demonstrated by comparing precipitation data observed within and around a large region of Texas where almost the whole area was irrigated. No significant difference was found, demonstrating that neither the amount nor the pattern of precipitation were modified by irrigation (Schickedanz and Ackermann 1977).

The impacts of irrigation on surface runoff can be observed both over the catchment and in the rivers. The amount and the rate of surface runoff increases due to the higher moisture content in the upper soil layers. The undesired consequences of the higher runoff rates are the increase of erosion potential, which may cause the degradation of soils by carrying away organic materials, and the increase of sediment transport in rivers.

The flow of rivers is also directly modified by irrigation. Part of this change is planned *a priori* when reservoirs are constructed to augment low river flows and to raise the amount of water available for irrigation in dry periods. The other impact is the considerable decrease in river discharge due to the high water consumption of irrigated areas. Irrigation is the type of water use where the difference between tha water intake and the effluent discharge is the highest; the difference may reach 70-80% of the intake, while the value is usually less than 10% in the case of industrial and community water supplies.

Another important change in rivers is caused by the possible increase of sediment load and the decrease in the capacity of sediment transport due to the decrease of discharge. The result of these effects is the rapid increase of siltation and the aggradation of river beds. Even more serious impacts on water quality are caused by the high salt content of effluents from irrigated fields. These changes are demonstrated in Fig.5.4, where characteristic parameters observed along the Rio Grande (USA) are summarized.

The most drastic change in the water regime due to irrigation is caused in the soil water zone, because the water distributed over the surface modifies completely the processes developing above the water-table. Infiltration is increased and the flux percolating downwards changes little with depth, because most of the pores are always saturated by irrigation water and therefore there is no free storage near the surface. Although the total amount of evapotranspiration is considerably higher than in the case of dry farming, any contribution from groundwater to evaporation will decrease because the gradient of capillary tension is lowered due to the

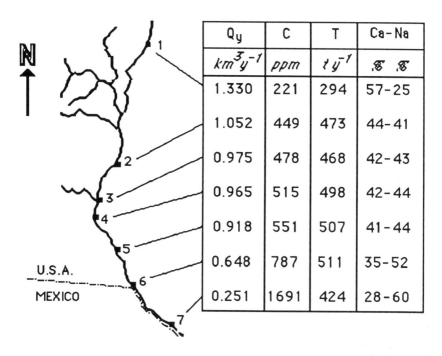

Q_y	C	T	Ca–Na
km^3y^{-1}	ppm	$t\,y^{-1}$	,% ,%
1.330	221	294	57–25
1.052	449	473	44–41
0.975	478	468	42–43
0.965	515	498	42–44
0.918	551	507	41–44
0.648	787	511	35–52
0.251	1691	424	28–60

Fig.5.4 Changes in flow and water quality along the Rio Grande:
Q_y = mean annual discharge; C = concentration of dissolved solids;
T = total dissolved solid transport; Ca-Na = calcium and sodium
as percentages of total cations (from Hotes and Pearson 1977).

113

continuously high moisture content in the upper layers of the soil profile. A new balance characterizes the average hydrological conditions in the soil water zone as modified by irrigation (Peczely 1977). The equilibrium level of the water-table may be raised close to the surface, causing a high probability of waterlogging if the irrigation system is not supplemented with efficient drainage.

The water-table is generally raised below and around irrigated fields due to the increase of infiltration, and this rise initiates or increases horizontal groundwater flow. Where this flow is towards neighboring non-irrigated lands, it may result in the development of areas where the excess water is discharged by capillary rise and evapotranspiration. The irrigated soils are leached, the salts are transported horizontally by groundwater flow, and then accumulate under the dryland areas. The accumulation of a high amount of sodium in the topsoil was reported in the lower- lying areas of the Murray valley (New South Wales) when irrigation was introduced on the higher terrain (Pels and Stannard 1977), and similarly an increase of areas affected by alkalinization was observed in the San Lorenzo irrigation project in Peru (Cornejo et al 1979).

Interventions from forest management. The type of vegetative cover influences considerably the development of the response of catchments to precipitation. The largest differences occur perhaps between forests and cultivated lands, and hence any change due to forest management (deforestation, thinning, clear cutting, reforestation, introduction of exotic species etc) may have considerable impacts on local hydrological phenomena.

The first difference is caused by the greater interception by forest than by grasslands or cultivated crops. The surface litter protects the soil against the splashing effects of raindrops, and the effect of the surface mulch of decomposing vegetation is to increase infiltration relative to surface runoff. Evapotranspiration is also increased, not only because of the direct re-evaporation of intercepted water from the canopy, but also because of the higher consumptive use of trees and the greater amount of water available in the root zone, which is much deeper than that of crops or pastures. The impact of the development of forest on evapotranspiration is demonstrated by the increase of annual evapotranspiration from a forest planted in the early 1950s (Fig.5.5). It will be noted that the groundwater recharge becomes more negative, implying that in this area the forest is making use of groundwater from another area (see text associated with Fig.6.3).

As a consequence of the influence of forest cover on the various elements of the water balance, the general regime of surface runoff from forested catchments differs considerably from the catchment response of agricultural areas. The most striking difference is the great decrease in flood discharge. To demonstrate this

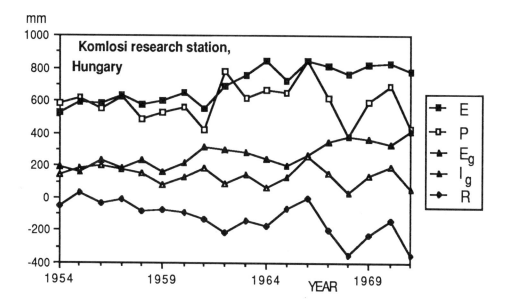

Fig.5.5 Components of the annual water balance for a growing forest: E = evapotranspiration; P = precipitation; E_g = upward (evaporation) flux from groundwater; I_g = downward (accretion) flux to groundwater; R = net groundwater recharge. (After Major 1975).

influence, the flood runoff depth has been plotted against rainfall for three representative basins in Mauritius (Fig. 5.6). The area of the catchments was practically identical and the climatic conditions were similar. In one basin (Bateau) the forest remained unchanged, while the other two areas (Gontran and Vacoas) were deforested. It can be seen that the flood volume was practically doubled by cutting out the trees. Similar increases in peak discharge were observed in several other experimental catchments (Parfait and Lallmahomed 1980).

Observations in humid areas also indicate generally higher base flow from forested basins due to the greater infiltration and storage capacity, although the numerical characterization of this difference is very difficult, partly because of the strong random variation of the discharge and partly because of the high uncertainty associated with the determination of base flow. The evaluation of data collected in representative basins, however, indicates that the multiannual average runoff is almost independent of the proportion of forests within the catchment (usually only a slight decrease with increasing forest area is observable). This indirectly indicates that the large decrease in flood volume is partly compensated by an increase in base flow.

115

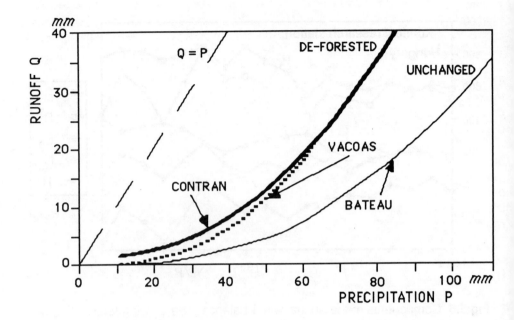

Fig.5.6 Effect of forest clearing on runoff for three experimental catchments in Mauritius (from Parfait and Lallmahomed 1980).

In low rainfall areas, the situation with regard to the effect of forest on baseflow is quite different. In Australia, the general result of replacing deep-rooted native forests and woodland with pastures, or other shallow-rooted vegetation, has been to cause an increase in groundwater recharge, often resulting in increased saline base flow in streams and salinization of lower land areas (Peck 1983).

In deforested areas, the loss of the protection for the soil provided by the forest means that the increasing intensity of surface runoff causes an extremely high increase of erosion. For example, in Brazil, 5-6 $t\ km^{-2}$ total solid discharge was measured from very steep deforested experimental areas during heavy rainfalls, while in the control area (where the forest remained intact) the value was 0-2$t\ km^{-2}$ (Bordas and Canali 1980).

It is very difficult to extend observations made in small experimental basins to large catchments, because the space scale strongly influences the hydrological variables. At the outlet of a large basin the discharge data provide integrated information on the behaviour of the catchment. The structural variation of the

116

surface (the proportions of forest, cultivated area and grassland, and the variability of slope conditions) eliminates the extreme impacts of special local conditions. However, the two strongest influences of forest (the decrease of both the flood volume and soil erosion) can be verified from the data of large rivers.

The data which have been analyzed indicate some aspects of the relation between forests and the local water balance. It can be stated in general that the water regime of forested catchments is more favorable from the point of view of water resources development (decrease of floods and erosion, increase of base flow in humid areas, only a slight decrease of average runoff) than that in cultivated or grassland areas. It is necessary, therefore, to predict all the expected impacts of forest management and particularly those of planned changes in land use such as deforestation. On the other hand, reforestation can be used in some situations to improve the environmental conditions and to produce a more favourable water regime.

Interventions from urban activities

Extent of urban areas. For many centuries people have been living together in cities. In many cases their choice of a certain location was based on the presence of water: for navigation and trade, for domestic and craft use, for food production (fish and crops), or even in defence of the inhabitants against enemies. The presence of urban areas affects the hydrology of local and adjacent regions in a number of ways, such as groundwater withdrawal for water supply, surface and groundwater pollution, increased precipitation downwind of cities, and floods.

Various facets of the urban hydrologic cycle are being studied in urban water balances and budgets. These studies must also include the man-affected flows of water due to systems for water supply and wastewater disposal. This complicates the approach to hydrological processes, compared with that which can be followed in rural areas.

Although the hydrology of urban areas with their numerous man-made features may seem superficially to be simpler than that of natural areas, a closer comparison of the hydrological systems leads to an opposite conclusion. This can be explained by the fact that the urban hydrological system comprises all the basic elements of the natural system plus many additional man-made features, some of which serve for manipulation of the urban water system to man's advantage.

Although the reasons for establishment of the first historical communities are not clearly understood, urbanization, which is characterized by a large concentration of people in relatively small areas, is now recognized as an inevitable historic process (Lazaro 1979). In 1965, about one third of the world's population lived in cities. By the end of this century, about one half of the world's population will live in cities, and in more developed regions about 80% of the population will be concentrated in urban areas (Lindh 1979). Such trends are reflected in the increasing number of large cities. In the early nineteenth century, there was only one city, London, with one million inhabitants. Estimates for 1985 indicated 270 urban areas with more than one million inhabitants and 35 areas with more than five million inhabitants (Lindh 1985). Further rapid growth of large cities is expected in the next 15 years, as indicated in Table 5.1.

Table 5.1

TEN LARGEST CITIES OF THE WORLD IN 1980 AND 2000 (after Lindh 1985)

Rank	1980	Population (Mp)	2000	Population (Mp)
1	New York	20.4	Mexico City	31.0
2	Tokyo	20.0	Sao Paolo	25.8
3	Mexico City	15.0	Tokyo	24.2
4	Sao Paolo	13.5	New York	22.8
5	Shanghai	13.4	Shanghai	22.7
6	Los Angeles	11.7	Beijing	19.9
7	Beijing	10.7	Rio de Janeiro	19.0
8	Rio de Janeiro	10.7	Bombay	17.1
9	London	10.2	Chicago	17.1
10	Buenos Aires	10.1	Calcutta	16.7

High concentrations of population in urban areas have several implications for water resources. The urban population acts as a driving force which changes the landscape and hydrological cycle of the area and it further requires certain services, such as water supply, flood protection and drainage, and water-based recreation. Because of the large populations of urban areas, any urban water resource projects have a potential to benefit a large population, but failures in design will on the other hand lead to large damages.

The data in Table 5.1 show the dynamic development of urban population and urban areas. Some areas will double their population in a short period of 20 years. The dynamic character of urban areas leads to non-stationarity of hydrological data. Lindh (1983) states that 80% of the world's available water resources is used to irrigate food crops, much of which is consumed in cities. In this sense, cities make use of the water resources of quite distant areas. However, they have a more direct impact on the hydrological conditions and water resources of adjacent areas, and these areas will therefore also be given some attention in the subsequent parts of this chapter.

Urban populations require certain water-related services for their existence and general well-being, including water supply and wastewater disposal, and flood protection and drainage.

Water supply and wastewater disposal. Urban populations place high demands on water supply for domestic, industrial, commercial and institutional purposes. Demand rates vary from very low values of $20\text{-}25 l\, p^{-1} d^{-1}$ to over $600\, l\, p^{-1} d^{-1}$ in large industrial cities (Hengeveld and De Vocht 1982). Depending on the density of population, water demand rates per unit area may be as high as $176\, m^3 ha^{-1} d^{-1}$ (World Bank 1978).

The high water supply demands of urban populations have to be met by withdrawals from local sources and imports from other areas. Local sources of potable water rarely suffice to meet population demands, and in any case often become polluted. Imports of water from distant areas are economically and technologically feasible; for example, Los Angeles transports water from areas as far away as $400 km$ (Pereira 1974). Imports of water from other watersheds affect their water budgets and ecology, so that the hydrological effects of urban areas extend well beyond their boundaries.

Most of the urban water use is non-consumptive (except possibly in arid zone cities, where large volumes may be used for watering of gardens and parks), and about 65% to 75% of the supply is returned to the surface or subsurface hydrological system. This returned water is polluted to some extent, and its discharge has large impacts downstream or on the local groundwater. Careless disposal of organic wastes may cause the spread of disease due to the presence of pathogens in drinking water (Van Burkalow 1982). To avoid such occurrences, it is necessary to provide adequate collection, treatment, and disposal of wastewater in urban areas.

Waterborne sewage disposal has evolved over a long period from simple pit latrines to septic tanks and to modern systems in which wastes from individual households are collected by sewers and transported to a central or regional plant for treatment and eventual disposal. Such advanced systems may be relatively costly and not

119

suitable for all conditions. It has been noted that there is a whole variety of sanitation technologies which, although less convenient, are less costly and fully adequate for use in developing countries (Hengeveld and De Vocht 1982). Regardless of the technological approach taken, it is imperative to provide adequate urban sanitation in order to avoid disease and concomitant human suffering and medical costs.

Rapid removal of rainwater and flood control. All urban areas can be generally characterized by large changes from the pre-urban surface drainage. There is an opportunity to manipulate these changes to man's advantage and, in this sense, flood protection and drainage should be considered as services to be provided in properly functioning urban areas (McPherson and Zuidema 1977). Approaches to surface drainage vary from very simple systems to rather complex schemes. A brief description follows of two basic systems, the separate system serving only surface drainage and the combined system conveying both stormwater and sewage.

A separate storm drainage system provides only for transport of surface runoff. The general objectives are to provide protection against large floods and to reduce local flooding (water ponding) and inconvenience. Recognizing these two principal objectives, the drainage system is sometimes divided into two interdependent and interconnected subsystems. The major system is designed for long return periods, up to 100 years. The minor drainage system reduces the incidence of local flooding and inconvenience. It includes smaller drainage channels, ditches, gutters, and storm sewers. It is usually designed for short return periods, from one to five years.

As will be apparent, there are various structural approaches to drainage practice. Drainage schemes employing surface channels are cheaper but possibly less convenient and more susceptible to illicit waste disposal than underground sewers. Open-channnel drainage systems predominate in developing countries, particularly in sloping areas, as a result of the much lower costs of open-channel drainage in comparison with those of underground sewers. In a combined sewerage system, a single conveyance system is used to transport sewage and surface runoff. This is more economical than the separate system, but the collection efficiency is lower because of possible sewage overflows during wet weather when the system capacity is exceeded and the difficulty of ensuring self-cleansing velocities in large pipes during dry weather flow . It should be noted that besides the fully separate or combined systems, there are numerous systems falling between these two extremes.

The provision of drainage in urban areas generally accelerates the concentration of overland flow and its transport from the area. Recognizing the increase in runoff volume due to changes in land cover, the accelerated transport contributes to increased magnitude and peakiness of urban runoff hydrographs.

In many metropolitan areas of the world, flood control and proper drainage are inseparable factors. The type of land use will affect the design criteria for flood control and drainage, and also the measures to be taken. It is clear that each situation is different and that climatic conditions do not play a primary role.

An example from Bangkok may illustrate this (Buning 1985). When the Bangkok metropolitan administration had decided to give the highest priority to protection against flooding and to the provision of proper drainage for an area of nearly $100km^2$, many constraints were met during the phase of project preparation. Some of these constraints were natural, and some were in the form of design guidelines requiring the minimization of land acquisition and resettlement, low overall costs and low-cost designs, and minimization of negative environmental and aesthetic impacts. The criteria for flood control included extreme hourly water levels with a return period of 100 years. The drainage system was designed for a 3-hour rainfall with a return period of 2 years, land use in the year 2000, and maximum 2-year water levels in the main and secondary drainage system not to exceed ground level. The design also assumed continuation of the present rate of land subsidence.

Landscape manipulation and diversity of urban areas. To accommodate large populations and to provide the services required by them, urbanizing areas undergo large changes in their physiography and attain certain physiographic features. It should be recognized that although urban areas occupy only a small fraction of the total land area, about 2% (Kuprianov 1974), their influence extends well beyond their boundaries, particularly in terms of water quality considerations.

Concentration of people in urban areas leads to dramatic changes in land surface cover. In particular, the permeability of the surface is dramatically reduced, due to extensive use of impervious cover and compaction of the top layer of the soil. Extensive impervious surfaces are typical of urban centres throughout the world. Such surfaces comprise roofs, roads and streets, sidewalks, and parking areas. The extent of the impervious area in an urban catchment is commonly described by the catchment imperviousness, defined as the ratio of the impervious area to the total catchment area. From the hydrological viewpoint, it is sometimes desirable to introduce another classification for impervious areas by distinguishing between directly-connected and non-connected areas, the former draining directly into impervious conveyance elements connected to sewer inlets, and the latter draining on to pervious areas.

Depending on the climate and socio-economic conditions, the use of man-made impervious materials in urban areas may be relatively low, with some streets and walkways not being paved. From the hydrological point of view, the absence of impervious cover is not significant, because soils in these areas become compacted and exhibit low infiltration capacities similar to those of impervious surfaces (Ando et al 1984). Urban land cover with inhibited infiltration contributes to high surface runoff rates and to accumulation and washoff of various pollutants.

Another feature of urban land is the stripping of natural cover and exposure of bare soils during the period of development. Such exposed surfaces are then subject to severe erosion. For example, Miller (1977) noted that such areas around the city of Baltimore averaged 19km^2 at any one time and the off-site movement of soil reached several kilograms per square metre of land.

Another important feature of urban areas is the high drainage density arising from the presence of a man-made drainage system.

The special physiography of urban areas and the need to provide services to the urban population contribute to a characteristic climate and relatively complex hydrological conditions in urban areas.

The diversity within an urban area is primarily determined by land use and the age of various subareas. Urban land is commonly classified into a number of land use categories. In a simple classification, five categories are recognized - residential, institutional, commercial, industrial, and open spaces. Each category could be further subdivided according to the type and density of development. The main differences between the various land use types are the imperviousness and environmental quality. Residential land use is usually classified according to population density as single-family, multiple-family, and high density (apartments) land use. The imperviousness of residential areas may vary from as low as 15% in suburbs to 85% in the case of apartment buildings. Residential areas are relatively clean if solid waste and wastewater collection are provided.

Institutional areas house office buildings, hospitals, schools, and similar institutions. Generally, they are characterized by an intermediate imperviousness and relative cleanliness arising from regular maintenance and limited land use activities.

Commercial areas vary from small neighbourhood plazas to large regional centers. They are highly impervious because of the large extent of rooftops and parking areas. Imperviousness values reaching 100% are not atypical for commercial areas.

From the environmental point of view, these areas generate high volumes of traffic with concomitant pollution. On the other hand, regular cleaning and maintenance reduces their polluting impact (Marsalek 1984).

Industrial areas are sometimes further classified according to the type of industry, which strongly affects their environmental quality and imperviousness. Modern high-technology industry provides a very clean environment as opposed to the heavily polluted environments of the steel or chemical industries. The imperviousness of industrial areas ranges from intermediate to high values.

Open spaces include various forms of recreational areas, parks, green belts and playgrounds. Although these spaces are part of the urban area, they have retained some of their natural character. They are characterized by predominantly natural land cover with high permeability, low runoff potential, and relative cleanliness.

Another factor influencing urban land character is the age of the developed area, which is reflected in building practices and possibly in the condition of the area. For example, older residential areas tend to have large lots with a lower overall imperviousness. Recently built areas are characterized by higher imperviousness, but some of them incorporate runoff controls. The age of the area may also affect its condition and the state of repair. Older, less attractive areas may lack proper maintenance which leads to higher pollution of surface runoff. It has been noted that poorly maintained roads generate higher runoff pollution (Sartor et al 1974).

The physiographic as well as thermal changes of urban areas may generate climatic effects. The climate of large metropolitan urban areas differs somewhat from that of the adjacent natural areas. Although these differences are often considered to be of secondary importance, they have been well documented (Landsberg 1981) and are summarized below for completeness. Urban areas exhibit 5-25 times higher contamination of the air (Landsberg 1981), which then affects other climatic factors. In particular, urban areas exhibit higher cloudiness (5-10% more), more fog (by 30 to 100%), more rainy days (by 5-20%), more precipitation (by 5-16%), and higher mean temperatures (by 0.5-2 ^{o}C) (Kuprianov 1974; Landsberg 1981). On the other hand, the global radiation and the relative humidity are reduced in urban areas by 5-20% and 2-8%, respectively.

Consequential effects on hydrological phenomena. Urban areas are characterized by a special hydrological system which is more complex than that of natural areas (Fig.5.7). The urban hydrological cycle includes all the features of the natural system plus many man-made intervention measures reflecting changes in the natural system. These additional features include water imports for water supply or low-flow augmentation, surface water and groundwater withdrawals, surface and

123

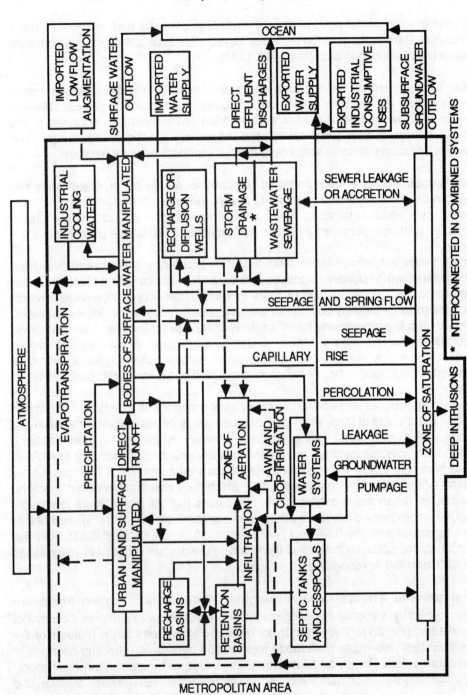

Fig. 5.7 Urban hydrological system (after McPherson and Schneider 1974)

subsurface drainage, and wastewater systems. Each of the above features has a number of typical elements which represent various measures for storage, transport, and treatment of water. Although Fig.5.7 focusses on physical features of the system, each element of the physical system has characteristic implications for the quality of water.

Many of the characteristics of urban areas discussed earlier create an opportunity for human manipulation of the hydrological cycle. The concentration of development in a relatively small area of high value makes it economically feasible to plan, design, and implement various water management measures, particularly in newly developing areas. The main goal of such measures is to reverse the adverse effects of urbanization on water resources. A brief summary of these measures follows.

Control measures are commonly applied to runoff from urban areas. The first group of measures uses runoff controls at the source either by reducing runoff generation (by means of permeable surfaces, enhanced infiltration and groundwater recharge), or by runoff detention. Control measures applied further downstream are more or less limited to runoff retardation by various forms of storage, improvement of its quality by treatment, and various forms of suitable disposal (e.g. in infiltration basins, or by enhanced evaporation) (Alley 1977).

Impacts on runoff. The hydrological impact of the various land use categories is evidenced particularly in differences in runoff, varying with the imperviousness.Typical runoff coefficients for various surfaces and a 5 to 10 year design return period, as found in catchment studies and applied in drainage design, vary from 0.7-0.95 for asphalt and concrete pavements and steep roofs, to 0.1-0.2 for lawns, gardens and playgrounds (Unesco 1987).

The major input to any drainage design is rainfall intensity and its areal variability. Sewers are designed for rainfalls of short duration. The intensities of tropical rainfalls are extremely high. For example, the intensity of tropical rainfall with a duration of 15 minutes and a 1 year frequency is 2.5 to 4 times higher than that of rainfall of similar duration and frequency in Western Europe. For 1 hour rainfall, the factor can be as high as 3 to 6. Combined sewer systems are not practical in tropical areas, as the large diameter sewers required would have to be laid on relatively steep gradients to create self-cleansing velocities during dry weather flow. In tropical climates with a seasonal rainfall, and also in arid or semi-arid climates with only a few rainfalls during the year, separate systems should be used to avoid solidification of sediment in sewers during dry periods. In the other climatological zones both separate and combined systems may be equally applicable.

The various approaches to drainage prectice mainly concern surface channels and closed conduits. The choice between the two depends on different factors. In climates with high rainfall depths and high rainfall intensities, and in sloping regions, the conduit system may be generally limited to institutional and commercial districts and built-up areas with high imperviousness, with open systems in residential areas in order to reduce costs. In general, this is the situation of metropolitan areas in developing countries. In the temperate climates of Europe and North America the closed system (either separate storm drainage or a combined sewerage system) is usual.

Impacts on groundwater. The main function of subsurface drainage systems in urban areas is to lower shallow water-tables to an acceptable level. Specially in areas which are fairly flat, this method of groundwater control is attractive due to high effectiveness, low installation costs (especially in newly built areas) and reasonable costs of maintenance. Drainage criteria have been developed for this purpose (Unesco 1987).

In any case, a careful design is needed, based on information about the construction of buildings and roads, and on soil properties and hydrology. Unforeseen complications, like land subsidence and the drying out of wooden foundation piles of old houses (as in the centre of Amsterdam) have to be avoided.

Another example shows the importance of knowledge of the physical components in an urban area (topography, hydrogeology, soil conditions, land use and size of the area). In one of the fast growing coastal cities in Western Asia, desalination of sea water is used for both water supply and wastewater. Due to the large volume of water available, the septic tanks were overloaded and the groundwater level rose by some metres within 2 -3 years. In some very crowded residential areas, where the hydraulic conductivity of the sandy subsoil was considerably lower than at other places, an unexpected drainage problem has arisen, specially along roads and streets. Sufficient knowledge of hydrogeological and soil conditions might have avoided this situation, although no data for a comparable situation existed elswhere.

Although groundwater is generally an important source of potable water, its value in urban areas is somewhat limited. This is caused by two factors: firstly, depletion of urban groundwater aquifers by excessive withdrawals and reduced recharge and, secondly, pollution of groundwater.

Increases in the imperviousness of land cover in urban areas reduce infiltration of water and recharge of groundwater aquifers (Sulam 1979). Reduced recharge of aquifers combined with large withdrawals results in the lowering of the water-table.

126

This then limits the usefulness of groundwater as a source of potable water, may adversely affect its quality, and cause additional problems by land subsidence (Poland 1984).

Another problem encountered in urban areas is the pollution of groundwater. Such pollution is caused by percolation of polluted surface water, seepage from sewers, and underground wastewater disposal, particularly in deep wells. In this connection, the devastating effect of toxic contaminant dumps on groundwater should be mentioned. In the United States alone there are 30 000 - 50 000 storage sites containing hazardous wastes and several thousands of these pose definite risks to human and environmental health, mainly through contamination of groundwater (Lennett 1980).

The level of pollution of stormwater is relatively low and the recharge of aquifers with stormwater may be acceptable in many areas. Malmqvist and Hard (1981) reported only minor changes in groundwater quality arising from stormwater infiltration.

It appears that groundwater in urban areas can serve as a source of potable water only if it is properly protected and not over-exploited. These restrictions require optimal and usually detailed information on the hydrological features of the area and on human activities, such as excavations, piling, or diversion of waterways, which might affect groundwater flow and urban water resources.

Impacts on streams and rivers. The impact of urbanization on water quantity stems from changes in the hydrological cycle in urban areas and particularly from an increase in and acceleration of surface runoff, and reduction in infiltration of rainwater into the ground. Increased and accelerated runoff lead to increased peak and mean flows in streams and rivers in urbanizing catchments. The effect of urbanization on peak flows has been widely documented and shown to depend on the ratio of the urbanized area to the total river basin area. For example, Andrews (1962) noted that addition of $60 km^2$ of impermeable surfaces, such as roads and rooftops, in the Thames river basin above Teddington Weir did not affect the river regime. This impervious area, however, represented only 0.5% of the total basin area. In basins with higher proportions of urbanized land, large increases in flood flows have been found. Riordan et al (1978) surveyed the literature on this subject and found the increase to be up to ten times.

Increased volumes of runoff contribute to increased mean flows which may have some water quality implications, specially with regard to river bed and bank erosion, and sediment tranport.

Reduced infiltration and diversion of wastewater effluents lead to reduction of base flows in urban streams and rivers. This has been documented on Long Island, New York (Simmons and Reynolds 1982) where 12% reduction in the total base flow volume was observed in streams in urbanized unsewered catchments and much greater reductions were found in sewered catchments. Such reductions may contribute to water quality problems in receiving waters.

Water quality in urban streams and rivers is strongly affected by the urban areas. Such effects include degradation of water quality due to discharges of wastewater, municipal effluents, combined sewer overflows, and stormwater.

Municipal effluents and wastewater include treated or untreated discharges of domestic, commercial, and industrial wastewaters. Their composition varies considerably depending on the source and the level of treatment. Generally, untreated or poorly treated effluents may be strong sources of pathogens, nutrients, solids, and some toxic contaminants. Where solid waste collection and disposal are not provided, such wastes may find their way into water bodies and contribute to their pollution.

Both stormwater and combined sewer overflows exert oxygen demands which may contribute to oxygen depletion through degradation of discharged organic matter and sediment in the receiving water. Both types of discharge also contain nutrients which contribute to eutrophication of receiving waters. Stormwater and overflows may also carry toxic substances, though at relatively low levels. Finally, stormwater discharges are a strong source of suspended solids, and overflows are a strong source of bacteria and pathogens. Bacterial and pathogenic contamination, caused mostly by sewage overflows and to a lesser extent by stormwater, is the most important impact which adversely affects water supply, water-based recreation, fishing and shellfish culture (Hvitved-Jacobsen 1986).

In the overall assessment, urban streams and rivers exhibit high variations in discharges with high peaks and low base flows. Extremely high discharges then contribute to increased incidence of flooding with concomitant threat to human life and large flood damages. The water in such streams tends to be polluted, certainly downstream of urban sources, and this limits its suitability for downstream domestic water supply without costly treatment. The polluted character of urban streams also limits their value for recreation and fishing. Some benefits may be derived from aesthetic aspects of such streams and rivers, water transportation, and water recreation without body contact.

Water and socio-economic development. In the overall assessment, urban areas have to rely on water imports from other areas. There is a strong conflict between high water supply demands and supply limited by quantity and/or quality.

This conflict needs to be remedied by proper planning and management of urban water resources and their use. The first remedial step is the conservation of water quantity and quality. Examples from highly developed countries (Hengeveld and De Vocht 1982) indicate that water demand rates can be controlled at intermediate levels. An attempt should be made to meet these reduced demands from local sources, which need to be protected in terms of quality and enhanced by recharging. Any remaining water shortages are then met by imports from other areas .

In urbanized areas attention has to be given to all relevant aspects which might affect the decision-making porocess, such as technical aspects, demographic trends, socio-economic perspectives for the region or city, financial restrictions and administrative rules. However, decision makers may encounter some problems when they have to give weight to different interests, which may relate to a range of return periods (from one year up to ten years and more). Some ten years ago it was concluded (Green et al1975) that planning and design of urban water projects were frequently determined by financial resources rather than by considerations of optimal design.

Undoubtedly the collection of information is necessary in preparing alternative solutions for further development of urban areas, for example for the planning of water resources development. In that case a further analysis of the socio-economic features of the area is required, because water demands are strongly related to the population growth, municipal facilities, industrial activities (possible changes in spatial extent and water consumption), agricultural needs etc.

In this respect it is worthwile to bear in mind that there are differences between developing and developed countries with regard to the rapidity of urban growth and the kind of urban development (population density per hectare, extent of industries and services). These differences will also extend to water demands and to the water-related measures which have to be planned and managed.

In the framework of an IHP project on urban hydrology, an international workshop was held on the socio-economic aspects of urban hydrology in 1976. The report of that conference (Lindh 1979) introduces the relationship between urban hydrology and human well-being, and underlines the importance of social indicators. However, a common set of indicators to measure human well-being that would be applicable on a world-wide basis, could not be devised. It is therefore important to recognise that some flexibility should be built into development plans, in order to facilitate changes during the process, if desired.

There are other advantages of a flexible approach: it will give an opportunity to introduce phases into the process of development and execution; and it may lead to differentiated solutions for smaller areas within the bounds of an overall infrastructure, e.g. for drinking water supply. An example of a proposed plan of supported activities in urban water resources development in the area of Surabaya, Indonesia, is given in Notodihardjo and Zuidema (1982).

6. Techniques for inter-regional comparison

Introduction

The basic idea of comparative hydrology is that the hydrological processes developing in various regions may differ considerably from each other and, therefore, different approaches should be applied to determine the hydrological variables in basins under different geographical conditions. The purpose of comparative hydrology is:

-   to delineate regions where sufficient hydrological similarity can be assumed to justify the application of techniques for characterizing hydrological processes;
-   to compare the hydrological processes occurring in different regions in order to determine how the various elements of the water balance depend on geographical conditions;
-   to summarize the various methods suitable for the simulation of hydrological processes and to quantify hydrological variables in different regions under different conditions.

To achieve these objectives, quantitative comparisons must be made between the hydrological processes developing in different regiona. Since the inputs are variable both in time and in space, direct comparisons of data sets cannot be made, but representative hydrological variables can be selected, and the characterization of the regions based on the range of these variables. This chapter gives some examples of this type of comparison.

Precipitation depth, duration and area

The most visible climatic differences between various regions are demonstrated by the amount and seasonal fluctuation of precipitation and evaporation as shown in the climate diagrams attached to the map at the front of this book. The characterization of a region requires, however, not only the description of average or median monthly or annual conditions, and departures from the median as shown by selected deciles, but also the quantification of extreme individual hydrological events, the occurrence of which is expected with a given probability. Such extreme events are usually characterized by the interrelations between the area, the depth and the duration of precipitation.

The most simple method of determining the catchment response (i.e. the expected volume and time distribution of runoff at the catchment outlet) to a given precipitation is the so-called rational equation (Chow 1964)

$$Q(t) = C \, A \, i \, (t - \tau) \tag{6.1}$$

This assumes that the discharge $Q(t)$ in the river as a function of time is proportional to the area A of the catchment and to the intensity of precipitation $i(t-\tau)$. The coefficient of proportionality is the runoff coefficient C. In the equation the delay between the rain and the development of discharge is considered as a constant time of concentration τ, equal to the travelling time of water from the most remote part of the catchment to its outlet.

When larger basins are investigated, the area has to be divided into subcatchments and the discharge hydrograph can be determined by superimposing the flows reaching the outlet, allowing now for the differences between the travel time to the outlet related to the various subcatchments:

$$Q(t) = \sum_{i=1}^{n} C_i \, A_i \, i \, (t - \tau_i) \tag{6.2}$$

The same basic principle, of interrelating rainfall intensity and discharge, and superimposing the runoff responses to precipitation events separated in space and time, is also used in more sophisticated models, such as the unit hydrograph. These techniques therefore require information on the probable extreme intensity of precipitation as a function of its duration and the size of the area.

If rainfall intensity were independent of its duration, then the structure of the rational formula indicates that peak discharge would occur when the whole of the catchment is contributing, i.e. when the rainfall duration T is greater than the longest travel time τ_i. However, as intensity with a given probability decreases with increasing duration, the peak discharge may in fact occur before all the catchment is contributing; this possibility is normally checked by testing some durations shorter than the whole concentration time of the catchment.

The above discussion gives the rationale for studies of the relation between rainfall intensity and duration, for a given probability of occurence. The most simple form of this relationship is the Montanari equation:

$$P = a\,T^b$$

or $$i = a\,T^{b-1}$$ (6.3)

where P is the rainfall depth with duration T, and a and b are constants characterizing the local conditions, which depend on the probability of the event. The process of calculating a and b starts by selecting the annual maxima of depth P for a given duration T. An empirical probability distribution can be constructed from this set of data, and fitting a theoretical distribution function allows the estimation of the depth expected with a given probability (Fig. 6.1; Bastug 1984).

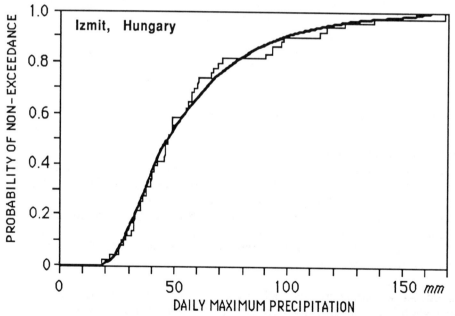

Fig.6.1 Maximum daily precipitation at Izmit, Hungary for the period 1929-1979

133

Repeating the analysis with other values of rainfall duration, pairs of values of P and T can be determined for any selected probability. These points plot approximately as a straight line on log-log paper, which verifies the form of (6.3). Fig. 6.2 shows the graphs determined in this way for two stations in Hungary, four in Sri Lanka, one in Kampuchea and one in Kuwait. For the construction of the graphs, the annual maximum intensity with a given probability was estimated by fitting a three-parameter gamma function to the empirical probability distribution. For Sri Lanka the original data were not available and the graphs show published probability values calculated by using the Gumbel distribution (Baghirathan and Shaw 1978).

The following points from Fig. 6.2 should be noted:-

- For some stations the depth-duration relationship is a straight line on log-log paper, while in other cases the lines are broken;
- The values of the parameter b, calculated from the ranges where the lines are parallel, can be regarded as regional characteristics (b = 0.21 in Phnom Phen; 0.27 in Sri Lanka; 0.30 in Hungary; and 0.60 in Kuwait);
- Four stations from one region of Sri Lanka (basins draining to the northeast shore of the island) are shown, to demonstrate that the value of b calculated in the range where the lines are parallel to each other, is a constant characterizing the region. They also show that breaks in slope may occur upwards, indicating that rainfall with a long duration may cause critical floods under special conditions.

Groundwater-atmosphere interactions

Although the flux of the vertical transport to and from groundwater is relatively small, the area of the surface involved is extremely large and therefore the total amount of water involved in interactions between the atmosphere and groundwater systems is a decisive component of the groundwater balance. The processes influencing this vertical transport are also important in the development of surface runoff, and particularly in the variation of runoff with catchment area. Hence the character of groundwater-atmosphere interactions has to be taken into account in any attempt to delineate hydrological regions.

Groundwater recharge depends on the hydrological events acting above the terrain (precipitation, snow melting, evaporation), on the vegetation covering the surface (a decisive factor in regard to interception and transpiration) and on the transport processes developing in the soil water zone. Both the methodological approach

134

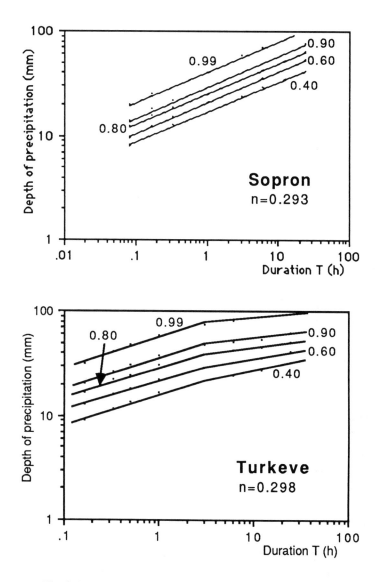

Fig.6.2a. Depth-duration-probability relationships for precipitation at two locations in Hungary.

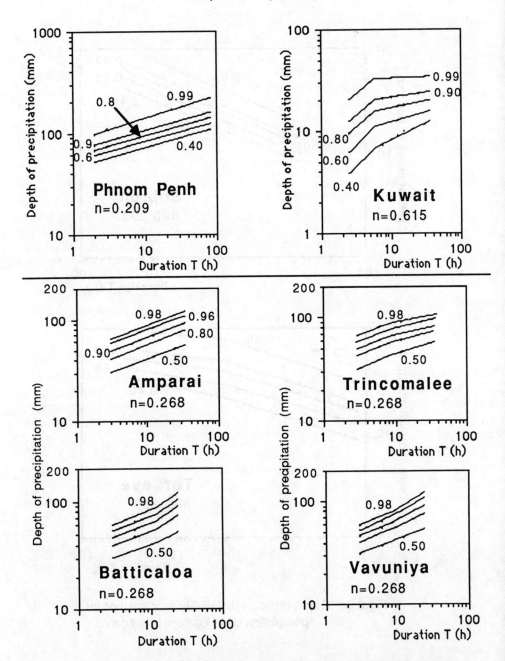

Fig.6.2b. Depth-duration-probability relationships for precipitation at Phnom Penh, Kuwait, and four locations in Sri Lanka.

approach and the data required to analyze these processes depend on the time unit selected for the study. Three hierarchic levels can be distinguished:

- when the purpose is the determination of the annual or multiannual average of groundwater recharge, a water balance approach, requiring only general hydrological data, can usually be applied. In low rainfall areas, however, the recharge is so low (often 1 $mm\ y^{-1}$ or less) that measurement errors make water balance techniques impractical, and recharge is best estimated by measurements with natural isotopes (Allison 1987);
- at an intermediate level, direct observations (regular measurement of the water content and the tension in the soil water zone, or the change in the water-table in a set of observation wells) are needed to estimate seasonal fluctuations in groundwater recharge. A reasonable time unit for this mezzo scale is between several days and one month;
- the most detailed information on the interactions can be gained by hydro-dynamic simulation (Nielsen et al 1986) of the processes in the soil water zone. The time unit of the model should be shorter than one day (several minutes are used in some cases) to facilitate the detailed description of the vertical water transport.

As was explained earlier, large space scales should be used for the regionalization of hydrological variables, to eliminate the random effects of local conditions. This implies the application of long time units, so that only the water balance approach is suitable for characterization of atmosphere-groundwater interactions in comparative hydrology. The water balance equation for the groundwater can be simplified considerably if limits for the balance time interval are selected appropriately.

For a sufficiently long period and assumed stationary conditions in the groundwater system, changes in groundwater storage are negligible in comparison with the fluxes to and from the groundwater body. The groundwater balance can then be written:

$$R = Q_i - Q_o \qquad (6.4)$$

where R is the recharge, and Q_i and Q_o are the lateral inflow and outflow of the groundwater body.

To analyze the variation of groundwater recharge with the average depth of the water-table in different regions, it is convenient to divide it into two elements, an accretion flux I_g from the soil water zone to the water-table, and an evaporation flux

E_g from the groundwater to the unsaturated zone:

$$R = I_g - E_g \qquad (6.5)$$

The value of the accretion flux I_g is always less than the surface infiltration I, and the difference increases with an increase in the average depth of the water-table and with an increase in the storage capacity of the unsaturated soil water zone. The general form of the variation with water-table depth is shown in Fig.6.3a. In arid regions there will be a limiting depth below which there is no accretion to groundwater, except in extreme precipitation events. In humid regions, there will be a depth below which the accretion to groundwater is independent of the depth of the water-table.

The total evaporation E originates from the evaporation of intercepted water E_i and the evapotranspiration E_s of water from below the ground surface:

$$E = E_i + E_s \qquad (6.6)$$

The first component is involved in the above-ground water balance

$$P = E_i + Q_s + I \qquad (6.7)$$

where P is the average rainfall and Q_s the surface runoff. The second evapotranspiration component E_s is composed of water consumed from the soil water zone and the evaporation flux E_g from the groundwater. The variation of these quantities with water-table depth is shown in Fig.6.3b . The ratio E_g / E_s is 1 when the water-table is at the surface and decreases slowly while the water-table remains in the root zone. It then decreases rapidly to zero at the depth below which there is no evaporation flux from the groundwater, a situation common in semi-arid and arid climates.

Using (6.5) , the variation of groundwater recharge R with water-table depth can be obtained from the difference between the curves for I_g and E_g , as shown in Fig.6.3c. This characteristic curve for the groundwater recharge has two important points:

      (i) the intersection with the axis at an equilibrium level d_{eq}
      (ii) a maximum value at a water-table depth d_{max}

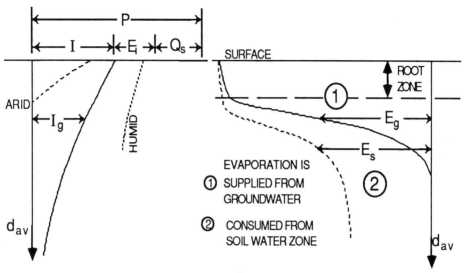

a. RELATION BETWEEN GROUNDWATER ACCRETION I_g AND AVERAGE WATER-TABLE DEPTH d_{av}

b. RELATION BETWEEN EVAPOTRANSPIRATION FROM GROUNDWATER AND AVERAGE WATER-TABLE DEPTH

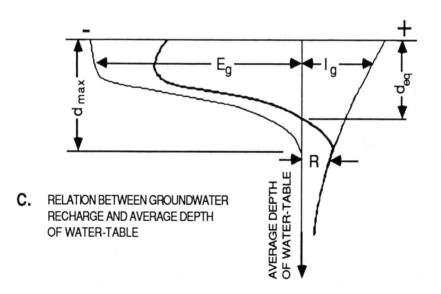

C. RELATION BETWEEN GROUNDWATER RECHARGE AND AVERAGE DEPTH OF WATER-TABLE

Fig.6.3 Characteristic curves for groundwater accretion, evaporation and recharge, as functions of average depth of water-table.

When the average water-table is above the equilibrium level, there is negative recharge. This condition characterizes "dry drainage" areas, when the lateral groundwater inflow Q_i is greater than the outflow Q_o . Below the equilibrium level there is positive recharge, and the lateral groundwater outflow exceeds the inflow. Between the levels d_{eq} and d_{max} , the recharge increases as the depth to the water-table increases, but below d_{max} the recharge falls to a fairly constant value with further increases in the water-table depth.

As the characteristic curve includes all the uncertainties in the groundwater accretion and evaporation curves, it cannot be used for quantification of groundwater recharge. However even this qualitative description of the vertical water transport provides useful information for several different studies, e.g. the delineation of recharge and drainage areas, rough estimation of the long-term average of groundwater recharge, characterization of the migration of salts above the water-table , and prediction of expected changes in the regime of soil water and groundwater due to human activities.

The general form and the main parameters of the characteristic curve provide some integrated information on the climatic, surface and soil conditions of a region. It can therefore be used to demonstrate the differences occurring in the interactions between the atmosphere and groundwater in hydrologically different regions.

Regional differences in river flow regimes

Duration curves characterizing multiannual average conditions are often applied to describe the water regime of rivers. The duration curve is the accumulated form of the empirical frequency distribution, indicating the number of days, relative to the total length of the study period, when the river discharge (or water level) was less than a given value (Fig.6.4). The number of days on the horizontal axis indicates the period when the discharge (or water level) was less than the value in question.

The duration curve is presented either in graphical form, or some characteristic points are listed in a table. For example, in Japan the discharges (or water levels) with durations of 35, 95, 185, 275 and 355 days are published in addition to the annual maximum and minimum. The geographical conditions of the Pacific coast region of Japan (i.e. relatively short rivers differing in geology and climate but with catchment areas of the same order of magnitude) facilitated a statistical analysis of the impacts of some influencing factors on the regime of the rivers, using the characteristic discharges of the duration curve (Musiake et al 1975; Kinoshita et al 1986).

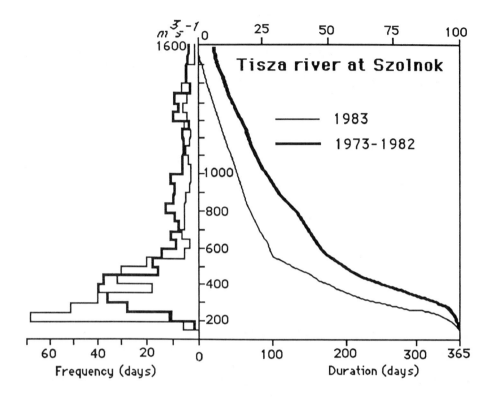

Fig.6.4. The duration curves for river flow on the right are the accumulated form of the frequency data on the left.

There are 124 catchments where the characteristic values of the discharge duration curves were calculated from the existing records. The region was divided into six climatic subregions (Fig.6.5) and the catchments were also grouped into six classes according to their geological structure (quaternary volcanic rocks; tertiary volcanic rocks; granite; tertiary formations; mesozoic formations; paleozoic formations). The climatic and geological conditions were quantified, and a regression analysis was performed to investigate the influence of climate and geology on the various characteristic discharges. The results (Fig.6.5) indicate that the large discharges are mostly determined by the climate, while geology becomes dominant in the range of low discharges.

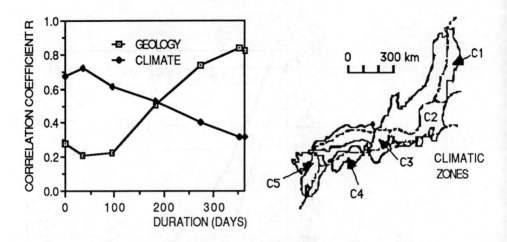

Fig.6.5 Influence of geology and climate on discharges with
different exceedance durations, for 124 catchments
in Japan (after Kinoshita et al 1986).

The basic concept of the duration curve can also be applied to describe seasonal fluctuations in the water regime. The duration curve is determined for each month separately, so that 100% on the horizontal axis represents 31, 30 or 28 days according to the number of days in the particular month. To represent the results in one integrated graph, the hydrological variable can be plotted on the vertical axis, the months on the horizontal one, and the probability used as a parameter. The figure constructed in this way is called a duration surface, and its character is demonstrated by Fig.6.6, which was constructed from the water level data recorded on the Danube at Budapest and the Tisza River at Szolnok, Hungary.

The duration surface is a useful tool for comparative hydrology, as it characterizes the influence of regional factors on the water regime of rivers. The comparison of the two duration surfaces presented in Fig.6.6 is an example of this application. The graphs illustrate the water regime of the two main rivers in the Carpathian basin. Although the distance between the two gaging stations is only 100*km*, the two duration surfaces differ considerably from one another, because of the different geographical conditions in the two catchments.

The main water source for the catchment of the Danube is the high mountain ranges of the Alps, with relatively large areas above the permanent snow line. The accumulation and melting of snow due to temperature fluctuations is the decisive factor in the development of the water regime. Hence, a flood exceeding a given

142

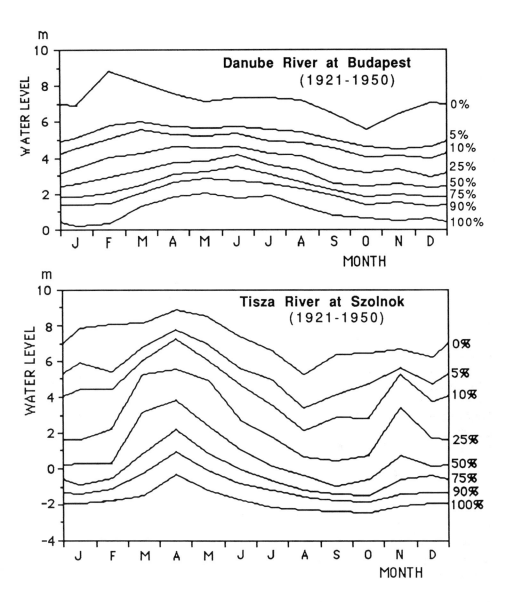

Fig.6.6 Duration surfaces characterizing different river water level regimes.

level at the gaging station is expected with almost equal probability in each month between February and August. The other consequence of the mountainous character of the catchment is that a low-water period develops in January and February due to snow accumulation, resulting in smaller discharges than in September and October, which are usually the driest months in this region of Europe.

The northeast boundary of the catchment of the Tisza river is the Carpathian mountains, where only a few peaks reach a level of 2000 m and even these peaks remain below the permanent snow line, so that the accumulation and melting of snow influences the water regime only in winter. The highest discharges due to snow melting are expected at the beginning of spring (March-April). The largest floods occur in this period and the baseflow discharge also reaches its maximum in April. The curves representing the water level with a given probability in the various months follow a very similar pattern, indicating a gradual decrease of discharge which reaches a minimum at the end of summer.

As clearly shown by these examples, the climatic, morphological and geological conditions have a decisive influence on the water regime of large rivers. The impact of climate can be observed mostly in the development of high discharges, while the base flow is determined mainly by the storage capacity of the catchment, which depends on the geological structure of the area. The morphology and the altitude of the catchment, in the example of the high mountains, may alter the role of accumulation and melting of snow, the consequence of which is the development of a water regime different from that characterizing the lower lying parts of the region. These and other similar influences should be considered when the principles of hydrological analysis are applied to the various regions in comparative hydrology.

Section B

Areas with catchment response

7. Sloping land with snow and ice

Introduction

The hydrological conditions and processes in mountain areas are characterized by an exreme spatial variability in the horizontal and vertical dimensions, which can be observed over a wide range of scales. In the micro-scale range of a few meters, great differences in the surface processes occur as a result of the micro-relief, slope angle and exposure, and their differential effects on net radiation, surface temperatures, wind conditions, precipitation, snow coverage, soil conditions and vegetation. In the meso-scale range of $100m$ to a few $1000m$, the same differential effects occur as in the micro-scale range; in addition, the vertical gradients of temperature, air humidity, precipitation and even of radiation components produce large vertical gradients in climate, which cause corresponding vertical gradients in vegetation, soil conditions, and the hydrological processes. Within a small basin, the melting of snow and ice and the condensation of water vapor on snow and ice surfaces can occur at the same time as evaporation from the bare soil, evapotranspiration from alpine vegetation or differential heating in the bare rock.

In the macro-scale range of a few kilometers to $100km$, the orographic aerodynamic influence of mountains on the motion of air masses and the large-scale distribution of precipitation is a dominant hydrological aspect in all mountain areas. High mountain ridges are known as weather dividers. These small to large-scale spatial differences in precipitation are at the same time related to spatial variations in cloudiness, radiation conditions, air temperature and humidity and to the resulting hydrological conditions. The orographic effect in mountain areas can produce dry semi-arid areas in well-shaded deep valleys with abundant precipitation in the higher elevation belts.

The water resources and runoff regimes of mountain areas above the forest line are strongly characterized by the climatic conditions described above. An important aspect is the low potential for storage and retention of water in snow and ice-free areas, because of the high energy of the relief and the limited presence of vegetation, soils and aquifers in certain areas to retain water. Runoff is therefore frequently a fast response to rainfall, which causes a high potential for flood runoff and erosional processes in those areas. Seasonal and long-term storage of solid precipitation in the snow covers and glaciers forms, on the other hand, an extremely important part of mountain hydrology with respect to flood retention, and above all in its seasonal and long-term natural regulatory effect on river flow and available water resources. Mountain regions are also very sensitive to climatic variations and changes which cause significant variations in the hydrologically relevant basin characteristics, i.e. variations of glaciers and vegetation (such as the treeline). The great variation of climate, vegetation, subsurface and hydrological processes within an elevation range of some $1000m$ is one of the fascinations of mountain regions - a range of variation corresponding to several $1000km$ in a change of latitude.

Hydrological processes related to climate

Relating hydrological processes to the climatic elements is an essential part of understanding the distribution of water in time and space on the earth's surface. There are large vertical gradients in the components of the energy balance of the surface, i.e. the net radiation, sensible heat flux, soil heat flux, latent heat flux of evaporation or condensation, and the heat used for melting ice or snow. Atmospheric motion and the laws of thermodynamics govern the atmospheric temperature and humidity distribution. In the discussion of the particular aspects of mountain hydrology, the vertical distributions of air temperature, air humidity and precipitation are of central importance.

Air temperature and air humidity. The vertical dry and moist adiabatic motions in the troposphere produce average temperature lapse rates of between 0.4 and 0.7 °C decrease of temperature for each $100m$ of elevation. In a similar way, specific air humidity (or vapor pressure) decreases with elevation. Both are decisive elements in the generation of clouds and precipitation, in the melt and evaporation processes, and in the conditions for vegetation. In the context of mountain hydrology, the following controlling functions of air temperature and specific humidity are of concern:

147

The air temperature controls:

- the elevation belts of the boundary conditions for vegetation
- the elevation where the transition from liquid to solid precipitation occurs
- the elevation up to which melting conditions occur on glaciers and snow surfaces
- the sensible heat flux as a component of the melting processes.

In an analogous manner the altitudinal distribution of specific humidity (vapor pressure) controls the atmospheric conditions for evapotranspiration and condensation at the earth's surface. This includes the control of the elevation zone on ice and snow surfaces, where the latent heat flux to the surface is reversed from condensation in the lower regions to evaporation in the upper regions.

The direction of the latent heat flux is of particular interest in snow and glacier hydrology because of its large specific energy turnover. The same amount of energy necessary to evaporate 1mm water equivalent of ice or snow is enough to produce 8.5mm of melt water. On the other hand, if conditions of condensation are prevalent, latent energy is released and contributes to melting.

These processes cause part of the variations in mass turnover of snow and glaciers in different climatic regions. Under maritime climate conditions, high precipitation and maximum accumulation rates occur in the firn basins (areas of snow partially consoliodated by thawing and freezing, but not yet converted to glacier ice), and high melt rates in the ablation areas, frequently supported by latent heat from condensation. This is, for example, frequently observed on the west coast of Norway and in the Pacific northwestern coast mountain ranges of North America, both in the west wind zone of the general atmospheric circulation. The European Alps are under less maritime conditions as far as precipitation and latent heat fluxes are concerned. Condensation conditions are predominant in the lower parts of the glaciers, while the upper parts are generally under evaporation conditions. In the subtropical zones and in continental areas, the mass turnover in mountain glacier areas is comparatively low; examples are found in the Andes of Chile and in the high mountains of Central Asia. A comparative study of hydro-glaciological conditions was undertaken for the European Alps and the Tianshan in China by Kang Ersi (1985).

Snow cover. The heat balance of the earth's surface is strongly influenced by its albedo (reflectivity in the shortwave part of solar radiation). Because the albedo of snow-covered areas is so different from that of other land surfaces, another particular aspect of many high mountain regions is the large variations in the snow coverage, in both time and space. The variations of mass balances of glaciers have been found to be significantly related to snowfall events during the ablation periods, when the effect of the high albedo of new snow is particularly strong in reducing the melt rates. With the same input of energy from solar incoming radiation, the melt rate at a snow-free glacier surface (albedo 0.3) may be two to three times that of a glacier covered with new snow (albedo 0.8). Summer snowfalls on glaciers have been recognized as main factors in the year-to-year variations of glacier mass budgets which have a direct effect on the runoff regime. The snow cover must therefore be regarded as a hydrologically significant land surface parameter.

Precipitation. The spatial distribution of precipitation in high mountain areas shows great horizontal and vertical variations. The orographic effect of mountain ridges on advective precipitation processes is twofold: on the weather side, the lifting of air masses is forced and results in increased precipitation. On the lee side of the mountains the air masses have less specific humidity after their loss of water during precipitation on the weather side. In addition, the downward component of the streamlines on the lee side causes adiabatic warming of the air masses and dissolution of the clouds, the well-known "foehn" effect.

Advection of air masses is a main condition for the increase of precipitation with elevation. Although the specific humidity decreases with elevation, the increase of air stream velocity with elevation is generally enough, together with the orographically forced lifting, to bring about the observed maximum precipitation depths in many high mountain regions. The question of the existence of an elevation zone of maximum precipitation is frequently discussed. Since there are no regions with sufficient and reliable precipitation measurements in the high mountains, there is still great uncertainty in every region. Computations of the atmospheric water vapor advection for the European Alps have shown an increase of precipitation potential up to the highest mountain ridges with elevations around 4000m (Havlik 1969).

Convective precipitation shows a different behavior in relation to elevation, because of different orographic interactions. It is frequently connected to the daily variations of the valley wind systems and to the forcing of thermal instabilities in mountain areas. In general the increase of precipitation with elevation is much less, and the maximum is lower. In the mountain areas of the tropical zones the wind velocity frequently decreases above the level of the friction layer at an elevation of about 1000m. This causes a tropical-convective type of vertical distribution of precipitation (Weischet 1965,1969; Lauer 1976). Another type of precipitation-elevation relation is observed in the trade wind zones, where the maximum of precipitation occurs near the elevation of the trade wind inversion. Lauscher (1976) and Barry and Chorley (1976) attempted to outline the different types of vertical

149

distribution of precipitation. Figs.7.1 to 7.3 give a survey of typical precipitation-elevation relationships in some of the world's mountain regions. Since the systematic error in precipitation measurements increases with elevation (increasing wind speed and increasing proportion of snow in total precipitation), the real precipitation depths at high elevations tend to be larger.

Evaporation. Of all the hydrological elements, areal evaporation is the one with the greatest uncertainty in our knowledge. Large spatial variations in the topography, surface and soil conditions, heat balance and water availability are the difficulties in using point measurements as a basis for areal computations. Therefore, where discharge records are available the water balance method is still the main source of our knowledge on evaporation from mountain basins, despite its weak basis in high mountain precipitation.

In a comparative treatment it is useful to recall the changes with elevation of the conditions for evaporation:

(a)   the disappearance of trees at a certain elevation will greatly reduce evapo-transpiration and evaporation of interception in mountain areas;

(b)   the percentage of partial areas with no vegetation and with littleor no soil increases with elevation,so reducing the availability of water for evaporation;

(c)   increasing snow cover duration with elevation increases the areas of potential evaporation. On the other hand, high albedo (low net radiation) and the upper limit to melting temperature of 0 °C set limits in energy input and in the surface water vapor pressure (maximum possible saturation vapor pressure 6.11 *mb* at a melting surface). This can even cause predominance of condensation at lower elevations where the atmospheric water vapor pressure may frequently be higher than 6.11 *mb* during the melting season.

It is generally accepted that the annual mean areal evapotranspiration and evaporation decrease with elevation. The average vertical gradients so far published are, however, very uncertain. Examples of evaporation-altitude gradients for mean annual values from different regions in the European Alps range from 71 to 365 *mm* decrease per 1000 *m* elevation increase (Lang 1981). In any case, we should expect that:

(a)   the tree line is reflected as a discontinuity in average evaporation-altitude relationships;

(b)   seasonal snow coverage and glacier areas are different in their surface

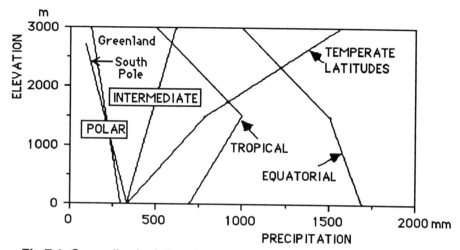

Fig.7.1 Generalized relations between mean annual precipitation and elevation (after Lauscher 1976).

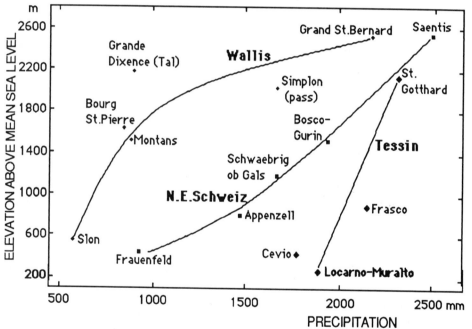

Fig.7.2 Relation between mean annual precipitation and average catchment elevation for regions in the Swiss Alps (from Swiss Meteorological Institute 1978).

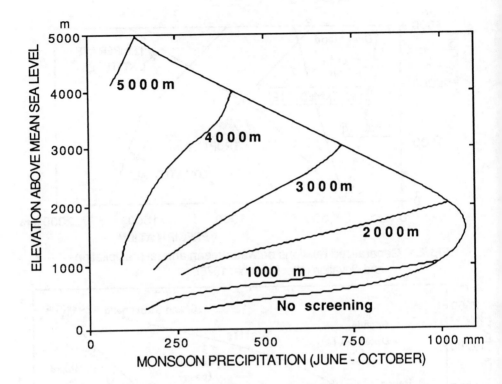

Fig.7.3 Relation between monsoon precipitation and average catchment elevation in the Pakistan Himalayas, behind screening mountain ridges of 1000 - 5000*m* average elevation (from Kushid Alam1973).

processes from the other surfaces, and therefore different in evaporation. Fig.7.4 gives an idea of the complex course of evaporation during the seasons in the Alps.

Water resources and runoff

Mountain areas are generally known as areas of rich water resources because of increased precipitation and, to a lesser extent, low evaporation. In certain regions and under a stationary climate, annual specific runoff follows on the average more or less the distribution of precipitation with elevation. Climatic variations at different time scales cause glacier mass balance fluctuations (storage changes) and the corresponding glacier runoff variations and trends.

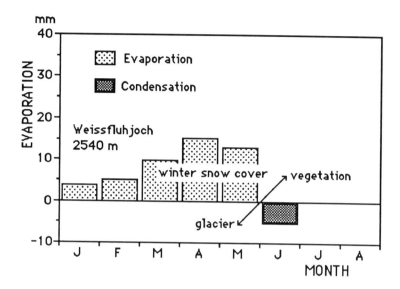

Fig.7.4 The seasonal variation of evaporation in the Alps, at elevations between 2000 and 2500 m above sea level (from Lang 1981).

In parts of the subtropics and in all higher latitude mountains, the seasonal course of radiation conditions and air temperature causes a distinct two seasonal runoff regime:

(a)   a cold winter season with low river flow, and at high altitudes frequently no flow at all: the accumulation of solid precipitation forms the winter snow cover;

(b)   a summer melting season with high river flow mainly from snow melt and glacier melt, sometimes combined with, or at different times from, flow from rainfall.

These seasonal cycles of natural storage in winter and release in summer are of great economic and ecological importance in semi-arid and arid regions.

Characteristics of snow melt and glacier runoff.   The regulating effect of snowmelt and glacier melt runoff in river flow is a distinctive characteristic of river basins which extend over a large elevation range. High melt rates generally occur during warm weather periods, with high incoming solar radiation at the same time as

153

lower elevation basins without snow and glaciers show maximum evaporation rates. During cloudy periods with rainfall, the melt rates are low or even stop because of reduced energy input. In cases of snowfall on glaciers, the high albedo of new snow can moreover reduce ablation rates for several days. The regulating effect of snow and glaciers in the flow regime of river basins is an important aspect of water resource management. For example, summer runoff variability, related to the area of glaciers in the catchment, shows a significant minimum at a glacier area of 40% (Fig.7.5). Lai Zuming (1982) reported the annual flow of rivers with and without glaciers in Northwestern China (Gansu Corridor) and their different deviation from average flow in dry and wet years (Table 7.1). The deviations are much smaller in river basins with contributions from snow and glacier melt.

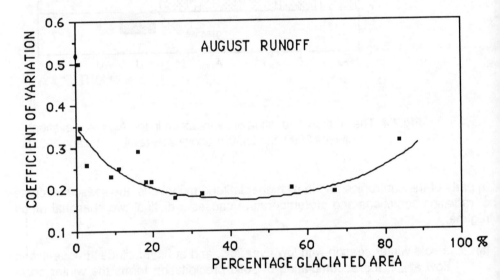

Fig.7.5 Relation between coefficient of variation (CV) of August runoff and percentage of glaciated area for alpine river basins, 14 in the Swiss Alps (0-67% glaciated, period 1968-1982) and 1 in the Austrian Alps (84% glaciated, period 1974-1983).

Fig.7.6 shows the contrasting reactions in the variations of summer flows, between a catchment with a precipitation regime and one with a glacier regime 120*km* away. Fig.7.7 shows a typical example of the seasonal flow regime of glaciated basins, together with the cumulative mass balance (storage change). There was very little precipitation in 1975/76 and therefore very little accumulation, which caused the mass balance to become negative as early as July. In 1976/77 there was ongoing

Table 7.1

STREAM FLOW DEVIATION (%) FROM AVERAGE FLOW IN NORTHWEST CHINA
(after Lai Zuming 1982)

River Basin	Condition	Dry year 1962	Wet year 1967
Xida	no glaciers	-41	+39
Gulang	no glaciers	-21	+53
Dang	with ice and snowmelt	0	-3.6
Changma	with ice and snowmelt	-9.5	+6

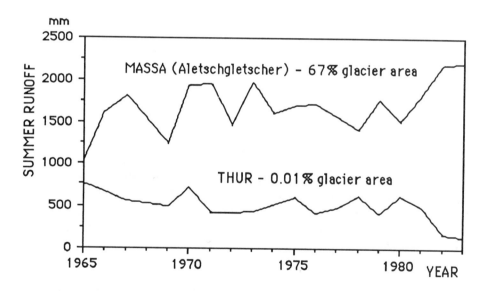

Fig.7.6 Mean summer runoff (May to September)
for two alpine river basins (from Lang 1984).

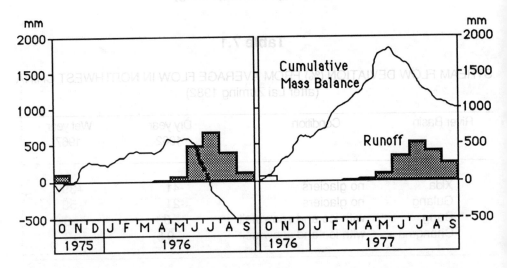

Fig.7.7 Cumulative mass balance and runoff for the Aletschgletscher river basin at Massa/Blatten (194.7km^2, 66% glacier area).

and partly heavy mass accumulation from October to the end of May, and the ablation period was interrupted by two accumulation events. This resulted in one of the hughest positive annual mass balances since records began in 1922 (Kasser et al !983).

The strong seasonal distribution of flow into a "dry" winter season and a high flow season in the melting period, shown in Fig.7.7, is at first sight comparable to monsoon flow regimes, with the exception of the snow storage and the regular diurnal cycles of melt water flow as shown in Fig.7.8, which are controlled by the diurnal cycles of surface energy balance. The base flow increases from day to day, mainly fed by melt water with long flow times from the accumulation area of the glacier. The sudden floods on 10/11 July were caused by carving glacier ice damming the river and followed by a sudden release (Emmenegger and Spreafico 1979).

Variation of mean annual runoff with elevation. The increase in mean annual runoff with increasing mean elevation of basins is the result of increasing precipitation with elevation, and to a smaller extent of decreasing evaporation with elevation. L'vovich (1979) presented a comprehensive treatment of runoff conditions in relation to elevation for various mountain regions of the world. For the Alps, for example, his graphs show precipitation, groundwater runoff, surface runoff, total runoff, evapotranspiration and total surface wetting. The groundwater

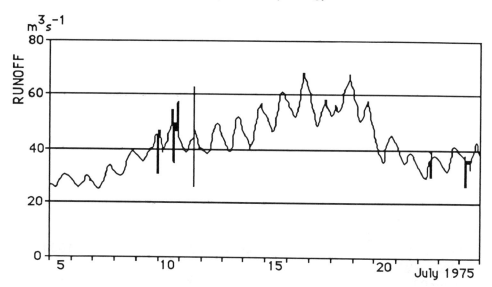

Fig.7.8 Diurnal variation of runoff from the Aletschgletscher river basin at Massa/Blatten (195km^2, 67% glacier area) during a typical melt period in July 1975, with high temperatures and high net radiation.

runoff (base flow) increases slightly with elevation, reaching a maximum near the tree line. This obviously reflects the favorable contitions of infiltration in the alpine forest belts and in the pleistocene (colluvial) and glacial deposits. On the other hand, a strong increase of surface runoff above the tree line is indicated, as a result of less permeable grassland soils and an increasing percentage of impermeable rocky area. In cold and rather dry mountain areas permafrost may become an important catchment characteristic as a barrier against infiltration.

The increase of runoff and water resources with elevation is of the utmost importance for all arid and semi-arid regions adjacent to mountain areas. In these regions, runoff is usually formed in the higher elevation zones only. Fig.7.9 gives as an example the runoff conditions of some mountain basins in Central Asia. In general, there is an appreciable reduction of runoff from west to east; this is in line with the distribution of rainfall (L'vovich 1979). An extreme example of aridity, even in the high mountains, occurs in South America, where for hundreds of kilometers, not a single perennial river is found running to the Pacific Ocean south of the River Loa in the Atacama desert. The arid zone there extends to the high mountains of the Andes. Dreyer et al (1982) have further analyzed the effect of mean catchment elevation on average annual runoff. Table 7.2 compares the variation of runoff with elevation under dry and wet conditions. The values for the

European Alps are taken from Baumgartner et al (1983).

Table 7.2

VARIATION OF MEAN ANNUAL RUNOFF WITH ELEVATION

Region	Runoff increase (*mm* per 1000*m* elevation)		Average annual runoff (*mm*)
Karakoram			
Northern slopes	approx	75	60*
Southern slopes	approx	200*	500*
Columbia River Basin			
Monash Mountains	approx	1800*	1000
European Alps			
Average		750	910
Minimum in southwestern region (KK5.2)		460	460
Maximum in central region (KK 4.1)		1060	1120

* values estimated by the author from the graphs given by Dreyer et al (1982)

Particular aspects of floods from mountain areas. The geomorphological conditions of mountain areas, such as steep slopes, lack of soil and vegetation, impermeable partial areas of rock, and permafrost, generally favor the generation of floods from storm rains. Often these events cause heavy erosion and sediment transport including mud flow and, as a consequence, sedimentation of the bedload downstream. On the other hand, there are mountain areas, particularly in middle and high latitudes, where part of the heavy precipitation is accumulated and retained as snow in the upper elevation zones. In these areas flood forecasting always needs reliable records and forecasts of temperature lapse rates in addition to precipitation.

There are three processes which are of particular concern in the formation of extreme floods in basins with snow cover and glaciers:

(a)   maximum melt water runoff in combination with excessive rainfall: there is a certain probability that the daily maximum melt water flow in a river basin in the afternoon/evening hours will coincide with a flood caused by a convective rainfall;

(b)   ice- or moraine-dammed lakes are occasionally formed in the marginal

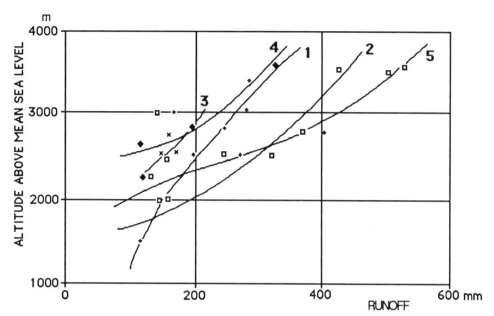

Fig.7.9 Relation between mean annual runoff and average catchment elevation for central Asia. Key: 1. Iren-Khabyrga range; 2. Urumchi area; 3. Bogdashan' range; 4. Koksha-Altau range; 5.Kahalyktau range (from L'vovich 1979).

zones of glaciers. It can happen that rivers are blocked by advancing or surging glaciers, or by snow avalanches and landslides. Sudden outbursts of such water bodies represent great hazards in various high mountain regions. The size of such flood events can by far exceed that of other weather induced short-term river flow extremes. Observations and discussions of such events have been reported by Hoinkes (1969), Rothlisberger (1981) and Haeberli (1983) for the European Alps, by Hewitt (1982) for the Karakoram Himalaya and by Vuichard and Zimmermann (1987) for the Nepalese Himalaya. See also Young (1980). Post and Mayo (1971) presented observations from Alaska;

(c)   in glaciers, large volumes of subglacial or intraglacial meltwater reservoirs may accumulate and burst out in catastrophic flood flows. On a world-wide scale, the size of such reservoirs and the frequency and magnitude of floods increase as a rule with the intensity of glacierization (Rothlisberger and Lang 1986). The most spectacular outbursts, however, seem to be related to volcanic activity (Bjornsson 1975,1976; Meier 1983). Iceland is

159

the classical country for this type of outburst flood, called "jokulhlaups".

At the same time, all these extreme floods cause extreme erosion.

Flood frequency analysis for river basins with glaciers has to take into account the possibility of different flood causes. In a statistical analysis, it may therefore be necessary to analyze the different samples separately. In Fig.7.10 the frequency distribution of annual maximum discharge of the Aletschgletscher basin is given as an example. Purely meltwater-induced floods are clearly limited by the maximum of possible heat input, which can be reliably estimated for a given climatic region. There is evidence that the three largest floods of the sample are combined rain/melt events which belong to a different statistical population.

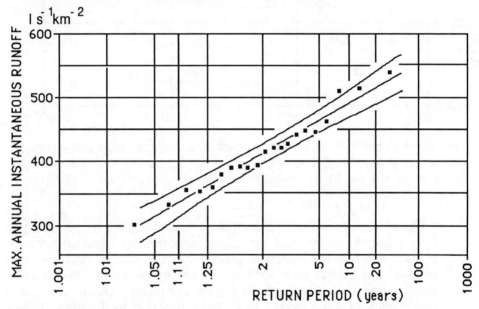

Fig.7.10 Probability plot of maximum annual instantaneous runoff for the Aletschgletscher river basin at Massa/Blatten (195km^2, 67% glacier area), for the period 1965-1985; with fitted straight line and 80% confidence intervals. (Data source: Hydrologisches Jahrbuch der Schweiz).

Long-term variations of water resources. Mountain regions are characterized by strong vertical gradients in climate and hydrological processes. This produces a high sensitivity of water resources to climate variations. Above all, it is air temperature that controls the elevation belts of vegetation and tree line and, together with precipitation, the climatic snow line and the extent of glacierization. Glacier variations have received much attention, particularly during and after the little ice age during which glaciers retreated. The world-wide retreat of mountain glaciers in this century has provided additional water flow in those basins as a result of the annual mass loss of glaciers. The predominance of annual negative mass balances stopped in the European Alps in 1965 when most glaciers started to have positive mass balances, mainly caused by wet-cool summer conditions (Kasser 1981; Patzelt 1973). Fig.7.11 presents the runoff in the basin of Aletschgletscher (195 km^2, including 67% glacier area in 1983) for the period of 1922/23 to 1984/85, together with the specific glacier mass budget. The additional contribution of meltwater to the runoff in the period of strong glacier retreat is clearly reflected in the high average runoff, particularly between 1940 and 1950. The low runoff for 1975 to 1980 is connected to the significant storage of water into positive glacier mass balance.

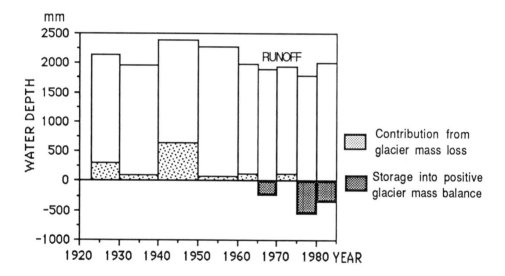

Fig.7.11 Runoff and mass balance in the Aletschgletscher river basin at Massa/ Blatten (195km^2, 67% glacier area). Data source Aellen (1985) and pers.comm..

In addition to the immediate effect on water flow of annual storage changes in glaciers, there is a long-term trend of glacier area variations as a result of mass balance behavior and ice mechanics. The pronounced shrinkage of glaciers in this century has above all reduced the lower parts of glacier tongues, where the greatest melt rates occur. An analysis of the long-term records of the Rhone river catchment at Porte du Scex ($5220km^2$) revealed a clear declining trend in summer runoff as a result of the reduction in glacier area (Kasser 1973). The catchment has an altitude range from 373 to $4634m$, and 13.6% of its area is covered by glaciers. In the period 1916-1968, this glacier area decreased by 19%, causing a reduction of mean summer (April-September) runoff of $14mm$, or 16% of the mean value of $887mm$.

Implications for water management

The foregoing has outlined the salient characteristics of mountain hydrology. The impact of mountain hydrology on the exploitation of water resources is summarized in the following points:

- mountain regions are land areas with precipitation amounts and river flow above average. Their water resources are of the utmost importance; in semi-arid and arid areas it can be the only water available. In many sub-tropical and all higher latitudes, the winter season shows very low river flow and the warm season is the meltwater flow season.
- The occurrence of seasonal snow cover frequently offers favorable conditions for natural storage of water which becomes available in the warm seasons. If the snow cover water equivalent is surveyed, seasonal river flow forecasting has a chance of being quite successful, which is helpful in optimizing water resource management.
- Specific types of extreme outburst floods from glacier areas are a hazard potential to be carefully investigated in any project on glacier-fed streams.
- Mountains above the tree line belong to the remote areas of the earth where observation networks and observation data are sparse. Satellite information is a useful tool here, particularly for snow cover assessment and forecasting (Rango et al 1977; Tarar 1982; Bagchi 1982).
- Glaciers are long-term natural water storage reservoirs. In river basins with contributions from glaciers, the river flow has less seasonal and year to year variations because of the regulatory effect of glacier meltwater runoff. This is of importance in the development of any water exploitation scheme such as irrigation and hydro-electric power reservoirs, etc. On the other hand, the long-term and secular trends of glacier runoff in connection with climatic variations must also be considered.

8. Humid temperate sloping land

Introduction

On a global scale, humid temperate areas are favored environments exposed, as their name implies, to extremes neither of temperature nor of precipitation. The name also emphasizes that the areas are regions of water surplus, but hidden behind this general image of moderation and plenty there are great variations. Humid temperate areas have been well studied both at the regional scale, where they have the densest hydrometric network, and at the research level where many hydrological and environmental concepts were first developed.

For the purposes of this chapter, humid temperate regions are defined as areas with annual potential evaporation between 500 and 1000*mm* and an excess of annual precipitation over evapotranspiration. The area covered is shown in Fig.8.1. In general, the southern limit (in the northern hemisphere) is determined by the point at which evapotranspiration exceeds rainfall, whilst the poleward boundary is fixed by the 500*mm* potential evaporation isoline.

Humid temperate areas, specially in Europe and northeast Asia, have a long history of economic and agricultural development. Consequently, land uses and environmental processes have changed dramatically over time, with particularly large changes following pre-agricultural forest clearance, the first period of industrial development in the 18th and19th centuries, and more recently since the second world war. This last period has seen not just major increases in industrialization but also an intensification of agriculture, with changes in farm structure and inputs. Human influences on hydrological processes are widespread and locally very significant, and high demand from all sectors of the economy frequently imposes strong pressures on the apparently abundant water resources.

This chapter attempts to review hydrological processes and regimes in humid temperate areas, to outline present and future trends in the exploitation of water resources, and to summarize the range and degree of human influence on hydrological regimes. First, however, it is necessary to set these processes and pressures in context by summarizing the key climatological, physical and human characteristics of the region.

Humid temperate environments

Climate. Humid temperate areas may be divided into regions dominated by maritime influences and those with a more continental regime. In general, maritime areas have a less extreme and variable climate.

Maritime humid temperate climates are found in Europe as far east (approximately) as central Poland, in southwest South America, the Pacific northwest coast of North America, Japan, New Zealand and Tasmania. The key climatic control in such areas is the continuous westerly movement of humid air masses, and weather patterns dominated by sequences of depressions. The ameliorating influence of the oceans is reflected in the low range in mean monthly temperatures, characteristically from 8 to 17 ºC. Precipitation is chiefly frontal, with convective storms also producing rainfall in summer, resulting in relatively constant precipitation through the year.

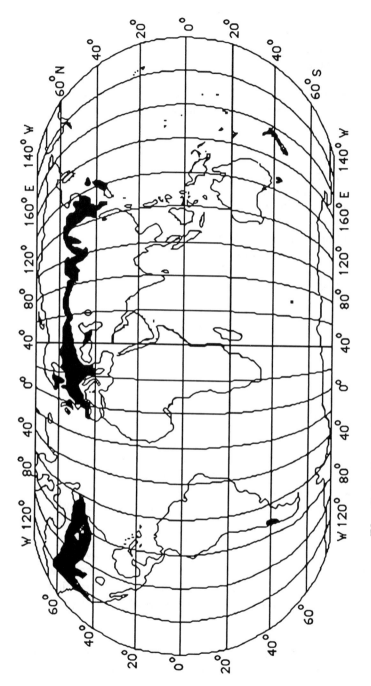

Fig.8.1 Generalized map of the humid temperate region

Winter precipitation tends to be higher in more coastal situations, due to the increased vigor of winter depressions, and orographic effects produce higher precipitation over west-facing uplands. Over much of the area, only a small proportion of precipitation falls as snow.

Continental humid temperate areas are found in eastern Europe, northeast North America, and northeast Asia and the Korean peninsula. These areas are dominated by anticyclonic conditions in both winter and summer, and consequently temperature ranges are much greater (mean monthly temperatures typically between -20 and 18 ºC). Most precipitation falls in summer during convective storms and a very high proportion of winter precipitation falls as snow (Unesco 1978 b). Unlike other continental humid temperate areas, the Korean peninsula is also affected by monsoons in autumn.

Annual precipitation totals increase with altitude and proximity to the sea. The highest rainfalls (over 2000*mm*) occur in uplands in western and central Europe and more generally in Chile, New Zealand, Japan and the Pacific Northwest. Rainfall less than 800*mm* is received in more inland areas in western Europe, and lower totals, less than 600*mm* , are characteristic of most continental humid temperate areas. The variation in annual totals between years is less than in more arid environments, and the coefficient of variation of annual rainfall is typically between 10 and 25% (Unesco 1978 b).

By definition, annual potential evaporation varies between 500 and 1000*mm*. There is relatively little variation over space (most of France has potential evaporation between 700 and 800*mm*, for example, while over much of the rest of temperate Europe it is between 600 and 700*mm* (Unesco 1978 b)), with variations closely related to temperature and therefore latitude and altitude, and more locally to wind exposure. Although the annual total is less than annual precipitation (again by definition), in virtually all humid temperate areas potential evaporation exceeds precipitation for part of the year. Fig.8.2 shows monthly totals for three sites. The pattern shown by the Dutch record is most representative and the French data illustrate the lower evaporative demand in upland areas. The Republic of Korea record (based only on one year) is more unusual in that the deficit occurs in winter, reflecting higher winter evapotranspiration and very heavy rainfalls during the autumn monsoon season. Potential evaporation varies less between years than annual rainfall (a standard deviation of 44.4*mm* at Winterswijk in the Netherlands compared with standard deviation of annual rainfall of 173*mm* (Buishand and Velds 1980).

Actual evaporation is less than potential, but the difference reduces as rainfall increases. For instance, a comparison of rainfall minus runoff (assumed to

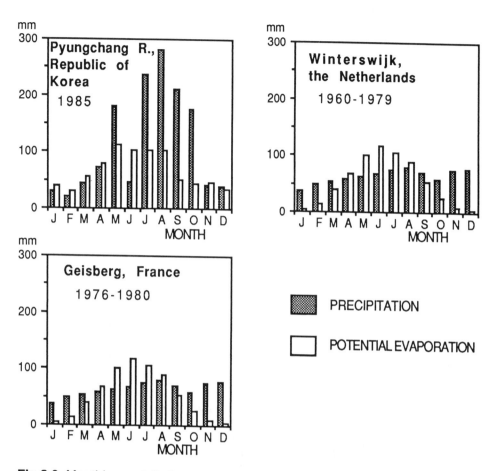

Fig.8.2 Monthly precipitation and potential evaporation (data from Ministry of Construction (1985-1986); Buishand and Velds (1980); Humbert (1982).

represent actual evaporation) with potential evaporation using data from many catchments in Britain (Institute of Hydrology 1980) produced the relationship shown in Fig.8.3. Actual evaporation varies much less over space than potential evaporation, and the annual value over most of humid temperate Europe is between 500 and 600*mm* (Unesco 1978 b). As climate becomes more continental, actual evaporation becomes more concentrated in the summer period and there may be no losses in winter.

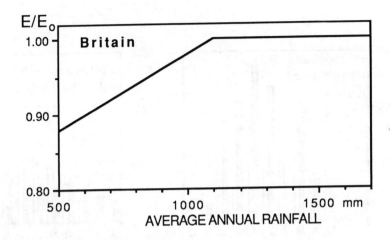

Fig.8.3 Approximate relationship between actual evaporation E and potential evaporation E_O in Britain (from Institute of Hydrology 1980).

Physiography and soils. Climate fundamentally determines the type, intensity and rate of geomorphological processes, but processes and land forms are influenced also by land cover and underlying structure. Humid temperate environments are characterized by low to moderate process intensities, and the classical landform model is of smooth, soil-covered slopes, gentle gradients and leisurely rivers. This pattern is disrupted by areas of steep relief - the European Alps and Carpathians, the New Zealand Alps, the Rocky Mountains in the Pacific Northwest, for example - where processes and land forms can be much more extreme. Humid temperate areas have many inherited land forms, most of which were fashioned during the colder conditions prevailing during the Pleistocene glaciations.

Soil types vary with past and present climate, the character of the underlying rock, relief and hydrology, time and land use. The great variability in soil characteristics is shown on the Unesco/FAO world soil map (FAO 1974), and Fig.8.4 represents an attempt to generalize relationships in humid temperate areas between relief, hydrology and soils. Although based on soils in England and Wales, this generalization is probably applicable at least to other maritime humid temperate environments. In general, lowland soils formed over sedimentary rocks are deep and well structured (although often clayey), with gleying in areas with gentler slopes. Resistant rocks in upland areas with harsher climates tend to produce thinner, more acid, soils subject to leaching and podsolization, while soils over chalk and limestone are thin and more calcareous.

168

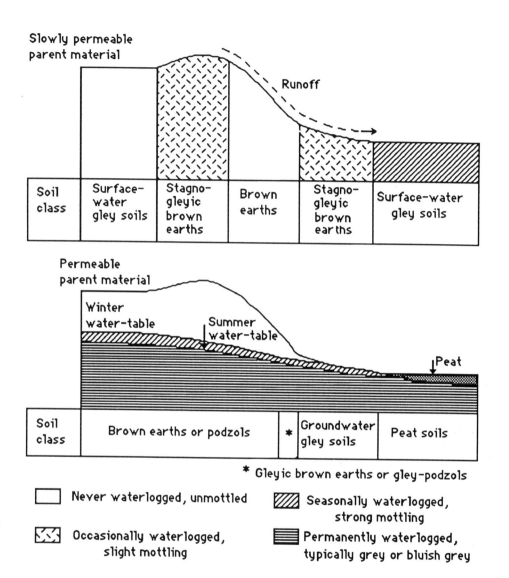

Fig.8.4 Relations between soil, relief and hydrological regime
in England and Wales (from Hodge et al 1984).

Land use. The natural vegetation of humid temperate areas is broad-leaf or
mixed coniferous and broad-leaf forest. However, and particularly in Europe, there

have been major and significant changes in land cover over several thousand years. Table 8.1 shows the percentage of land under four different uses for several humid temperate countries, and it is immediately clear that agriculture - cropland plus pasture - is generally the dominant use. The category "other land" includes land unsuitable for agriculture or forestry, land used on a temporary basis and urban and industrial land. In humid temperate areas very little land is too poor for continued use (the poorest land is used for forestry) so urban and industrial uses dominate the "other" category. Since the mid 19th century there has been a continued increase in urbanization, and this is taking place particularly at the expense of cropland (WRI/IIED 1986).

Population densities in Europe and the Koreas are among the highest in the world, reflecting long histories of economic and agricultural development. These

Table 8.1

LAND USE IN SELECTED HUMID TEMPERATE COUNTRIES, 1981-1983.
(Source: WRI/IIED 1986)

	Cropland %	Arable and meadows %	Forest %	Other land %	Pop. density $(p\,km^2)$	% Urban
Austria	19	25	39	17	89.3	56.1
Belgium	25	21	21	33	298.5	89.2
Bulgaria	38	18	35	9	83.1	68.6
Czechoslovakia	41	13	37	9	122.4	66.3
Denmark	62	6	12	20	119.4	85.9
France	34	23	27	16	99.8	77.2
German Dem. Rep.	47	12	28	13	153.6	78.2
Fed. Rep. Germany	31	19	30	20	116.1	57.0
Ireland	14	70	5	11	34.9	57.0
Netherlands	25	34	9	32	388.8	92.5
Poland	49	13	29	9	120.1	59.2
Romania	46	19	28	8	97.1	54.8
Switzerland	10	40	26	23	152.3	60.4
U.K.	29	48	9	14	227.3	91.7
New Zealand	2	54	37	7	12.2	83.7
PDR Korea	19	0	74	6	166.6	63.8
Rep. Korea	22	1	67	11	415.0	65.3

countries also tend to have a high proportion of their population in urban areas and industrial or sevice-based economies. Consequently pressures on land and water resources are most likely to come from the urban and industrial sectors.

Water balance components and regimes

Runoff generation. River flows in humid temperate areas derive from three sources, namely groundwater, subsurface flow and overland flow (including direct channel precipitation). The relative importance of these sources varies over both time and space, and a key concept central to the understanding of the dynamics of humid temperate runoff is that of the "variable source area". This concept was introduced by Hewlett and Hibbert (1967), and basically states that only a temporally variable portion of the catchment contributes to storm runoff. The key elements are shown in Fig.8.5, and Dunne (1978) provides a comprehensive review.

Natural infiltration capacities are high in humid temperate areas due to the dense vegetation cover, and Hortonian overland flow (where rainfall intensities exceed infiltration capacities) is therefore rare. Infiltrated water flows downslope (as "throughflow" or "interflow") along less permeable horizons or in pipes or macropores within the soil (Jones 1987), and saturated zones build up at the bases of slopes and in hillslope concavities. Water moves through the saturated zone more quickly than through unsaturated soil, and the channel flow hydrograph is therefore strongly influenced by the growth, contraction and location of saturated ("contributing") areas. Water may also be forced ("shunted") out of the bottom of a saturated zone by fresh water arriving at the top (Anderson and Burt 1982). As rain continues, the saturated zone may reach the ground surface. Subsurface flow will emerge as return flow and be combined with direct precipitation to generate saturated overland flow.

The relative importance of these processes varies between and within catchments. With deep permeable soils (and specially with steep slopes), subsurface flow dominates the response hydrograph, and although this may be rapid if shunting occurs or there is a dense macropore network, can produce peaks following perhaps several days after rainfall (Weyman 1974; Burt and Butcher 1985). If relatively impermeable horizons lie close to the surface and slopes are gentle, overland flow may be more important (Dunne and Black 1970), although Bonell et al (1984) observed that overland flow in a forested catchment in Luxembourg was severely limited by high hydraulic resistance.

171

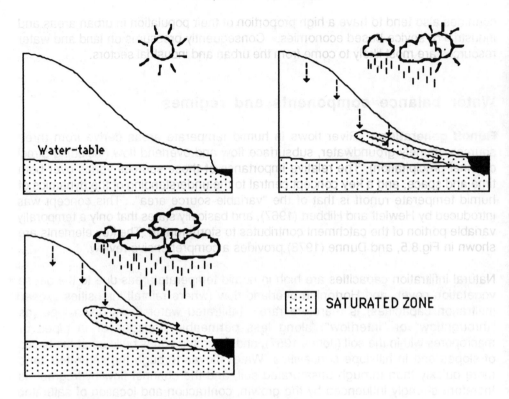

Water-table

SATURATED ZONE

Fig.8.5 Idealized summary of flow generation processes
in humid temperate areas.

There are similar variations in the source of base flow, which depends not only on geology but also on topography. The steeper the catchment the greater the proportion of total storage that is provided by the soil rather than groundwater storage, and in such catchments throughflow provides most of the base flow (Hewlett pers. comm. in Ward (1984)). In flatter catchments or catchments underlain by major aquifers groundwater will dominate base flow.

In practice it is difficult to separate the contributions from the different sources (although separation based on chemical composition has been widely used (Sklash and Farvolden 1982)), so a distinction is made between "quick flow" (the flow under the storm hydrograph) and "delayed flow", "base flow" or even "non-separated flow". Storm event data from Britain (Boorman 1985) show that on average approximately 40% of precipitation in a rainfall event goes to quick response runoff, although there is a great variation between catchments and events. Average catchment percentage runoffs vary from only 1.1% on one chalk catchment to 77.2% on a clay catchment, and runoff for a catchment in southwest England varied between flow events from 8.6 to 61%. Takei et al (1975) found that less than 30% of event rainfall went to quick flow in two small granitic basins in upland Japan. More simply, the base flow index (based on the separation of a daily hydrograph) measures the proportion of the total runoff that can be regarded as base flow (Institute of Hydrology 1980). Values of the base flow index are related to geology and soil type, and catchments in western Europe and New Zealand have values over a wide range from 0.15 to 0.99. The higher values are found in catchments with more permeable soils overlying extensive aquifers.

It is clear from the above that only certain parts of a catchment will contribute to the storm flow hydrograph. These contributing source areas will vary between and within rainfall events, and depend on topography, soil characteristics, the state of the vegetation (plant growth affects soil structure) and antecedent moisture conditions. The location of areas generating storm runoff is of practical significance for catchment management, as such areas are sources not only for water but also for sediment, and their development or alteration may have dramatic consequences. It is therefore necessary to be able to identify contributing areas in the field: Dunne et al (1975) describe some methods based on topography, soil and vegetation characteristics and Van de Griend and Engman (1985) review remote sensing techniques.

Annual water balances. The volume of runoff in a year depends primarily on the volume of precipitation, evaporative demand and soil and rock types. The highest volumes are found in western and mountainous areas, and can easily exceed 2000 *mm*. Low runoff volumes are found not only where precipitation is low but also where the underlying rock is permeable and much precipitation goes to groundwater storage. A large part of north central France south of Paris, for example, is underlain by a permeable chalk aquifer and has annual runoff volumes less than 100 *mm*; one catchment of 203 *km²* (the Val de Bonce at Montboissier) yields only 8 *mm* of annual runoff on average.

Graphs showing the relationship between annual rainfall and runoff for each year are frequently used to describe regime characteristics. The form of the relationship depends on basin properties, and there is a wide variation in the correlation between rainfall and runoff due to differences in the variability between years in evaporative demand and to the time lag between rainfall and response. Fig.8.6 shows relationships for four European catchments listed in Table 8.2, and indicates the variations found; there is a good correlation on the Beult in England, where there is a rapid response to rainfall inputs, and a poor correlation on the Pang which is fed by a chalk aquifer.

173

The long-term difference between precipitation and runoff varies little over humid temperate areas and variations depend primarily on the underlying characteristics of the catchment. Annual values in Western Europe vary from 300 to over 900mm with an average of approximately 470mm, but values for half of the stations sampled were within 410 and 560mm. Because the difference varies less over space than the precipitation, the runoff coefficient (ratio of runoff to rainfall) varies considerably. Runoff can be well over 90% of precipitation in high rainfall impermeable areas and less than 10% in some chalk catchments, although values of between 20 and 35% are, however, more typical in Western Europe. The difference between precipitation and runoff is proportionately higher in more continental areas as summer evaporative demand increases (Unesco 1978 b), and is less in wetter humid temperate environments such as the Korean peninsula, Japan and New Zealand (Unesco 1978 b). For example, Takei et al quote runoff coefficients of 51% and 71% for two upland granitic catchments in Japan.

Table 8.2

LOCATION AND AREA OF EXAMPLE CATCHMENTS

River and gaging station	Area (km^2)	Location
Beult at Stile Bridge	277.1	Southeast England
Kym at Meagre Farm	137.5	East England
Pang at Pangbourne	170.9	Southeast England
Severn at Plynlimon	8.8	Central Wales
Ammer at Oberammergau	114.0	Bavaria, Fed. Rep. of Germany
Grosser Regen at Zweisel	177.0	Lower Danube, Fed. Rep. Germany
Lange Bramke at Oberharz	0.8	Northeast Fed. Rep. of Germany
Reiche Ebrach at Herrnsdorf	269.0	Central Fed. Rep. of Germany
Schussen at Magenhaus	246.0	Bavaria, Fed. Rep. of Germany
Wurm at Randerath	305.0	Northwest Fed. Rep. of Germany
Zusam at Pfaffenhofen	505.0	Bavaria, Fed. Rep. of Germany
Gardon de Mialet at Roucan	239.0	South France
Layon at St George	250.0	West France
Maumont at La Chanourdie	162.0	Southwest France
Orgeval at Le Thiel	104.0	North central France
Vesanan at Halabeck	4.7	Southern Sweden
Lutschine at Gsteig	379.0	Central Switzerland
Yeongsan at Naju	2060.0	Southwest Rep. of Korea

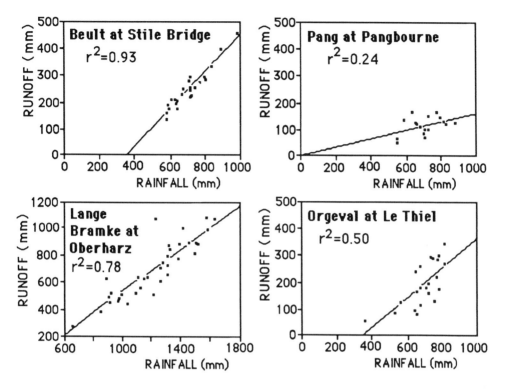

Fig.8.6 Relations between annual rainfall and runoff, for four of the catchments listed in Table 8.2.

Annual precipitation, runoff and their difference vary from year to year (as shown in Fig.8.6), but variability in humid temperate areas is less than in drier environments (Unesco 1978 b). The coefficient of variation (CV) measures this variability, and an average CV of annual runoff of 0.26 was calculated from 382 catchments with more than 20 years of record in Western Europe; 90% of the individual catchment values lay between 0.16 and 0.4, and there is evidence that the CV of annual runoff falls as average annual runoff increases. Less data are available on the relative variability of annual precipitation, runoff and their difference, but Table 8.3 shows the CV of these quantities for 9 catchments in Western Europe. Annual runoff varies more than annual rainfall, and Dacharry (1974) found in the upper Loire catchment that the less the amount of storage in the catchment and the greater the amount of evapotranspiration relative to precipitation, the greater the relative variability of annual runoff. The difference between precipitation and runoff varies less than annual runoff, but the relationship with variations in annual rainfall is less clear. The

175

variability depends not just on rainfall but also on the variability in evaporative demand, and the relative differences shown in Table 8.3 may be due to different variations in summer temperatures.

Table 8.3

COEFFICIENTS OF VARIATION OF ANNUAL RAINFALL, RUNOFF AND THEIR DIFFERENCE, FOR SELECTED CATCHMENTS IN WESTERN EUROPE

| | Coefficient of variation | | | No. of |
	Rain	Runoff	Difference	years
Borne at Chadrac Borne (France)*	0.135	0.283	0.131	45
Loire at Bas-en-Basset (France)*	0.135	0.330	0.173	40
Lignon at Pont de Lignon (France)*	0.157	0.428	0.276	25
Ance du Nord at Laprat (France)*	0.163	0.288	0.175	45
Harpers Bk at Old Mill Bridge (England)	0.142	0.319	0.150	46
Beult at Stile Bridge (England)	0.145	0.299	0.082	23
Exe at Thorverton (England)	0.126	0.187	0.084	29
Lange Bramke at Oberharz (F.R. Germany)	0.194	0.305	0.213	35
Orgeval at Le Thiel (France)	0.151	0.428	0.150	22

* from Dacharry (1974)

River flow regimes. There have been several attempts to classify global flow regimes into a series of groups, with the best known being that of Parde (1947). At smaller scales Grimm (1968) defined typical flow regimes in Europe and Aschwanden and Weingartner (1985) have defined regimes in Switzerland. However, it is difficult to classify what is essentially a continuous process into discrete groups, and a better indication of flow regimes in humid temperate areas may be obtained by considering a few selected catchments. Accordingly, histograms showing the variation in flow between months for eight small catchments in Western Europe (listed in Table 8.2) are given in Fig.8.7. The actual pattern of flow during the year depends on both precipitation and evapotranspiration, and on the soil and geological characteristics of the catchment.

Most of the regimes shown in Fig.8.7 are characterized by maximum flow in winter, with the relative dominance of winter flow depending on the distribution of precipitation. In the west of Europe precipitation is highest in winter and evapotranspiration is highest in summer, leading to a larger difference between summer and winter flows. In more continental areas most precipitation falls in

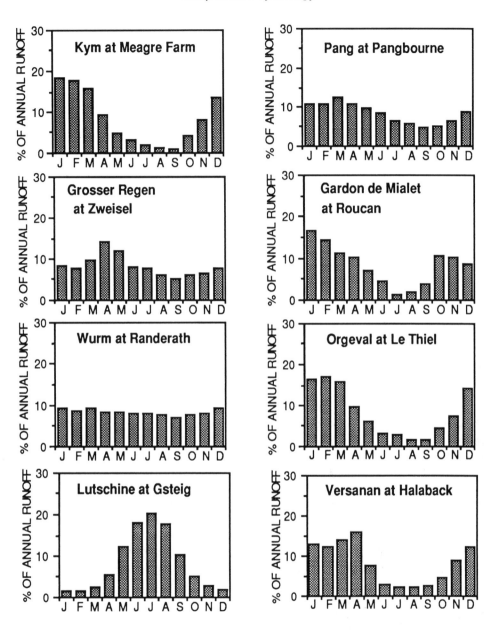

Fig.8.7 Monthly runoff histograms, showing average monthly runoff as a percentage of average annual runoff, for eight of the catchments listed in Table 8.2.

summer when evapotranspiration is also highest, and there is thus less difference between summer and winter flows. In mountainous and continental areas snowmelt has an important influence on seasonal flow regimes, as shown in the histograms for the Lutschine, Grosser Regen and Versanan catchments.

The importance of geology in influencing seasonal runoff patterns can be seen by comparing the histogram for the Pang catchment - underlain by chalk - and the Kym catchment which lies on impermeable clay. The two catchments are exposed to similar maritime humid temperate climates, but the difference in regime is striking: the greater infiltration into and storage capacity of the chalk means that high flows are sustained through summer, and the peak flow occurs in spring after the period of highest precipitation.

An indication of the short-term variation in flows is given in Fig.8.8, which shows annual hydrographs for six of the catchments listed in Table 8.2. The contrast between the slowly-responding chalk Pang and the flashy clay Kym catchments is emphasized, as is the difference between the Pang and the Wurm in northern Germany. These two catchments have similar monthly flow regimes, but from the annual hydrographs it is clear that this is not due to similar geologies, and therefore the similar monthly regimes may result from rather different combinations of inputs and catchment types.

Low flows. The low flow behavior of rivers in any climatic region may be characterized by a range of techniques for describing the properties of the flow hydrograph. They may be grouped into three different classes: the first analyzes the statistics of discharge values, the second the durations that the flow is above or below a given value, and the third the relationships between storage volumes and reservoir yield. Fig.8.9 shows an example of the second of these which illustrates the flow duration curves for six rivers. The importance of geology in controlling runoff patterns has already been noted, and this process is most marked at extreme low flows. For example, the 95 percentile discharge (the flow exceeded on average in all but 18 days in each year) is 35% of the average flow on the aquifer-fed Pang river compared with 3% of the average flow from the clay Kym. Fig.8.10 shows the 10 day annual minimum series for four catchments with different hydrogeologies. The strong control of catchment geology is evident.

Table 8.4 shows the mean and range of the 95 percentile discharge and mean annual 10 day minimum for seven areas in Europe. There is a high correlation between the two low flow statistics (Institute of Hydrology 1980) and the mean and range are similar in most regions. The Rhone basin has the lowest values of low flow statistics, reflecting the very low rainfall in the summer months and low groundwater storage to sustain summer flows. Within any region the hydrogeology

178

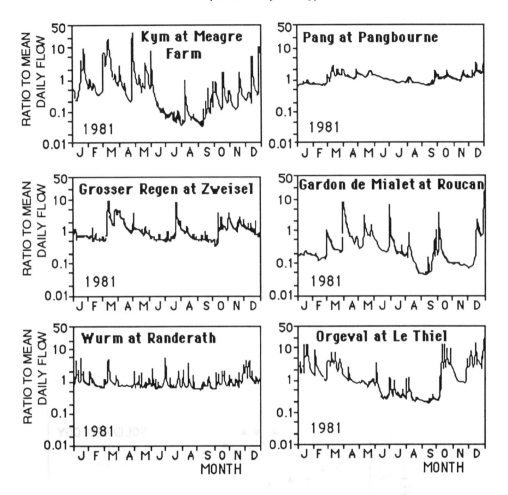

Fig.8.8 Daily flow hydrographs for six of the catchments listed in Table 8.2.

is the main influence on low flows, although very high drought flows may be found on impermeable lithologies where lake storage dominates the outflow hydrograph. In order to develop low flow estimation techniques for ungaged sites it is necessary to index the role of catchment geology, and in Britain (Institute of Hydrology 1980) this was achieved by calculating a base flow index using a simple base flow separation procedure applied to daily mean flow series. The index relates closely to both low flow measures and catchment geology, and is currently being computed over a large area in northwest Europe as part of IHP Project 6.1(Flow regimes from experimental and network data). Similar values to those in Europe have been computed in New Zealand (National Water and Soil Conservation Authority 1984).

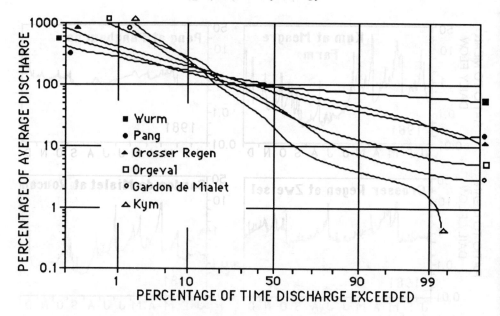

Fig.8.9 Flow duration curves for six of the catchments listed in Table 8.2.

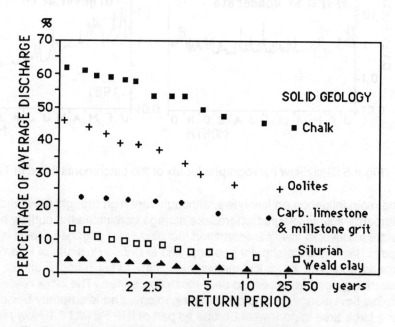

Fig.8.10 Annual minimum 10-day flows with different
geology in Britain (from Gustard 1983).

Table 8.4

LOW FLOW STATISTICS FROM WESTERN EUROPE

Location	No. of catchments	Q95[*] Mean	Range	MAM(10)[#] Mean	Range
Rhine basin (F. R. Germany)	134	21.1	2.6 - 54.6	24.0	3.3 - 61.6
Seine (Belgium/France)	93	23.5	1.1 - 69.1	29.1	1.7 - 88.1
Upper Danube	28	31.8	12.0 - 61.1	32.4	11.1 - 60.1
Rhone basin	29	8.0	1.2 - 21.5	8.8	1.6 - 22.3
Bretagne, Loire, S.W. France	103	13.0	0.1 - 43.8	14.1	0.8 - 49.5
Denmark	20	24.0	2.0 - 67.1	26.4	3.4 - 68.8
S.E. England	94	21.9	0.0 - 74.8	23.2	0.0 - 79.0

[*] 95 percentile discharge expressed as % of mean flow
[#] mean of 10 day annual minimum discharges as % of mean flow

Other studies have related low flows to other catchment characteristics, including stream order and basin area (Yoon 1975) and catchment geology (Musiake et al 1975).

Although low flow statistics provide a useful summary of the regime they do not convey information about the spatial or temporal distribution of flows; these can be illustrated by examining historical droughts. However, there may be difficulties in the definition of a drought, and Beran and Rodier (1985) distinguished six types of drought according to their duration, season of year and severity. Table 8.5 summarizes the most notable droughts in western Europe between 1970 and 1979.

The decade was a period of exceptionally low flows in every country in northwest Europe with the most extreme and widespread drought occurring in 1975 and 1976 (Doornkamp et al 1980). More recently the 1984 drought, although less widespread and of shorter duration, was more severe in Wales and western England (Marsh and Lees 1985). Recent droughts have not been confined to Europe: the period 1976-1977 was one of exceptional low flows in parts of the Pacific Northwest coast of North America.

Comparative hydrology

Table 8.5

SUMMARY OF DROUGHTS IN WESTERN EUROPE 1970-1979
(Source: Beran and Rodier1985)

1971   Exceptionally dry in most European countries; lowest levels on Rhine since 1818.

1972   European territories of USSR; lowest flows of previous 5-80 years.

1973   Winter 1972-Spring 1973; driest in 200 years in eastern U.K., second driest year on record in Czechoslovakia.

1974   Unprecendented 9 week drought in parts of Sweden, Denmark and the Netherlands.

1975   Widespread European drought.

1976   Continuation of below average rainfall over much of Europe. U.K.16 month rainfall lowest since 1727, Belgium driest since 1921,Denmark since records began in 1874, Netherlands February-August rainfall least for 125 years. Very severe drought in Czechoslovakia, East Germany and Hungary.

1978   September to November rainfall in S.E. England least since 1752. Driest October and November on record in parts of western France.

Floods. The distinction between maritime and continental humid temperate climates is reflected in patterns of flood behavior. In maritime areas the largest annual flood usually occurs in autumn or winter following prolonged heavy cyclonic rainfall (with a snowmelt contribution in some years), but many large floods have occurred after intense thunderstorms in summer and early autumn. In more continental areas there may be two flood seasons: one in the spring caused by snowmelt (often in combination with some rain), and one in autumn or early winter.

The average magnitude of floods is measured by the arithmetic mean of the series of annual maximum floods. This is fundamentally controlled by catchment area, but investigations using data from western Europe have shown that channel slope, catchment rainfall and soil characteristics are also important. Analysis of British data, for example, yielded an equation predicting the mean annual flood from catchment area, slope, soil type, drainage density, storm rainfall and lake storage (NERC 1975), and studies in New Zealand related the mean annual flood to catchment area and rainfall (Beable and McKerchar 1982).

182

The variability of annual floods at a site is indexed by the coefficient of variation. In western Europe this varies from 0.13 to 1.3, with most catchments having values between 0.29 and 0.7. Fig.8.11 shows frequency curves for several catchments (listed in Table 8.2) with different hydrological and catchment characteristics. The curves are general extreme value distributions fitted by the method of probability weighted moments, and are all standardized by the mean annual flood. The Reiche Ebrach, Maumont and Layon data show curves typical of lowland catchments with low annual rainfall: large floods are caused by summer thunderstorms, and several years have very low annual maxima. Lowland catchments with more maritime climates have flatter frequency curves, as shown by the Ammer and Zusam data.

Slightly steeper curves are found in upland catchments (the upper Severn, for example) although flatter curves may be observed in upland catchments with lakes or more significant snowmelt (as in the Schussen).

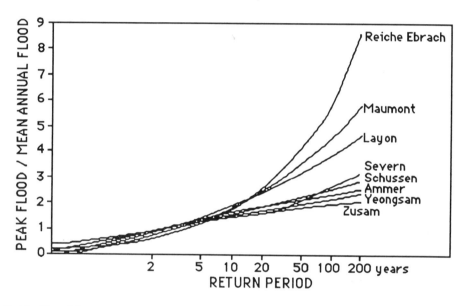

Fig.8.11 Flood frequency curves for eight of the catchments listed in Table 8.2.

Soil water regimes. As a generalization, soils in humid temperate areas dry during summer and become wetter - and perhaps waterlogged - during winter. Climate determines the large scale pattern in variations in soil water over time and through the soil profile, but depth to groundwater, soil type and land use all significantly affect the details of response to rainfall and evapotranspiration.

The closer the water-table is to the surface, the greater the influence that groundwater, rather than climate, will have on soil water content. Fig.8.12 illustrates this, showing soil water content at $0.9m$ depth varying much more in the Mahlkunzig with the deeper water-table. Soil water variations over time are greatest in looser textured, less well consolidated soils (such as those with high sand contents) due to the lower water tension and higher hydraulic conductivity. Profiles for two nearby sites in central England are given in Fig.8.13. The effects of land use are illustrated in Fig.8.14, which shows that soil water contents at a given depth are lower under forest.

The variability of climatic inputs - particularly precipitation - means that soil water contents and accumulated deficits vary between years. Fig.8.15 shows that both the duration and timing of soil water deficit differ between years, although as with short term variability, patterns depend on soil, groundwater and land use characteristics.

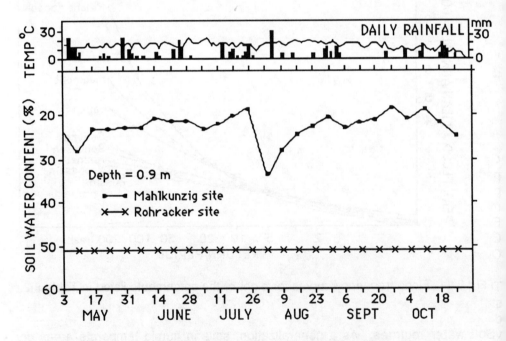

Fig.8.12 Effect of groundwater level on soil water (shallow water-table at Rohracker, deep water-table at Mahlkunzig). (Source: Morgenschweis 1980)

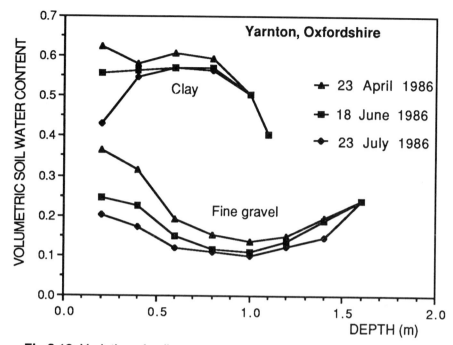

Fig.8.13 Variation of soil water with soil type, date and depth
(from Gardner 1986).

Groundwater resources and regimes.   The groundwater regime within a particular aquifer is governed by the volume and seasonality of inputs, aquifer permeability, the extent and permeability of cover, and the degree of groundwater management.   In general terms, the three most important aquifers in western Europe are Permo-Triassic sandstones (particularly Bunter and Keuper series), the Cretaceous chalks (particularly Upper and Middle Chalk) and Tertiary and Quaternary sands and gravels (Commission of the European Communities 1982).

The aquifer potential of the Permo-Triassic sandstones is related to their high storage capability and relatively high permeability, which may vary widely with the degree of cementation. Poorly cemented Bunter sandstones have typical porosity values of 30% in the British Midlands (Rodda et al 1976), but may decrease to 5% in highly cemented sandstones in southwest England, and porosities of 1% to 6% nearer faults have been reported for Bunter sandstone in the Federal Republic of Germany (Keller 1979). Typical transmissivity values for the Permo-Triassic sandstones vary with the relative importance of fissure and inter-granular flow (Rodda et al 1976).

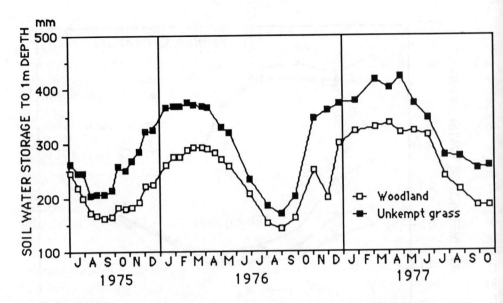

Fig.8.14 Effect of vegetation on soil water content (from Gardner 1981).

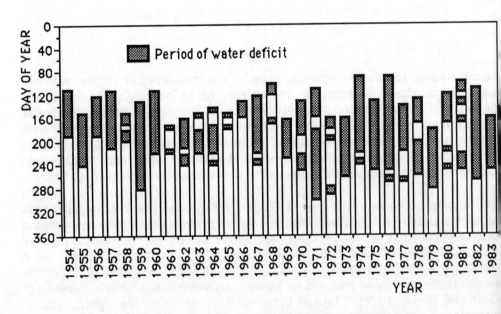

Fig.8.15 Periods with water deficits in a heavy clay soil under grass (from Bouma 1986).

Although the primary porosity of chalk is low, its aquifer potential arises from high secondary porosity and permeability due to fissures and joints. The porosity of Upper Chalk in southern England generally lies between 41 and 50% compared with 21-30% for Lower Chalk, but values for the Chalk in northern England are generally lower (Rodda et al 1976). Transmissivities of the Upper Chalk are broadly similar throughout Western Europe, but with wide local variations in relation to topography and structure.

Tertiary and Quaternary sands and gravels, of fluvio-glacial origin, constitute an important aquifer because of their low transmissivity rate and high storage capacity. Deposits, typically more than $20m$ thick but more than $200m$ thick in the upper Rhine fault, are most extensive in Belgium, Denmark, France, the Federal Republic of Germany and Ireland (Commission of the European Communities 1982). Coastal aquifers, however, are prone to saline intrusion.

Groundwater recharge is a function of the volume of residual rainfall, surface infiltration and geological percolation rates. Annual recharge has been estimated for the U.K. at 25-50mm on Boulder Clay in Essex, increasing to nearly 100mm in Norfolk and on the Chalk in S.E. England, rising locally to 350mm on Chalk on Salisbury Plain (Rodda et al 1976). Preuss (1975) found rates of 62-163mm y^{-1} for sand and gravel aquifers in northern Germany. It is possible to infer approximate recharge by annual hydrograph separation: analysis in western Europe reveals that approximately 80% of runoff (about 150-200mm) from the Paris Basin is derived from groundwater. Even higher proportions of runoff come from groundwater in more heavily fissured limestone areas.

Variations in both residual rainfall and infiltration rates with time mean that groundwater recharge may vary both annually and seasonally. In chalk in southern England, for example, there may be only125mm of percolation with 500mm of rainfall, but up to 500mm of percolation with 900mm (Rodda et al 1976). As factors affecting infiltration are seasonal, so mean groundwater storage in maritime areas follows an annual pattern of winter recharge and summer depletion in sympathy with soil water regimes. This seasonal fluctuation of late winter peaks and early autumn minima is summarized by the annual maximum, mean and minimum water-table levels for a chalk aquifer in southern England (Fig.8.16), although the magnitude of the variation of course varies greatly.

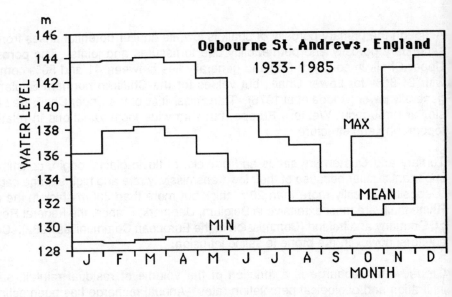

Fig.8.16 Mean, maximum and minimum groundwater levels in a chalk aquifer.

Sediment transport and water quality

Sediment transport.   Peak and mean suspended sediment transport in Europe are low on a global scale (Walling and Kleo 1979), although there are large variations between catchments. Loads decrease as basin size increases and increase with relief, but also depend on land cover, soil and rock characteristics and human interference which together influence material availability. In western Germany, for example, mean annual yields are mostly less than $50t\ km^{-2}$, but vary from 6 to $300t\ km^{-2}$. Walling and Webb (1981) found in Britain that annual yields could be up to $500t\ km^{-2}$ in wet upland areas and less than $1\ t\ km^{-2}$ in flatter, larger basins (or small basins with very resistant bedrock), but claimed a value of $50t\ km^{-2}$ was more typical. Sediment yields in other humid temperate areas can be higher than those in Europe: an estimate of $7000t\ y^{-1}km^{-2}$ has been made for the Waipaoa River in New Zealand (Fournier 1969). These areas are characterized by high rainfall amounts and intensities, steep relief and easily erodible surface materials. In areas of very steep relief, debris torrents may pose a significant hazard. Slaymaker et al (1987) describe torrents triggered by rockfalls and soil slides in the Cascade Range, British Columbia, one of which transported over $60\ 000m^3$ of material in a few hours following an intense rainstorm.

Bedload constitutes only a small proportion of total solid transport in humic temperate rivers. Gregory and Walling (1973), for example, found that bedload only accounted for 1.1 to 11% of total sediment load in some small catchments in southwest England. In steeper mountainous catchments, however, bedload may be a much higher proportion of total load.

Solute transport. Concentrations of dissolved material vary less over space and time than suspended sediment concentrations, and depend on the chemical composition of atmospheric inputs, the amount of precipitation that goes to runoff and, perhaps most importantly, the characteristics of the underlying soil and rock. Walling and Webb (1981) presented typical concentrations for rivers draining different rock types in England and Wales, showing for example that specific conductances of 40 000 -80 000μS m^{-1} can be expected from chalk catchments, while specific con- ductances of less than 4000 to 9000μ S m^{-1} are typical of more resistant granitic and metamorphic bedrocks.

Total solute loads depend not only on concentrations but also on runoff volumes, and areas with the greatest loads do not coincide with areas of greatest concentration (Walling and Webb 1981). Loads in Britain range from 10 to 400t $y^{-1}km^{-2}$, with the greatest values in central England; higher concentrations in the east coincide with lower runoff volumes.

In humid temperate environments more material is removed from a basin as solute load than in solid transport. Between 55 and 77% of total transported load in several small basins in southwest England was found to be in solute form (Gregory and Walling 1973), while Probst (1986) discovered that 68% of material transport in a basin in southwest France was accounted for by solute load. Solute load can in some areas be very much more important - between 93 and 95% in two Polish rivers (Jaworska 1968), for example - but as relief increases and stream velocities rise solid transport dominates.

Water use

By definition, humid temperate areas are well supplied with water, but this water is rarely found exactly where and when it is needed. Table 8.6 shows for many humid temperate countries (and also, for comparative purposes, some other countries) long term average total and per capita resources. Many countries in humid temperate areas are abundantly supplied with water, but others - such as Belgium, England and Wales and Poland - have relatively low resources per capita. This is generally not due simply to a lack of rainfall but also to high population densities (although Poland's problems are compounded by low rainfall). Table 8.6 also shows

189

Table 8.6

RESOURCE AVAILABILITY AND WITHDRAWALS IN SELECTED COUNTRIES
(Sources: WRI/IIED 1986; Van der Leeden 1975)

	Long-term average resources			1985 pop. (10^3p)	Resources per head $(l\,d^{-1})$	Withdrawals	
	$mm\,y^{-1}$	$(10^3$ $Ml\,d^{-1})$	%from upstream			$(10^3$ $Ml\,d^{-1})$	%of res.
Austria	1071	246.4	38.9	7487	32910	7.5	3.1
Belgium	379	34.2	31.3	9880	3464	22.6	66.1
Bulgaria	1774	539.4	90.9	9220	58500	?	?
Czechoslovakia	703	246.4	68.9	15648	15746	13.7	5.6
Denmark	300	35.3	0.0	5144	6866	2.1	6.0
France	329	492.8	18.8	54608	9025	73.9	15.0
German Dem. Rep.	243	71.7	38.2	16642	4310	22.7	31.7
Fed. Rep.Germany	643	438.1	52.5	61106	7169	90.9	20.8
Hungary	1290	328.5	95.0	10797	30429	15.3	4.7
Ireland	424	119.6	0.0	3595	33281	4.6	3.8
Netherlands	2446	247.8	88.9	14506	17081	39.4	15.9
Poland	188	161.0	10.9	37556	4287	42.5	27.0
Romania	807	525.7	79.9	23065	22791	16.8	3.2
Switzerland	1220	136.9	16.0	6289	21767	6.8	5.0
England and Wales	452	187.9	0.0	49918	3763	26.3	14.0
European USSR	248	3616.7	15.4	199900	18189	?	?
New Zealand	1476	1086.9	0.0	3291	330273	2.8	0.2
Rep. of Korea	643	172.5	0.0	40872	4220	29.3	17.0
Kenya	25.4	40.5	?	20600	1967	?	?
Tunisia	20.4	9.2	?	7209	1272	2.9	32.0
Malaysia	1382	1248.5	?	15551	80282	25.8	2.0
Iraq	194	231.3	?	15676	14758	115.8	50.1
Sudan	7.4	50.7	?	21550	2350	49.7	98.1

withdrawals and the proportion of theoretically available water that is withdrawn, but it must be remembered of course that not all water withdrawn is consumed and the same water may be withdrawn several times. Nevertheless, the Table gives an indication of countries with the greatest pressures on water resources.

The gross figures shown in Table 8.6, however, conceal more than they reveal, and say nothing about more localized variations in resources and demand (Table 8.7 shows resources per person within England and Wales), about the quality of available water (90% of the water available in the Netherlands comes from the Rhine and the Meuse, and these rivers - particularly the Rhine - are heavily polluted), about variations in supply and demand during the year, or about stresses imposed in drought years. These issues will be considered more closely in the following sections which deal with water use by sector.

Table 8.7

WATER RESOURCE AVAILABILITY IN ENGLAND AND WALES
(Runoff data from U.K. Surface Water Archive)

Water authority	Long-term average resources $(Ml d^{-1})$	Resident pop. $(10^3 p)$	Resources per head $(l d^{-1})$
North West	30586	6866	4455
Northumbrian	11680	2619	4460
Severn-Trent	20698	8315	2489
Yorkshire	15157	4623	3279
Anglian	11235	5157	2179
Thames	7711	11565	667
Southern	10400	3944	2637
Wessex	9993	2340	4271
South West	22051	1442	15292
Welsh	51227	3047	16812

Domestic, commercial and industrial use.　　Domestic, commercial and some industrial users require high quality, clean water. In general, groundwater is of higher quality than surface water (which needs at least primary treatment to remove sediments), and is thus preferred for public supply needs. The proportion of water supplied from groundwater sources of course varies with geology, and in some areas (Wales, for example), resources of extractable groundwater are very small. Abstractions from groundwater in western Europe are shown in maps prepared by the Commission of the European Communities (1982). These show that over much of Europe less than 25mm is withdrawn each year, but that very much higher abstractions can be found around urban and industrial areas: over 500mm are withdrawn each year in the Ruhr area in the Federal Republic of Germany, for

191

Comparative hydrology

example. The maps also show that resources in some industrial areas are currently seriously over-exploited. As an alternative to groundwater, water is therefore taken either directly from rivers or from impounding reservoirs; in England and Wales all three sources contribute equally.

Other industrial users withdraw water themselves, and although some is clean most is not. Indeed, much water directly abstracted by industry is used for cooling (83% in the Netherlands (Colenbrander 1986)), and virtually all comes from surface water. Power stations are major withdrawers of water in humid temperate environments, and in both the Netherlands and the Federal Republic of Germany withdraw more than the rest of industry. Virtually all is used for cooling and is taken from surface waters. In many countries demand for cooling water is increasing, but in Britain it has fallen, reflecting a move towards coastal locations.

Table 8.8 shows change over time in supplies to different sectors in England and Wales. A steady increase in public water supply is apparent, which, as in other countries, is attributable to an increase in domestic consumption associated with increased use of such items as dishwashers and washing machines. Demand is forecast to rise still further (Archibald 1983), with projections depending on

Table 8.8

WITHDRAWALS BY SECTOR IN ENGLAND AND WALES
(Source: WAA 1987)

Year	Withdrawals ($Ml\,d^{-1}$)				
	Public supply	Direct to industry	Spray irrigation	Other agric.	Power station
1975	15360	6560	111	94	13714
1976	15009	6655	161	96	13211
1977	14747	6958	116	120	13406
1978	15828	6626	81	150	12539
1979	16267	6762	106	139	12710
1980	16186	5034	92	133	10474
1981	16105	4973	117	111	9284
1982	16331	4729	139	117	8931
1983	16224	4093	171	118	8254
1984	16402	3892	199	122	7052
1985	16641	3939	107	130	4369

assumptions made about toilet flushing (which accounts for 32% of domestic water use in England and Wales (WAA 1987)). In contrast, demand in Sweden has already levelled off, following the installation of modern equipment in most households.

Industrial consumption from public and particularly direct abstractions are declining (at least in relative terms) in most of western Europe (in Sweden and the Netherlands (Colenbrander 1986), for example). This is due both to a decline in traditional heavy water-using industries such as steel (a tonne of steel requires an average of 100 000 *l*, for example (WAA 1987)), and an increase in efficiency of water use. The ratio of water used to water abstracted (the "recycling ratio") in industry in the Federal Republic of Germany rose from 2.65 to 3.00 between 1975 and 1977, for example (DOCTER 1987).

Patterns of supply and demand obviously vary through the year, with greatest stresses in summer when flows in maritime areas tend to be least (Fig.8.7). In such areas, impounding schemes are therefore necessary to hold the winter surplus through to summer. Available water is easily sufficient to satisfy summer demand in most years, but occasionally there are shortages. Rainfall was well below average over much of northwest Europe between May 1975 and August 1976, for example, with England and Wales particularly affected. Many water supply reservoirs fell to very low levels (one in southwest England fell to only 8% of capacity (Gardiner 1980)), and restrictions on supplies were introduced. Domestic supplies were reduced by 5 to 10% over much of England, and by up to 25% in parts of Wales. In a more continental environment, maximum flows occur in spring and summer (Fig.8.7), and reservoir storage is therefore necessary to maintain supplies (and provide for hydro power) over winter.

The recent droughts in western Europe heightened awareness among the public and water resource planners of the variability of water resources and potential limits to the supply of clean water, even in humid temperate areas. They have also had a longer term effect on consumption, and the 1975-1976 drought is credited with encouraging industrialists to examine carefully their water usage.

Water and agriculture. Of all uses of water, agriculture is the one most strongly influenced by climate and the availability of water resources. Plants are the major consumers of water, and since most land in humid temperate areas is devoted to agriculture (including forestry) a large proportion of evapotranspiration can be considered to be agricultural consumption. Agriculture also uses a sizeable proportion of surface and groundwater resources, but in general in humid temperate areas a surplus of water is more of a problem to agriculture than a shortage.

193

Table 8.9 shows for several humid temperate countries the proportions of agricultural land irrigated and drained, and it is clear that drainage is much more extensive than irrigation. Field drainage is necessary when soil water contents are too high, as waterlogging inhibits plant growth, encourages disease and restricts access by heavy machinery. Most field drainage in Europe is concentrated around the Baltic, North and Irish Sea coasts, primarily for climatic reasons (Green 1980). In these areas drainage is necessary to reduce waterlogging before and after the growing season (there is a winter effective rainfall maximum in these temperate coastal areas) to allow the use of heavy machinery. In more central parts of Europe with continental climates characterized by a summer rainfall maximum, drainage is required to reduce waterlogging during the growing season. Fields may be drained by open ditches or buried drains, and the relative proportions vary greatly: virtually all field drainage in Hungary, for example, is by open drain while 90% of field drainage in Poland is by underground drains (Framji et al 1982). The large increase

Table 8.9

PROPORTION OF FARMLAND WITH FIELD DRAINAGE AND IRRIGATION
(Sources: WRI/IIED 1986; Framji et al 1982; Green 1980; Higgins et al1987)

Country	Field drainage	Irrigation	
	% of crop and pasture	% of crop and pasture	% of crop only
Austria	7.0	0.0	0.0
Belgium	25.0	0.0	0.0
Bulgaria	1.5	19.4	29.3
Czechoslovakia	14.0	2.3	3.0
Denmark	49.0	13.9	15.0
France	7.0	3.6	6.0
German Dem. Rep.	20.0	2.4	3.0
Fed. Rep. Germany	35.0	2.4	4.0
Hungary	63.0	2.4	3.0
Netherlands	45.0	21.1	59.4
Poland	23.0	0.8	1.0
Romania	?	16.2	24.7
Switzerland	16.0	1.2	6.0
England and Wales	50.0	1.1	1.6
New Zealand	?	1.3	51.1
PDR Korea	?	52.2	55.4
Rep. of Korea	?	46.8	47.0

in drainage all over Europe since the 1960's has been stimulated not just by the need for drainage and improving technologies but also, and more importantly, by such influences as government grant aid and agricultural subsidies. Framji et al (1982) reported large predicted increases in drainage in many countries, but the future rate of change in drainage will depend on government attitudes to food and agricultural production and surpluses.

Irrigation in humid temperate areas is necessary not to initiate agricultural production but to improve the quality and consistency of yields. Some irrigation of soft fruits and vegetables is also undertaken in spring to prevent frost damage (Framji et al 1982). Table 8.9 emphasizes the small proportions of agricultural land irrigated in humid temperate areas, although there are some important exceptions. A large proportion of farmland in the Korean peninsula is used for rice production in paddy fields, the Netherlands has a large area of horticultural and vegetable cultivation, and Romania and Bulgaria both have extensive areas with low rainfall during the growing season. Irrigation need varies considerably between years, and Fig.8.17 shows the theoretical frequency of need in England and Wales (defined as the frequency with which potential evaporation exceeds rainfall by more than $75mm$). In much of the U.K. irrigation is rarely required, but in some parts - specially in the drier east - it may be beneficial to crop growth nearly every year. A review of irrigation requirements in England and Wales (ACAH 1980) estimated that, in a dry year, approximately $86 \times 10^6 m^3$ of water would be demanded (equivalent to $235Ml\, d^{-1}$, although need is of course concentrated in summer) by farmers to irrigate approximately $125\ 000ha$, with the greatest area under irrigation for grassland, potatoes and vegetables. There is, however, a great variation in demand between years (as shown in Table 8.8) with the greatest withdrawals of course in dry years. Withdrawals for irrigation are less than abstractions by other users, but virtually all is consumed and is thus a loss. The greatest demand for irrigation water inevitably comes in areas with the greatest pressures on water resources. Over 50% of irrigation water in England and Wales, for example, is used in the east of England where rainfall - particularly in summer - is least.

The area of land under irrigation in Europe has increased over the last 20 years, although in some countries the rate of increase is declining as the most economically effective schemes are completed. Large increases in England and Wales occurred after the 1976 drought, and Evans (1983) expected a 2-3 fold increase by the end of the century. Czechoslovakia plans a doubling in the area under irrigation by 1990 (Framji et al 1982). Most irrigation schemes developed since the mid 1960's have been spray (or sprinkler) irrigation schemes, but earlier schemes have used flood or furrow irrigation. Many schemes have also been developed (in the Netherlands, for example) to maintain soil water and groundwater levels by controlling surface water levels.

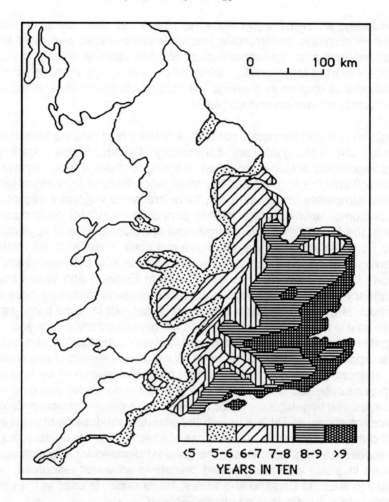

Fig.8.17 Frequency of irrigation need in England and Wales, expressed as the number of years in ten (after Pearl 1954).

Irrigation is not the only agricultural use of water in humid temperate areas and may not even be the most important, but is often the only use for which statistics are available. Irrigation uses only approximately 46% of agricultural withdrawals in the German Democratic Republic (Framji et al 1982), with the rest used predominantly by livestock. In England and Wales approximately $170 \times 10^6 m^3$ are used each year for feeding and cleaning livestock, specially dairy cattle (ACAH 1980), and this is twice as much as would be used for irrigation even in a dry year. Total consumption is less varied from year to year than abstractions for irrigation (depending more on

196

the number of livestock than rainfall deficits) and only modest growth is forecast (ACAH 1980). Water for livestock needs to be of high quality, and much (70% in England and Wales) is therefore taken from public supplies.

Implications for water resource management. The preceding sections have indicated that while humid temperate areas are characterized by large supplies of water, pressures on water resources can be very high and overall demand is increasing. It is particularly important to secure adequate supplies of good quality water, and subsequent sections will indicate the extent of human influences and the consequences of water use on water quality. Supply management involves the provision of the infrastructure - predominantly physical - to supply water to consumers, whereas demand management is concerned with providing the infrastructure - predominantly institutional, legal or operational - for adjusting demand for water. Both methods are used extensively in humid temperate areas.

The most obvious and common supply management technique is to provide reservoirs to maintain water supplies during the drier part of the year. The ready availability of water in humid temperate areas means that reservoirs do not need to store water between years, although some of the larger reservoirs in Britain are designed to accomodate two summers and one winter. Most early reservoirs were of the direct supply type where an aqueduct carried water direct from an upland reservoir to the demand centre. Recent developments have regulated the river to maintain flows at a downstream abstraction point by making occasional releases from an upstream regulating reservoir. This provides a much greater yield for the same reservoir storage. In some areas (e.g. London) pumped storage reservoirs are used to maintain supplies. In densely-populated areas, however, it can be difficult to find sites for reservoirs. The solution of Poland's water management problems, for example, is constrained by the lack of suitable reservoir sites; large areas of the country have gentle relief and permeable geology, high population densities and economic activity concentrated in river valleys (Laski 1977).

Other supply management techniques include aquifer recharge or groundwater augmentation, where aquifers are artificially recharged from surface water to enable extraction at a greater rate than natural recharge, and short-distance water transfer between basins. Large-scale water transfer schemes have not been implemented due to their high cost, although a "national water grid" for the U.K. has frequently been proposed to enable the transfer of supplies from water-rich to water-poor areas in drought years. Some water users use treated effluents, but this too is expensive and will only be economically justifiable if there is no other source of good quality water. Desalination of salt water is also expensive, so is not attempted in humid temperate areas.

Demand management techniques to manipulate the magnitude, location and timing of demand are widely used. These techniques include framing charging policies for consumers to encourage efficient and discourage wasteful use, and controlling direct withdrawals by means of a licensing system. Other techniques used to manipulate demand in drought periods range from temporary restrictions on water use to public education campaigns. Many water management authorities integrate supply and demand management techniques, and in the Thames basin in England the supply network is operated so that specified restrictions on use are tolerated with given frequencies (Jamieson and Nicolson 1984).

Water resource management in water-rich humid temperate areas is as critical as water management in drier environments, but the great variety of economic and political systems found in humid temperate areas, together with large variations in demand over space and time, mean that many different strategies are adopted. In centrally-planned economies, for example, water resource development is usually very closely linked with national economic development plans (see Laski (1977), for example), but in market economies the link is less explicit.

Hydrological impacts of human activities

Humid temperate areas are among the most densely populated parts of the world, and consequently there has been a long history of human modification of hydrological regimes. It is important to distinguish between direct effects (such as those due to river regulation and abstractions from rivers) which are often anticipated, and indirect consequences which may be unforeseen.

Direct effects on flow regimes. River impoundment and effluent returns have major impacts on flow regimes. The return of effluent water - whether from industrial or public withdrawals - can have a noticeable effect on low flows, and downstream of large urban areas such water can constitute the bulk of flow. River impoundment schemes, however, have a much more widespread effect on flow regimes. Upland and lowland rivers are impounded for a variety of purposes, including water supply, power generation and flood control, and Petts (1984) has indicated the scale of the continued increase in impoundment. It is of course impossible to generalize about the effects of impoundment as they depend on reservoir size and operating rules, but it is clear that few major rivers in humid temperate areas remain unaffected and many upland headwater streams have been impounded. The most common effects are changes in the distribution of flows through the year with higher low flows and lower total volumes of runoff, and a reduction in flood peak magnitude.

Indirect effects on hydrological regimes. Land use changes have a major and widespread influence on hydrological processes, and in humid temperate areas significant changes in land use have occurred over very long time periods. Limbrey (1983) reviews the evidence for changes in hydrological regimes following the clearance of natural forests in Bronze Age, Neolithic and even earlier times, and it can be argued that very few areas in humid temperate environments - particularly in Europe - can realistically be termed "natural"; human influences stretch back thousands of years.

Evaporation of intercepted water is greater from forest than grassland vegetation (although the difference depends on climate, and appears greatest in wet upland areas), and removal of forests consequently increases runoff. There have been few studies of deforestation in Europe, where much forest clearance was done before the development of hydrology, but studies in New Zealand using several experimental catchments (Scarf 1970) have shown that peaks of small floods can be over an order of magnitude higher after clearance of forest and scrub.

More recent changes in land use, and in particular afforestation, have been closely studied. Early studies in northern England showed that 290*mm* more water was evaporated from forests than grassland (Law 1956) - with major implications for forested reservoired catchments - and the subsequent debate led to the establishment of a major paired catchment experiment at Plynlimon in central Wales. This showed that runoff from the forested catchment was 15% less than runoff from the adjacent grassland catchment (Blackie and Newson 1986). Further experiments at Thetford Forest in drier lowland Britain showed, however, a lesser difference between forested and open catchments (Gash and Stewart 1977), attributed to lower interception losses.

Although in the long term, afforestation reduces runoff (at least in upland areas), several studies have shown that drainage prior to afforestation leads to short-term increases in total runoff and peak flows and reductions in response times. Robinson (1986), for example, found in a small catchment in upland England that pre-planting drainage led to increases in low flows and hence annual yields, that flood peaks were increased by approximately 20% (although the increase declined as the forest matured) and that lag times were reduced by a third.

In many parts of Europe there have been major changes in the structure of farms over the last twenty years in an attempt to increase efficiency and productivity. This restructuring (termed "remembrement" in France) usually takes the form of consolidation of fields into larger units, the removal of many field boundaries and the provision of new infrastructure, and can have locally significant effects on runoff. Bucher and Demuth (1985) compared the water balance components of

199

two small paired catchments near Freiburg in the Federal Republic of Germany, finding that the restructured catchment had an increase in runoff volume as well as a decrease in both evapotranspiration and soil water retention, leading to a reduction in groundwater recharge (Fig.8.18). This was attributed to soil consolidation and the development of asphalted roads and a drainage system.

It was noted earlier that a large and increasing proportion of farmland in humid temperate areas is drained. There has been a controversy over the effects of this drainage, with studies showing both an increase and a reduction in flows, and it appears that response depends on the pre-drainage characteristics of the soil and antecedent conditions. Drainage of a previously wet soil tends to reduce subsequent floods (as more water can be retained in the soil), while drainage of a drier permeable soil increases runoff by providing a faster escape route for rainfall (Robinson et al 1985). Similar results were found by Warmerdam (1982) in the Netherlands.

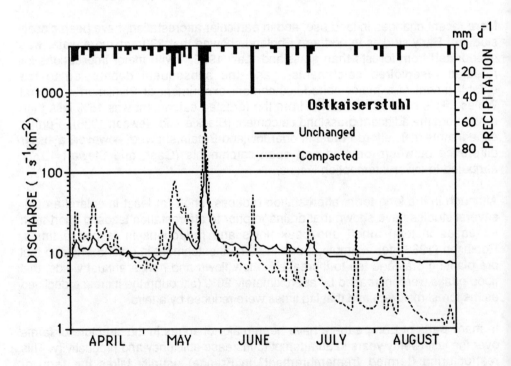

Fig.8.18 Effect of field compaction on runoff, southeast Federal
Republic ofGermany (from Bucher and Demuth 1985).

At the catchment scale it is difficult to distinguish the effects of field drainage from the effects of channel improvements which are often made at the same time. Bree and Cunnane (1980) studied several catchments in Ireland, and discovered that channel improvements increased downstream flood peaks, with an average increase of 60% in the 3-year flood. On a much larger scale, Engel (1977) showed how the hydrograph of a 1882 flood on the Rhine would change form with increasing river improvement; if the same flood occurred at the present time, the peak at Maxau would be nearly 40% higher and two days earlier.

The most extreme modification of land use since initial forest clearance is urbanization, and its effects are well known and well studied. In essence, urbanization involves the replacement of the natural drainage network with a denser, more permanent and efficient network and a change to an impermeable ground surface. The total volume of runoff (particularly storm runoff) is therefore increased, lag times between rain and flood are shortened, and low flows are reduced, but the effect of urbanization depends not just on its type and degree but also on the characteristics of the basin; the more permeable and slower responding the basin before urbanization, the greater the effect on flow regimes of development. Several studies have examined the magnitude of the effect of urbanization, and Fig.8.19 shows one hour unit hydrographs derived by Walling (1979 a) before and after urbanization. Packman (1980) found that the mean

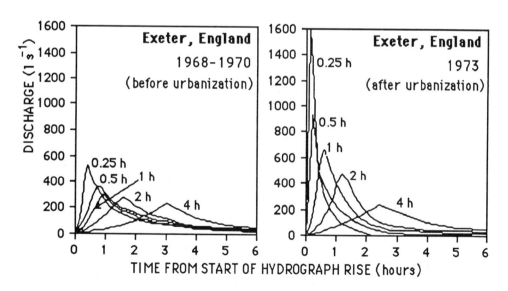

Fig.8.19 Unit hydrographs from a catchment in southwest England before and after urbanization (from Walling 1979 a).

annual flood in England and Wales could be between 3 and 7 times higher after urbanization, although the effect of urbanization reduces for larger floods. Ikuse et al (1975) showed that the total volume of runoff was higher (twice as high for 100*mm* of rainfall) in an urban catchment near Tokyo than in a nearby rural catchment, and peak flows were five times higher.

Other activities such as mining can also have dramatic influences on flow regimes. Clearance of large areas of topsoil and compaction by machinery may lead to increases in runoff, particularly in storm events, as would discharge of pumped mine drainage into river channels. In other cases the reduction in water-table levels associated with open cut mining and drainage of deep mines leads to a reduction in flows, and large urban and industrial withdrawals of groundwater can also lead to a lowering of the groundwater level. In still other regions water-tables are rising due to reduced industrial groundwater withdrawal. Marsh and Davies (1983) describe rising groundwater levels threatening cellars and basements in London.

Effects on sediment yields. Deforestation not only leads to higher runoff but also increases sediment yields. Annual yields of over 1000*t* km^{-2} are found over much of Rumania, for example, and are attributed to extensive deforestation in the 19th and early 20th centuries (Ichim et al 1984). Landslides contribute much of the sediment. In Luxembourg, Imeson et al (1984) measured specific sediment yields in a cultivated catchment four times higher than yields from a catchment under oak and beech forest. This was due not just to increases in exposed areas but also to the poorer soil structure in cultivated areas enabling greater subsurface erosion and transport.

Although sediment yields are lowest under forest, afforestation does not necessarily reduce sediment loads and output can be greatly increased in the short term by erosion of drainage ditches dug before planting. Robinson and Blyth (1982) measured sediment concentrations up to two orders of magnitude higher immediately after ditching in a small upland English catchment, although concentrations settled to approximately four times higher after a few years and will reduce further as ditches become vegetated. Nevertheless, Newson (1980) measured bedload yields 3.7 times as high in a forested as in a grassland catchment in Wales, more than 30 years after planting.

The greatest effects on sediment concentrations and yields, however, are produced by industrialization and the process of urbanization. Walling (1979 b), for example, found increases in sediment yields by factors of between 5 and 10 during urban construction, and although average loads from developed urban areas are low, concentrations in small areas during storms can be very high. Ellis (1985) measured sediment concentrations of up to 4500*mg* l^{-1} in a small basin in

London, with most being washed from roads. High sediment yields can be found downstream of mines and other mineral extraction sites. Concentrations between 40 000 and 60000$mg \, l^{-1}$ are found in a river draining tailings from an iron mine in southwest England, for example (Cominetti 1986), making it one of the most sediment-laden rivers in any humid temperate area.

Effects on water quality. The effect of forest clearance on stream water chemistry depends not just on atmospheric inputs and underlying geology, but also on the area and timing of clearance. In one experiment (Pierce et al 1970), woodland covering a catchment in the Hubbard Brook experimental area, New Hampshire, was cleared completely and left over winter, and in the following years nitrate concentrations were 50 times higher than in an uncleared control catchment. This large increase was caused by the decomposition of large amounts of plant material and the complete lack of nitrogen uptake by vegetation. Sulfate concentrations decreased, due to the inhibition of bacterial activity associated with high nitrate concentrations. The results of this study, however, reflect an extreme change in land use; only rarely is a complete catchment cleared, and early clearing for cultivation was probably on a much smaller scale. Hornbeck et al (1975) showed in the same area that staged clearing had a much less dramatic effect on water quality than complete clearing.

There is a great deal of evidence that agricultural activities other than forest clearing are in many areas having increasing effects on the chemistry of both surface water and groundwater. Perhaps the greatest concerns are expressed over increasing nitrate concentrations, and many studies of processes and rates of change have been undertaken (Roberts and Marsh 1987). There are many sources of nitrogen, including decomposition of crop residue and other vegetative matter, inputs from animal manure, inputs from synthetic fertilizers and atmospheric inputs (OECD 1986). Organic nitrogen is first mineralized to ammonia and then, if the soil is well oxygenated, converted to nitrates by nitrification. The relative importance of each source of nitrogen varies over space and time.

Following a review of several studies, Young (1986) estimated that plowing established grassland to initiate arable cultivation released between 200 and 400$kg \, ha^{-1}$ of nitrates for removal by subsurface flow or by leaching to groundwater. Plowing provides plant material for decomposition and by aerating the soil stimulates mineralization and nitrification. Young (1986) concluded that large scale plowing in the 1940's led to increases in nitrate transport to groundwater, and that the more recent increase in the use of synthetic fertilizers has meant further increases in nitrate concentrations. Peak concentrations of 22$mg \, l^{-1}$ were recorded during very low flows in eastern England in 1975/1976 (OECD 1986), but higher concentrations have been measured in the Girou basin in southwest France.

203

Probst (1986) showed that concentrations increased with discharge, and varied from 16$mg \, l^{-1}$ during low flows to 34$mg \, l^{-1}$ in storms. High and increasing nitrate concentrations in groundwater have been recorded in many countries (Roberts and Marsh 1987). Henin (1986), for example, showed that in 1969 none of the catchments in the Yonne basin in France had nitrate concentrations in excess of 44$mg \, l^{-1}$, while 21.1% had such concentrations in 1979. Concentrations and loads can be expected to increase as fertilizer inputs rise, with important consequences for public water supplies.

Other locally important agricultural influences on water chemistry include animal manure, and particularly pollution from liquid silage. This has a very high biological oxygen demand (BOD) - much higher than untreated sewage - and can have a dramatic effect on water quality when silage is washed from fields during rain or leaks from inadequate storage facilities. Modern agriculture increasingly uses pesticides and herbicides, and these too may have locally significant effects on water quality. Although there is a general lack of quantitative data on pesticides and water quality, several studies have shown that effects can be measured. In the U.K., for example, concentrations of up to 5$\mu g \, l^{-1}$ of phenoxyalkanoic acid have been recorded (Croll 1986), but long-term trends and residence times are unclear.

The impact of agriculture on water quality has been studied for a much shorter time than the more immediately obvious and longer established effects of urban and industrial areas. In contrast to agricultural pollution, inputs from urban and industrial pollutants come from point sources such as sewage treatment works, storm sewers and factory outfalls, and drainage from landfill sites. There is of course a great variety of impacts, with different activities contributing different substances to rivers and older industrial areas having more pollution. Mine drainage and water draining from surface spoil heaps, for example, can be very acid and loaded with heavy metals. Accidental spillages of chemicals and pollutants into rivers can have more dramatic effects on water quality and river biology, specially if the receiving waters were previously relatively clean. Major spills can kill much of the fish and plant life in a river and cause problems for water supply; abstractions from the Rhine in Netherlands were temporarily halted in 1987 after a major pollution incident in Basel, Switzerland, for example.

Runoff from urban areas can also have a significant effect on water quality, and indeed Ellis (1985) estimated that 35% of the annual pollutant load of British rivers was due to storm sewer flows which occur only between 2 and 10% of the time. Most chemicals come from road surfaces, and many are associated with fine particulate sediments. High concentrations are found not just of chemicals derived from rubber, bitumen, oil and grease but also of heavy metals. Following a review of a number of studies, Ellis (1985) estimated that mean storm concentrations of lead lay between 0.03 and 3.1 $mg\ l^{-1}$, although maximum concentrations could be much higher. The quality of storm sewer runoff is similar to the quality of overflows from combined storm and sewer systems, suggesting that improvement in the quality of urban discharges can only occur if storm runoff is also treated.

Finally, many transport authorities in humid temperate areas add salt to roads during cold periods in winter to prevent or remove hazards due to freezing. This salt is often washed off roads into streams, particularly if - as is common in areas with mild winters - forecast snow falls as rain, and can have major effects on water quality. Kunkle (1972), for example, found salt concentrations in a stream in Vermont draining a salted road 15 times higher than those in nearby streams.

Surface water acidification. In the last twenty years there has been increasing concern over the acidification of surface water. This was first noticed in southern Scandinavia, then reported in eastern Canada, and has now been observed in many parts of Europe. Although there is a good deal of uncertainty about the causes, the main culprit at present appears to be acid precipitation. Sulfur and nitrogen oxides emitted by the burning of fossil fuels can oxidize in the atmosphere, forming sulfuric and nitric acids which reach the ground by precipitation. The acids also release toxic cations, in particular aluminium, from the soil. As a result of acid precipitation, rivers become more acidic with consequent effects on fish and plant life and have higher concentrations of toxic aluminium. Snowmelt may produce intense flushes of acidity and aluminium. The response of a catchment to acid precipitation depends on land use and soil chemical characteristics (Cosby et al 1986). These factors vary across space, but in general more acidic soils - as commonly found in upland areas - with high aluminium contents are most prone to change.

The presence or absence of forests is a major control on response to acid precipitation. Runoff from coniferous forest is already acidic, because pine needle litter is highly acidic, but forests encourage the movement of acid from the atmosphere to runoff. Trees harvest more acid from the air (partly by dry deposition, or fallout directly on to leaves), concentrate acids by evaporating relatively more of rainfall and encourage chemical processes in the soil by slowing down flow routes to river channels. In the uplands of England and Wales, for example, pH values of water can be half a unit lower in forested catchments than in moorland catchments and aluminium concentrations can be two to three times higher, with larger differences during high flows (Hornung et al 1987).

Although most problems so far identified have been in upland areas characterised by acidic soils and dense forest cover (in Czechoslovakia, Germany, France, Poland and eastern Canada, for example), the problem of acidification is not necessarily

205

restricted to upland areas. Acid precipitation falls over large areas of both eastern and western Europe, although the predominantly westerly direction of airstreams means that western areas are less affected, and increasing afforestation might lead to major and widespread increases in surface water acidification, with subsequent implications for both flora and fauna and water supply.

Conclusions: past, present and future

Humid temperate environments are defined by their wetness and moderation, but there is a great deal of variability. Although climate exerts the greatest control over processes at the largest scale, at the local and even regional level features such as land form, geology and land use are more important. Several examples have shown not only that similar climatic inputs can have different outputs under different conditions, but also that similar outputs can arise from different inputs.

Human influences on hydrological processes, particularly in Europe and northeast Asia, have been widespread and of long duration. There is stratigraphic evidence, for example, of the effects of prehistoric forest clearance on flow regimes in northern Europe. Greater influence has been felt since industrialization, and both direct and indirect (often inadvertent) alterations to hydrological regimes are common.

In principle, there is enough water in humid temperate areas to satisfy current demands, but in practice there are large variations over space and time on pressures on good quality resources. Too much water is more often a problem for agriculture than too little (much more land is drained than irrigated), and irrigation is only applied to maintain the consistency of yields. In dry years, however, agricultural demands can place severe pressure on available resources. Industrial and domestic usage tends to be higher than agricultural usage, reflecting both climate and the state of economic development in humid temperate countries.

But what does the future hold? The evidence suggests that demand for water - particularly good quality water - will continue to increase, although industrial concerns appear to be moving towards more efficient resource use. This increased demand will require careful management, because although there is plenty of water in most humid temperate areas, there is often not enough clean water. Water resource management is therefore as much a matter of quality management as quantity management. Great advances have been made in recent years in cleaning up rivers - particularly the dirtiest rivers in Europe - but pollution from non-point sources such as agricultural nitrates and acid precipitation is currently the focus of concern. The pace and extent of change in resource use and human impact is influenced by economic and possibly political conditions, and these are difficult to predict. The effect of agriculture on demand and water quality in western Europe, for example, will depend on future changes in the EEC Common Agricultural Policy, which is currently charcterized by surplus production of specific products.

Perhaps the most important issue facing water resource management in humid temperate areas in the long term, however, is the prospect of climatic change. After several years of controversy, climatologists are now reaching a consensus that an increase in the level of carbon dioxide in the atmosphere, due to the burning of fossil fuels and, to a lesser extent, deforestation, is causing a global warming. Less clear is the effect this will have on water resources (Askew 1987).

Several studies (reviewed by Beran (1986)) suggest that global warming will lead in humid temperate areas to increased precipitation in winter and higher temperatures, and hence more evapotranspiration, in spring and summer, with a poleward migration of climate zones. The greatest impacts will therefore be in the southern (in the northern hemisphere) parts of humid temperate areas where evapotranspiration may exceed precipitation. Localized effects will depend on catchment characteristics, and Beran (1986) suggests that catchments with less storage will see a greater reduction in water resource availability. Not only will the pattern of resources be changed, but demand, particularly for agriculture, is also likely to increase. Although it may be possible to curb further increases in CO_2, the effects of past increases have not yet filtered through to water resources, and some change is inevitable. Water resource managers need therefore to be prepared for increased resource variability as well as increased demand.

9. Dry temperate sloping land

Occurrence on a global scale

The dry temperate zone is basically equivalent to the geobotanical steppe and prairie zone in Fig.1.4, labelled 70 on the ecohydrological map in Fig.1.3, but also includes some dry subtropical areas labelled 30 in southern Siberia. Basically, there are two main global regions belonging to this zone (Fig.9.1):

(a) the dry and partly fertile steppes of the USSR, including three main sub-regions: the European part, west of the Urals, northern Kazakhstan and western Siberia around Lake Baikal. The belt continues into Mongolia and north China, and includes the downstream Huang Ho basin and the Beijing area ;

(b) the northern part of the fertile Great Plains region in North America.

208

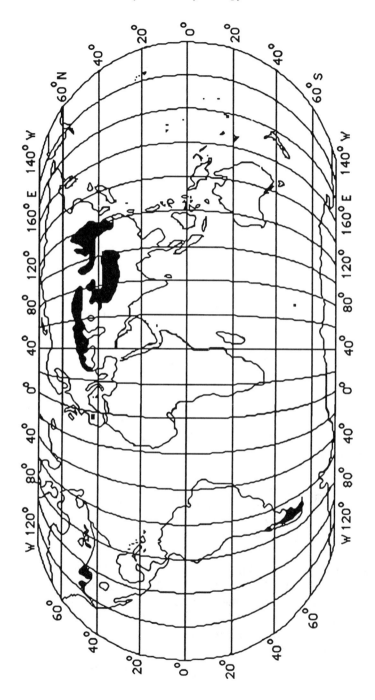

Fig.9.1 Generalized map of the dry temperate region

Comparative hydrology

On the other continents there are some minor areas of the dry temperate region, mainly Patagonia in South America and the coastal region in Australia between Adelaide and Melbourne.

The Asian steppe zone passes a sequence of main regions (Pokshishevsky 1974):

- the westernmost part cuts through the river basins of the Volga, Don, Dnieper and Dniester west of the Ural chain. This is a district where wheat is grown intensively;

- the slightly elevated arid steppes of Kazakhstan where agriculture based on wheat as the main crop is a dominating activity. This is the major virgin land development area of the USSR;

- the steppe zone of the West Siberian lowland where soils are excessively saline due to water shortage;

- the areas around Lake Baikal with its pronounced continental climate and incidental or discontinuous permafrost. The main activitites are industry and stock breeding;

- the Mongolian highland including the source region of the Selenga river, the main tributary to Lake Baikal;

- the northeastern part of the Huang Ho river basin in China where agriculture is intensive and loess soils extremely fertile, and Inner Mongolia including the Beijing and Tianjin areas. It includes the main wheat growing district of China, where more than 50% of the land is cultivated.

The North American plains region is composed of the upper part of the Missouri basin on the US side, passing through Montana, Wyoming and North Dakota, and on the Canadian side the southern part of the province of Saskatchewan, drained towards the northeast by the Saskatchewan river, a tributary of the Nelson river.

Thus, agriculture is a dominating activity in the dry temperate zone as long as the summer heat allows a growing season of adequate length. The dry climate makes supplementary irrigation extremely attractive as a means of increasing agricultural production. This has generated plans for large-scale water transfers both in USSR and China. Where the climate is too cold, as in the southern part of eastern Siberia, priority is given to meat and milk production. Thanks to the thin snow cover, pastures may continue to be available through the winter (Pokshishevsky 1974).

Water balance composition

General characteristics. In the dry temperate sloping lands, the annual evaporation demand is by definition 500-1000*mm*. Depending on the precipitation this will create a water deficit which may amount to 500*mm* (Unesco 1978 b). The precipitation stays below 800*mm* in the Asian region with only 200 - 400*mm* in Kazakhstan. In the Great Plains region it generally is 250-500*mm* (USWRC 1978) with a runoff generation of less than 25*mm*.

L'vovich (1979), in his comparative global analysis, characterizes the water balances of the two regions as shown in Table 9.1. Of an annual precipitation of the order of 500*mm*, only about 5-10% reaches the terrestrial water systems in aquifers and rivers and 90-95 % evaporates. There is a water deficit of at least 400*mm* and the actual evapotranspiration is only half or even less of the potential. The proportions between rapid flood flow and time-stable base flow are quite variable: from 3:1 up to 12:1, depending on the geological conditions and the continentality of the climate.

Table 9.1 allows a comparison between three of the sub-regions. When comparing Kazakhstan with the European sub-zone we can see that the smaller precipitation is mainly reflected in lower infiltration, supported by a larger share of snow and therefore smaller evapotranspiration, whereas the runoff is only slightly smaller. In the North American case, the evaporation demand is considerably higher due to climate differences. In spite of the rather similar precipitation the runoff is much higher in this sub-zone, mainly from higher flood flow due to lower infiltration. The annual biomass production in these two sub-zones is the same, in contrast with a considerably lower level in Kazakhstan, where 140-180 days are under snow cover. Although the groundwater recharge is quite low in the European part, it is even lower in the other two sub-regions.

Interdependencies. The principal interdependencies between the different water balance components have also been studied by L'vovich (1979). For each of the main eco-hydrological zones he found quite similar interrelation patterns for the different continents. Fig.9.2 shows the interrelations in the steppe and savanna zones between infiltration on the one hand and evaporation and groundwater recharge respectively on the other. In this diagram, the dry temperate zone or the steppe/prairie zone corresponds to the drier part. At greater depths of infiltration, the curves are transformed into relations valid for various forms of savannas under conditions of higher potential evaporation. From the hydrological aspect therefore the water balance components of the dry temperate zone interact in much the same way as in the dry savanna.

Table 9.1

ANNUAL WATER BALANCE COMPONENTS (*mm*) FOR THE STEPPE AND
PRAIRIE ZONES IN ASIA AND N. AMERICA (from L'vovich 1979).

	USSR		N. America
	European	Kazakhstan	
Precipitation	495	410	510
Total runoff	40	20	65
Groundwater component	9	2	5
Surface flow component	31	18	60
Infiltration	464	392	450
Actual evapotranspiration	455	390	445
Potential evaporation	890	860	1200
Biomass production $t\ ha^{-1}y^{-1}$	6-9	2-4	6-8
Groundwater recharge as fraction of infiltration	0.02	0.006	0.01

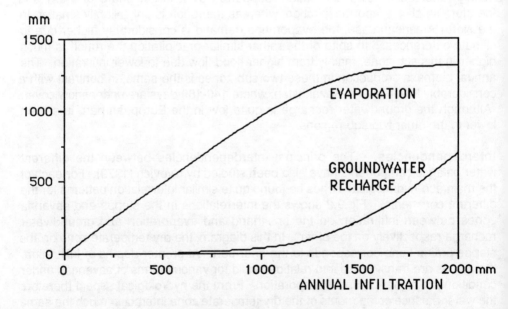

Fig.9.2 Water balance relations in the steppe and
savanna zones (from L'vovich 1979).

River regimes

L'vovich (1979) has also studied the river regimes on a global basis, codifying the main types of water composing the runoff and the season when the flood flow occurs. His study shows that the dry steppe in Kazakhstan has a special type of river regime: the streams do not flow for a large part of the year, and there is almost no base flow. The river flow is generated from melting snow, and the part due to rainfall is very low. Water flows in the streams only during a brief period in the spring. This regime is typical of extreme continental hydrological conditions and does not exist in other parts of the world.

L'vovich stressed that rivers supplied by snowmelt form important components of the human environment. The high regime contrast between different seasons complicates the use of the water resources, specially in the arid regions where much water is required for irrigation. This demand can be satisfied only if reservoirs can be built to even out the seasonal fluctuations. During intermittently dry years with little snow or extreme years when the soils do not freeze as usual, the snowmelt infiltrates and spring flow is small. The risk of such occurrences leads to radical transformations such as the Irtush-Karaganda Canal or the huge inter-regional pipeline from the Ishim river.

Rivers in eastern Siberia have more variable regimes. Generally runoff from rainfall is the dominating source of the water supply due to the influence of the Pacific monsoon, giving precipitation in the summer. The deep winter anticyclone causes the winter precipitation to remain low, leading to only a small or even negligible surge from snowmelt. In some regions snowmelt runoff may be the dominant component but the high flow may be partly carried over to the summer, merging with high flow from summer rains. An extremely small groundwater runoff is typical of the Transbaikal region, due not only to the aridity in general, but also to the common presence of perennially frozen ground.

In the easternmost part of the Asian sub-zone of dry temperate sloping land, comprising north China and the downstream part of the Huang Ho, the regime is influenced by the summer monsoon. It is however a rather diffuse monsoon type regime in that less than half of the annual flow passes during the monsoon season. There is a rather large groundwater component due to the deep and permeable soils. The low flow period in winter is short.

In the dry temperate zone in North America, rainfall dominates the runoff in the southern part with high flows in the summer, whereas the Canadian part of the prairie has snowmelt as the main water source, with high flows during the spring.

The same regime dominates the plains part of Patagonia in South America where the total runoff is very low, mainly occurring in the spring because of snow cover.

Water use and related problems

Seasonal water accessibility. In the semi-arid zone, local rivers are generally ephemeral, whereas larger rivers often carry water emerging from source areas outside the region itself. In the U.S. Great Plains region for instance, the major rivers are perennial and carry exogenous runoff generated in the Rocky Mountains to the west. The situation is different on the Canadian side. Being part of the upstream region of the Nelson river, the watercourses carry endogenous water only; the tendency is for 96% of the annual flow to occur during spring months, and local watercourses tend to dry up during summer and autumn (Unesco 1978 b). The large seasonality in river flow typical of the dry zone implies that reservoir storage is essential for adequate year-round water supply.

In general, groundwater conditions in the dry temperate zone may vary greatly due to differences in geology. In the Great Plains region, the major aquifers follow the major rivers and are replenished from perennial streams (USWRC 1978). Outside the river valleys, groundwater may then be an unreliable source of supply. In some cases, incautious use may lead to problems as is reported to be the case in Montana (USGS 1984), where uncontrolled flowing wells create concerns about potential groundwater mining of large yield aquifers in buried stream channel deposits.

In the densely populated North Plain region around the multimillion Chinese metropolises of Beijing and Tianjin, water supply poses major problems. Gustafsson (1984) reports that, with the limited surface water resources in this dry region, the water supply of Beijing has to be based largely on groundwater. Over 40 000 wells have been drilled, the majority being used for irrigation of $200000ha$ of fertile land. The intensive use of groundwater has led to over-exploitation: since 1970 the water-table has been sinking $1-1.5m\,y^{-1}$, producing a depression over $100\,000ha$.

The water supply is even more serious in the Tianjin area, where the overuse of groundwater has produced serious land subsidence. An immediate water shortage in 1981 had to be provisionally solved by diverting water through the Grand Canal from the Huang Ho. The necessary dredging, excavation and construction work on the canal system was carried out by 640 000 civilian workers within a three month period during the autumn harvest and sowing period. The long-term water need has now been solved by a $234km$ canal bringing $1000Mm^3y^{-1}$ from the north.

In North Dakota the aquifer system consists of several sandstone layers (USGS 1984). The water is however generally unsuitable for many uses because of salinity. In Wyoming, the main aquifer is a very thick and extensive sandstone, shale and coal bed. Locally the water has been polluted by leachates from mine spoils and oilfield holding ponds (USGS 1985).

Water use. The dominant water use sector in the dry temperate agricultural regions is generally agriculture calling for irrigation to increase agricultural productivity. In regions with large industrial activities, such as mining or coal production, industry may be the dominating sector.

Particular problems are caused by energy-minerals development implying large water demands not only for mineral extraction and processing, but also for transport of coal as a slurry in pipelines. In Wyoming, agricultural and industrial users compete for the available supplies while new storage and diversion structures are presently being planned (USGS 1984). In the USSR, the water needs for development of mining areas have generated water diversion projects, as will be discussed later.

Montana and Saskatchewan also have large strippable coal reserves now constituting widely used aquifers. Consequently, potential long-term and cumulative hydrological effects on water-tables and groundwater quality from coal mining activities is causing increasing concern (USGS 1984). The effect on the quality of drainage water may even cause bilateral problems between Montana and Saskatchewan due to trans-boundary water systems carrying the effects across the national border.

Mining activities may also call for river diversions when areas rich in mineral resources have to be drained. One example is the Coal Creek mine in Wyoming (Bowles et al 1985), where a local creek has to be carried around the area proposed for mining. Arising impact problems typical for this hydrological zone include the transformation of a playa area into a detention basin. This may create risks of dissolving salts left behind in the evaporating playa over a period of centuries, so degrading the water quality in the future.

Other water quality problems are those emerging in local ponds and strip mine ponds in the northern Great Plains region, discussed by Andersson and Hawkes (1985). They found pond water often to be slightly brackish or even saline due to the richness of the soils in alkali or sodium and the enrichment taking place mainly by evaporation. This limits the beneficial uses to which such ponds may be put, like irrigation and livestock watering. For example, lead concentrations may be too high.

215

Droughts. Typical of this region is the problems caused by droughts, one example being the 1976-77 drought in the Upper Missouri basin (Matthai 1979). Precipitation was only 25% of the long-term average, generating very low runoff and producing record low water levels in local reservoirs. Seventy-seven counties were designated as disaster areas. This was the most severe drought of the century and was explained by the low water content in the mountain snowpack and low soil water in both the topsoil and the subsoil. Droughts are common in the Northern Great Plains but vary in terms of severity and duration.

Consequent hydrological problems include low groundwater levels and in this case highly toxic concentrations of blue green algae, which led to the death of dogs and cattle. Emergency operations had to be organized for the city water supplies and there was a great risk of grassland fires during the summer season.

Water resource assessment problems

Determination of runoff. The sensitivity of runoff generation to fluctuations in evapotranspiration creates particular problems when determining runoff from ungaged catchments. Card (1979) has shown that models commonly used to calculate synthetic streamflow data in humid climates were not applicable in the semi-arid Canadian priaries. He found a relationship between basin area and mean annual runoff but no connection with median and extreme values of runoff. One particular difficulty encountered is the variability of the effective drainage area from year to year due to ever changing depression storages in rather flat environments.

The fact that not all surface runoff reaches the main stream in a prairie watershed, some of it being caught in depressions such as sloughs and lakes from which there is no drainage even under wet conditions, implies problems from a water assessment perspective. There is in other words a dead area to be taken into account when determining the drainage area of hydrometric river flow stations. In practice, this is done by distinguishing between gross and effective drainage areas in the prairie provinces (PFRA 1983).

McMahon (1979) has approached the predictability of hydrological behavior in the arid zone. He compared flow records and peak discharge series from six arid regions and determined hydrological characteristics. He concluded that hydrological characteristics cannot be extrapolated to arid zones and that arid zone streams are considerably more variable than those in the humid zone.

The fact that precipitation in semi-arid areas is characterized by high variability in both time and space has directed increasing interest to event-based models of

216

precipitation when modeling ephemeral streamflow (Duckstein et al 1979). The approach may be used to generate stochastic input to management models such as design of flood retarding structures, water supply system operation etc.

Extreme flows. Streamflow in dry regions may be increased by groundwater inflow in some river reaches and decreased by channel seepage and evapotranspiration in others (Riggs 1979). Thus low flow characteristics may change along a watercourse. Losses may be particularly large during seasons with high evapotranspiration losses.

In his study on streamflow in ephemeral streams, based on studies in the Cheyenne river basin in Wyoming and the lower Yellowstone river in Montana, Riggs suggested that channel width may be a better indicator of the n-year flood yield than drainage area. Evidently transmission losses along the river contribute to irregularities in the flood development process.

Maximum potential streamflow regulation. The potential level of effective development of surface water resources is different under humid and arid conditions. The fact that the coefficients of variation are considerably higher in the arid than the humid zone effectively means relatively larger reservoir capacities in the dry zones. McMahon (1979) in an intercontinental comparison concluded that in North America the maximum potential stream flow regulation is of the order of 40%, whereas it seems to be less than 10% in Australia due to the higher flow variability.

Hydrology and land use

Agriculture in water-short regions. As already indicated, the dry temperate zone is widely used for agriculture, but the insufficient rainfall gives farmers an incentive to increase their yields by supplementary irrigation. The feasibility of trying to make agriculture drought-proof by large-scale irrigation is therefore a topical issue at present in the Canadian parts of the Great Plains. The issue is however closely related to economic issues such as the cost of wheat on the world market. In the USSR, the steppe zone includes large parts of the main wheat growing region; the agriculture is mainly rain-fed. The traditional areas of large-scale irrigation are further south, but irrigation is under extension to more northern regions, provided enough water can be made available.

In China the dry temperate land includes the fertile Northeast Plain and the North China Plain, which provide a considerable share of the agricultural production of the country. The dry climate however represents a constraint by the threat of crop failure during dry yeare, creating great interest in soil and water conservation. At the same time, the non-cultivated grasslands represent an important future potential. With a cultivated area of only $0.1\,ha\ p^{-1}$ the task of improving land productivity is seen as a major component of socio-economic development (Gustafsson 1984).

Agricultural water management. In Canada, the prairie provinces have been agricultural lands since the immigration era. The soils are fertile but the length of the winter shortens the growing season leaving a limited crop choice. Careful management of the available water forms part of the present policy. In this region, snow is seen as the most manageable water resource, beside irrigation water, and an important issue for instance is to find ways of reducing the wind-generated snow transport out of a field. Methods for snow management are now being developed by the hydrologists at the University of Saskatchewan, the idea being to leave an adequate stubble height in the fall in order to keep the snow cover from blowing away.

At the same time, meltwater infiltration may be facilitated by mechanical treatment of the soil during fall (Gray pers. comm.). Important research areas in the region therefore include infiltration in frozen soils (Gray et al 1986). Furthermore, operational estimates of areal evapotranspiration are needed, which would allow microcomputer based on-farm water balance accounting on an operational basis (Morton et al 1985). It is also important to understand the influence of cloud cover on evapotranspiration when water is scarce in a farming district of great economic potential (Mudiare et al undated).

In spite of the gentle overall slope of the landscape of the Saskatchewan prairies there are also some typical flatland phenomena such as a dense pattern of potholes, so-called sloughs. The water quality in such natural potholes, which may be hydrologically complex, may be rather poor, making them unsuitable for human consumption. In these dry regions, water hauling is a traditional way of getting drinking water out on the farms. Rainwater harvesting in so-called dugouts or artificial sloughs is another way of water management.

Large-scale water transfers. The fact that land use, to such a high degree, depends on or is restricted by the renewable supply of water from aquifers and local rivers makes water resources development an important activity in the dry temperate zone, and still more so of course in the adjacent dry warm areas. There is for instance a considerable interest in increasing water security and agricultural yields in the large natural grassland areas of the northern hemisphere, specially in Eurasia.

218

Northern China. The dry climate, fertile soils, rapid agricultural development, dense population, large metropolitan areas and rapid industrial development in northern China generated discussions on water transfer from the water-rich south to the dry north as early as the 1950's. According to Liu et al (1985), three major schemes have been discussed: the west, middle and east routes. They all imply transfer from the Yangtze river: the west route up to the Huang Ho west of the bend, and in the other two alternatives passing beyond the Huang Ho up to the Beijing area (middle route) or to Tianjin (eastern route).

The analyses and preparations of the alternatives accelerated with the severe drought in 1972 and an engineering plan was put forward in 1977 for the east route. This scheme favors both water supply of the northern plains and navigation on the Grand Canal and has met generally positive attitudes. It includes enough water for 4.3*Mha* of irrigated land in addition to 2.7*km³* of water for industrial, mining, and municipal uses. One main disadvantage is however that pumping is essential along the Grand Canal which makes this alternative energy-consuming.

USSR. Interbasin flow diversion in the USSR started in the European region in order to improve water availability in water-short regions. Korzoun (1978) reports an actual level of total diversion of $40km^3y^{-1}$. A series of additional large-scale transfers are presently under discussion to meet future water needs in the nation. There are mainly three types of situations generating such plans (L'vovich 1979):

-   supplying the water necessary for developing mining resources and other industries in resource-rich but water-short areas. The Karaganda coal basin is an example, already supplied by water transfer from the Irtysh river;

-   production of farm crops in water-short areas. The Kara Kum canal in south USSR is an example, bringing water from the Amu Darya to the Caspian Sea;

-   major ship canals like the Volga-Don, the Volga-Baltic and the White Sea-Baltic canals. Some of these canals combine the functions of shipping and the transport of water (from the Volga to Moscow)

-   combating falls in the water level of the Caspian Sea, the Aral Sea and Lake Balkash; increased levels of water consumption in agriculture in the tributary basins have reduced the inflow to these lakes and radically changed their ecosystems (e.g. the famous sturgeon population).

A number of huge water diversion plans are under discussion for such purposes: transfer from the Pechora and Vychegda rivers to make up for expected shortages in the Volga; of the Siberian rivers Irttysh and Ob to Kazakhstan and central Asia; and other transfers to combat the water level decreases in the Aral Sea and Lake Balkash. The immense dimensions of some of these projects however tend to delay their realization. The interested reader is referred to Voropaev and Velikanov (1985) for details. As for the best way of managing decreasing water levels in arid lakes, there is also the possibility of stabilizing them at a lower water level as proposed by L'vovich (1979).

Man-induced streamflow changes. Streamflow changes induced by land use changes are particularly frequent and significant in arid regions. Szesztay (1979) has demontrated the influence of aridity, expressed by Budyko's RIA coefficient (see Chapter 1, p.17), on the runoff-forming fraction of the precipitation, i.e. the runoff coefficient. Fig.9.3 shows the relationship for some 40 river basins for both humid and arid conditions, based on data from Hungary. In humid regions, soil water is already available to satisfy the evaporative demand and the runoff forming fraction is determined by climate alone. This is expressed by the line

$$E_o / P = 1 - Q / P \qquad (9.1)$$

in the figure, which can be interpreted as the long-term water balance equation

$$P = Q + E \qquad (9.2)$$

with $E = E_0$ under conditions of minimal water stress.

The water deficit conditions in arid regions evidently make actual evapo-transpiration sensitive to man-influenced water availability in the plant root zone. Man-influenced streamflow changes can therefore only be understood, forecast and controlled in terms of the water balance equation.

Hydrological role of farming methods. Agricultural measures like tilling and harvesting affect the part of the terrestrial water cycle which controls the partitioning of rain and meltwater between overland flow and infiltration on the one hand and plant uptake (evapotranspiration) and groundwater recharge on the other. One should therefore expect a consequent modification of the local water balance.

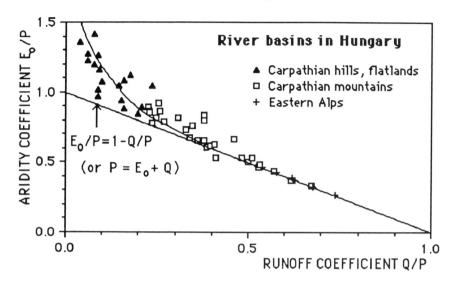

Fig.9.3 Relation between aridity and runoff coefficient
based on mean annual data (from Szesztay 1979).

The changes generated by cropping of the virgin steppe and its further development in the present century have been studied by L'vovich (1979) on the basis of comparative studies between virgin steppe and research fields. Fig.9.4 illustrates his reconstruction of the water balance in the central chernozem region over the last thousand years, assuming rainfall to have remained unchanged. The graph shows that the plowing of the steppe starting 1000 years ago brought about a successive reduction of infiltration into the soil, increased the surface runoff and consequently increased both flood flows in streams and erosion. The fallow lands of the lea system contributed to this development.

In the mid 20th century, there was a further transformation with the total adoption of fall plowing instead of the traditional spring plowing, increased depth and better plowing dates, means of reducing overland flow where plowing is insufficient, extensive use of fertilizers, and adoption of protective forest belts. As agriculture facilitated infiltration and plant uptake of soil water, flood flow was successively reduced, and infiltration and evapotranspiration increased, thereby decreasing the total stream flow.

Integrated land and water management. The part of the Huang Ho basin in China in this zone presents particular problems due to the large erodibility of the loess soils once deposited there. The Huang Ho used to be known as "China's

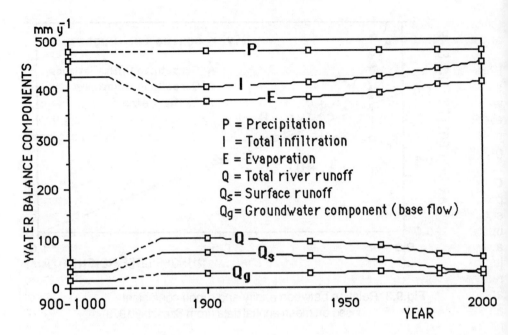

Fig.9.4 Reconstruction of water balance changes in the central chernozem region of the USSR (from L'vovich 1979).

Sorrow " (Gustafsson 1985) - not only is it the world's most silt-laden river but big floods carried down from the highlands of Tibet and from torrential rains on the nearby mountains have caused the river to change its lower course 26 times of which nine were particularly serious. The silt loads complicate the use of both reservoirs and dikes, intended to reduce the flood risks, which tend to accelerate through the rapid accumulation of silt in the lower reach and in the reservoirs.

The silt mainly originates from extensive areas of barren loess land along the middle reach of the river. Erosion was severely aggravated by the call for intensified agriculture as part of the socio-economic development (cff Mao: "Make grain the key link", Mosely 1985). Consequently soil conservation is a fundamental measure in the long-term solution of the Huang Ho's problems.

Indeed, soil conservation has a double purpose; not only does it reduce the erosion risk but it may also increase infiltration, and therefore contribute to conserving the precious water. Translating information on renewable water supply and population density given by Gong and Mou (1985) for the Wuding tributary basin in the middle reach gives a water competition level of close to 2000 *p/flow unit* (cf Ch. 1). This gives an idea of the level of water stress in this densely populated and water-short area.

Both Gustafsson (1985) and Gong and Mou (1985) describe the experiences of some of the numerous farmer brigades in their development of integrated soil and water conservation on the local scale. The main components of these works, according to Gong and Mou, are check dams to catch the silt load of local rivers, reservoirs, flood diversion canals, warping areas, irrigation, and artificial recharge of groundwater.

Conflicting land use interests. In the Huang Ho basin, the upstream flood production, midstream silt production, and the downstream vulnerability to both sedimentation and flooding, which reinforce each other, has created a need for upstream-downstream interactions in basin-wide planning (Mosely 1985). There is a congruence in interests between the local desire in the middle reaches to improve agricultural production by soil and water conservation, and the lower reaches desire to be relieved of the burden of sediment. In the lower reaches there are however contradictions between the need to set aside land for flood control and other uses of land, triggered by the high population density. The fact is that the designated Beijing flood detention area, which is to be opened should a very large flood peak occur, has not been used during the last 30 years. The land has instead been used for agricultural activities and farmers have even built houses in the basin. Another example is $2000km^2$ of the river floodplains in the lower Huang Ho, where unofficial tillage by a population of more than one million has taken place. To protect their fields, they have built inner sets of dikes which however tend to endanger flood control work in general.

10. Humid warm sloping land

Occurrence on a global scale

Many meteorologists consider a main feature of the warm humid regions to be the equatorial trough, a belt of low pressure in the central tropics. Because of the convergence of air flows into the equatorial trough it is referred to as the Inter

Tropical Convergence Zone (ITCZ). Changes in the position of the sun through the year are mainly responsible for ITCZ fluctuation. The centre of the ITCZ tends to move up to 10-20° North of the equator and its actual distance from the equator varies with the longitude. In the Atlantic and eastern Pacific the equatorial trough is located near the equator or to the north of it; the greatest movement of the ITCZ occurs over Africa. Conditions favorable for the formation of various types of wet/dry river regimes are at the rims of the ITCZ.

Monsoons have another impact on the river regimes and water balance in the humid warm regions. They are usually associated only with the Asian continent, but their actual zone of influence is considerable, because any wind of biannual periodicity and stability of direction can be referred to as a monsoon. Some monsoons show a cycle of three to ten days of strong wind alternating with equal periods of weaker air flow.

Deviations from more or less regular circulation are frequently observed in the tropics. In general they are called tropical disturbances and they can occur over an area of considerable size. Generally recognized disturbances include hurricanes, wave disturbances and linear systems. Hurricanes, also known in the tropics as cyclones and typhoons, resulting from synoptic disturbances in the equatorial trough or at medium latitudes, are easily recognisable on aerial photographs. They have a somewhat small diameter, with a minimum pressure of less than $990 mb$, and wind circulation speeds of $32 m s^{-1}$ or more. Low pressure systems with low wind speeds are referred to as tropical depressions.

Fig.10.1 shows the humid warm areas and the areas affected by tropical hurricanes of varying frequencies. It can be seen that the American humid warm area includes the Amazon basin shared by large portions of Brazil, Bolivia, Equador, Peru, Colombia and Venezuela. The area extends northwards into Mexico, the chain of islands in the Caribbean, and south-central U.S.A.. In Africa, the areas included are centred on the Congo Basin, covering Zaire, Congo and the Central African Republic, and extend westwards along the Guinea Coast of West Africa. Southwards and eastwards, the humid warm areas continue towards Zimbabwe while they also occur in the Seychelles, Mauritius and the northern half of Madagascar excluding its northern tip. Some of the areas shown for both the Americas and Africa on this small scale map however include areas which more properly belong to sub-humid climates with seasons or months when $E_0/P>1$. The Indo-Malayan humid area extends from western India and Sri Lanka to Thailand, Indo-China and the Philippines,as well as through Malaysia to Indonesia and Papua New Guinea. It is also found in the northwestern and eastern Ghats and in the lower part of the Himalayas, Khasia hills and Azani (Richards 1964; Oyebande 1981). The region continues southwards in a narrow strip along the east coast of Australia.

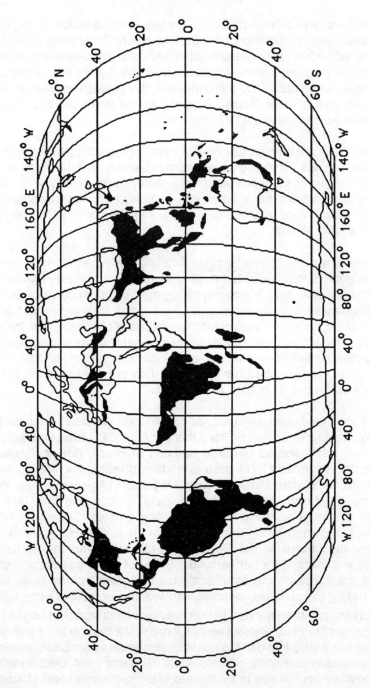

Fig. 10.1 Generalized map of the warm humid region

This is one hydrological environment where the hydrological system illustrated in Fig.2.2 finds an almost complete fulfillment, on account of the copious availability of moisture and heat energy throughout the year. The atmospheric, land and underground components are all very active. The major fluxes of flow rates within and between storages are apparently less episodic so that the storage components change frequently. Only the snow/ice box in Fig.2.2 is not applicable to the humid warm areas.

In general, the lateral movements of water are more important than the vertical ones. In particular, evaporation from bare soil is less pronounced in these humid warm areas of sloping land, but evaporation from canopy interception can be of very high magnitude. Litter also has significant effects on infiltration rates and capacity, and hence on overland flow.

Main water balance components

Precipitation. For a proper understanding of the distinctive hydrological features, it is necessary to refer to the two main sub-regions of the humid warm areas: one that is affected by tropical cyclones and one that is not.

(i) Sub-region not affected by tropical cyclones. This consists of two areas: one of continuously wet climate where precipitation is not seasonal, and the other which is not continuously wet but has seasonal precipitation. Each of the sub-divisions is affected by different rain-generating mechanisms. The continuously wet areas, for instance, which occur mainly within a relatively narrow band 3° to 5° north and south of the equator have convection as the main rain-generating mechanism. Rain falls in relatively short intense bursts of a few hours a day (WMO 1983, 1987). For example, long term rainfall distribution in the entire Lake Victoria catchment ($193,000 km^2$) in East Africa shows that no month of the year has less than 5% of the total annual amount:

J	F	M	A	M	J	J	A	S	O	N	D	Year
73	82	159	208	165	90	118	100	156	191	138	91	1398 (*mm*)

Most of the non-continuously wet areas on the other hand experience seasonal variations in precipitation which result primarily from the seasonal displacement of the sun relative to the earth and the related displacement of the ITCZ. Within the non-continuously wet areas, the areas not affected by trade winds or monsoons are very limited and include part of the south-western Amazon plain, a few small areas sheltered from the trade winds, and the humid tropics of West Africa. The two major rain-generating mechanisms in the areas are convection and convergence related

227

to the ITCZ. The length of the wet season and total annual precipitation decrease with increasing latitude and a clearer impression may be given of two rainy and two dry seasons of different intensities.

Fig.10.2 shows the seasonal frequency of storms, the distribution of total rainfall and storm duration, and the diurnal distribution of rainfall at Ibadan, some 100*km* away from the coast. The annual rainfall is about 1230*mm* and occurs during most of the year, but is concentrated in March-October, each month of which receives at

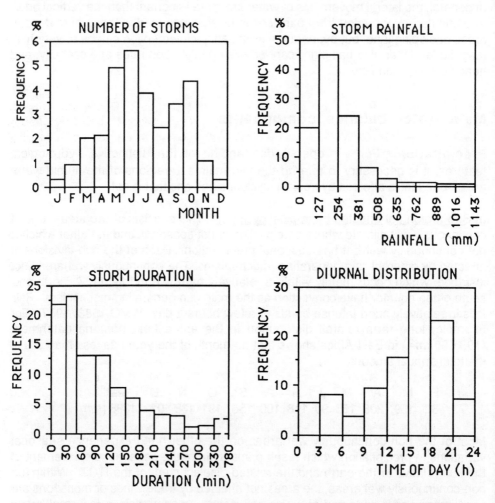

Fig.10.2 Storm rainfall characteristics at Ibadan, Nigeria (1960-1980)
(from Oguntoyinbo and Akintola 1983).

least 7% of the annual total. Of 470 storms between 1960 and 1980, only 2% had a rainfall exceeding 51 *mm.* Over 50% have durations less than one hour, and less than 20% last longer than three hours.

Areas affected by monsoons mainly occur in the vicinity of the Indian Ocean. Trade winds emanate from either oceanic or continental high pressure systems in the higher latitudes. They are usually stable initially, but as they descend, traverse long and progressively warmer ocean surfaces, and accumulate an increasing amount of moisture and may become unstable.

The monsoon trade winds are known for generation of precipitation lasting for several days over widespread areas. Seasonal variations of precipitation are larger than in the sub-areas described above.

(ii). <u>Areas affected by cyclones.</u> Tropical cyclones (also called hurricanes or typhoons in certain regions) are intense low pressure systems (WMO 1983, 1987). Gales and heavy rainfall may extend up to 500*km* from the center. Tropical cyclones originate in warm ocean areas with temperatures above 27 °C. The total rainfall in the locality of their passage is generally of the order of a few hundred *mm* and may last from three hours to three days.

Areas so affected are found at latitudes greater than 5° N and S and are usually limited to 22° N and S, although the damaging effects such as floods and storm surges may extend well outside this range of latitudes. However, the colder ocean currents and upwelling of the south Atlantic and south-eastern Pacific ocean preclude the formation of tropical cyclones in those areas.

In the Central American and the Caribbean region, the rainfall regime with rain falling throughout the year is still torrential, so that this feature combined with the morphometric characteristics leads to the occurrence of flash floods and devastating inundations. Torrential rains are expected every month, but the most intense and longest events are always associated with the passage of cyclones (Arenas 1983). Table 10.1 shows the frequency of torrential rains higher than 100 *mm* in 24 hours. Table 10.2 compares the intensity and duration of very heavy non-cyclonic and cyclonic rains for the region.

In Cuba and to a large extent Central America as a whole, the intensity of noncyclonic rains rarely exceeds 250 *mm h⁻¹* and decreases drastically after 150 minutes of continuous rain. On the other hand, in cyclonic rains longer durations and higher intensities are sustained. The depth-area-duration curve for western Cuba (Fig.10.3) also shows the widespread nature of high-intensity cyclonic rains.

229

Table 10.1

FREQUENCY OF TORRENTIAL* RAINS IN CUBA FOR THE PERIOD 1930-1970
(after Arenas 1983)

Month frequency	Storm duration in minutes									Total
	100-150	151-200	201-250	251-300	301-350	351-400	401-450	451-500	> 500	
Jan	8	5	-	-	-	-	-	-	-	13
Feb	8	4	1	-	-	-	-	-	-	13
March	13	3	1	-	-	-	-	-	-	17
April	34	3	1	-	-	-	-	-	-	38
May	100	33	6	5	2	1	1	-	-	148
June	140	35	15	12	4	3	-	-	-	209
July	42	7	1	-	-	-	-	-	-	50
Aug	72	10	5	3	5	3	1	1	1	101
Sept	95	25	8	11	3	2	-	-	-	145
Oct	162	66	17	5	5	1	1	1	1	259
Nov	52	18	7	1	-	-	1	-	-	79
Dec	24	3	1	-	-	-	-	-	-	28

*Higher than 100 *mm* in 24 *h*.

In parts of Queensland in northeast Australia, the summer months of December-March are the most hydrologically active period, during which 63% of the annual rainfall is concentrated, and intensities for individual storms range from 70 to 150$mm\,h^{-1}$. Daily totals in excess of 250mm are also common, resulting from well organised tropical cyclones which develop in the monsoonal trough (Bonell et al 1983).

Thus, in the humid tropics, the rain-generating mechanisms of convection, convergence, tropical cyclones, monsoon trade winds and orographic effects may, in various combinations, produce the complex rainfall and runoff conditions often encountered in the region. Only small river basins not more than a few thousand km^2 will have drainage areas located entirely within one single climate sub-zone; larger basins may be influenced by several climatic zones, including some outside

Table 10.2

PRECIPITATION INTENSITY ($mm\ h^{-1}$) FOR CYCLONIC AND NON-CYCLONIC
STORMS IN CUBA FOR THE PERIOD 1926-1982
(after Arenas 1983)

(a) Non-cyclonic important rainfalls

Zone	Period of continuous rainfall (*min*)						
	5	10	40	60	90	150	300
Western	180	165	138	109	82	50	-
Central	122	122	101	75	57	44	23
Eastern	163	162	91	82	65	45	-

(b) Heaviest tropical storms

Date	Period of continuous rainfall (*min*)										
	5	10	40	60	90	150	300	720	1440	2800	4320
20 Oct 1926	342	282	144	115	87	65	43	23	-	-	-
4-7 Oct 1963	132	114	84	72	66	55	48	37	29	26	21
15-16 Nov 1971	264	216	165	150	133	96	64	33	30	26	-
9-10 Sept 1979	96	84	69	64	60	53	45	35	21	21	-
2-3 June 1982	168	150	135	127	115	102	80	52	31	31	-
18-19 June 1982	156	156	123	125	105	95	78	50	29	-	-

the humid warm areas.

Table 10.3 indicates the variation in precipitation in the humid warm areas, and reflects fhe influence of the different rain-generating mechanisms.

Interception by forest. During a storm rainfall, the main effects of forests are exerted through interception of rainfall by the canopy (branches and leaves), absorption of rainfall by the litter layer, and the changes in soil permeability and storage capacity due to the activity of the root system.

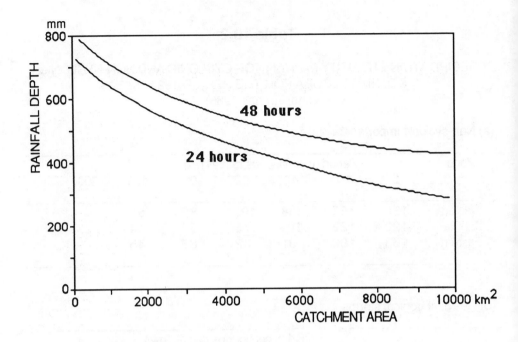

Fig.10.3 Depth-area-duration envelope curves for
cyclonic rainfall in western Cuba.

The obstacles presented by a tropical forest to the free fall of rain cause changes in the quantity, rate, and time of water delivery to the ground. Interception in the forest occurs at two levels within the forest cover: in the crown canopy and in the ground litter. Three main components of crown interception can be identified. These are the interception E_i (water retained by the crown surfaces and later evaporated), throughfall T (water dripping through and from the leaves to the ground surface), and stemflow S (water that trickles along twigs and branches and finally down to the ground surface via the main tree trunk). The interception E_i represents water which never enters the soil. The amount depends on the ability of the forest to collect and retain rainfall (the interception capacity), on storm intensity, and on the potential evaporation. The density, type and height of the canopy will affect the interception capacity, as does the age of trees.

Ground litter, like the crown canopy, detains and retains precipitation that reaches it. Some of the water may evaporate. The litter protects the underlying soil from the impact of rain drops.

Table 10.3

AVERAGE RAINFALL AND PRINCIPAL RAIN-GENERATING MECHANISMS
IN THE HUMID WARM AREAS

Area	Average annual rainfall (*mm*) and seasonal distribution	Principal rain-generating mechanisms
South-Central Java	4770	CY, MTW
Central Taiwan	2100-2500, April-Sept (80%)	CY, MTW
Hainan Island (China)	1800	CY, MTW
Mangalore (India)	3428, May-Oct (90%)	MTW, CN
Central Amazon (Manaus)	2000	CN, CE, OR
East Texas	1207	CY, CE
Tropical Coast of Queensland, Australia	4175	CY, OR, MTW
Papua New Guinea	2000-4000	CY, MTW, OR
Guyana	3200	CN, CE, OR, CY
Cross River Basin (S.E. Nigeria)	3320	CN, CE, OR
Lake Victoria Basin	1398, >5% for each month	CN, CE
Congo Basin (Zaire)	1530	CN, CE

CE= convergence/ITCZ; CN = convection; CY = tropical cyclone;
MTW = monsoon trade winds; OR = orographic.

In the tropical rain forest, interception values for heavy rainfalls are less than for light falls. Observations in peninsular Malaysia indicate that with a total annual rainfall of about 2500 *mm*, the interception loss amounts to 450-500 *mm* or 18-20%. Most of the rainfall reaches the ground surface by the process of throughfall (Lockwood 1976). In Nigeria, Lawson et al (1981) obtained the results in Table 10.4 from 30 storms each with rainfall in excess of 5 *mm*, sampled in a tropical forest in Ibadan during the May-September wet season in 1979. The value of interception loss ranged from 13 to 19% for monthly rainfall of at least 100 *mm* and averaged 17%. The average throughfall was 73%. Table 10.5 shows the values of T, S and I for tropical forest locations in India, Western Java, Cote d'Ivoire, Zaire, Puerto Rico and Kenya. The effect of rainfall intensity on interception in a tropical forest in Tanzania is illustrated in Table 10.6. For gross storm amounts equal to or greater than 10*mm*, interception ranges from 13.5 to 18%. Forest type and rainfall intensity thus appear

233

Table 10.4

RAINFALL AND INTERCEPTION COMPONENTS OVER FOREST AT IITA CATCHMENT (IBADAN, NIGERIA), 1979 (from Lawson et al 1981)

	May	June	July	August	Sept	Season
Number of storms sampled	5	6	7	5	7	30
Total rainfall P (*mm*)	89.4	107.8	238.8	76.3	123.8	645.1
Throughfall T (*mm*)	68.4	85.0	170.0	50.8	98.8	473.0
Stemflow S (*mm*)	9.5	8.9	29.8	6.9	9.3	64.4
Interception I (*mm*)	11.5	13.9	39.0	18.6	24.7	107.7
T/P (%)	77	79	71	67	74	73
S/P (%)	11	8	12	9	7	10
I/P (%)	13	13	16	24	19	17

Table 10.5

THROUGHFALL, STEMFLOW AND INTERCEPTION FROM RAINFALL IN SELECTED TROPICAL FORESTS (after Oyebande 1987)

Forest Type	Location	Gross rainfall (*mm*)	Percentage of total rainfall		
			Through-fall (T)	Stem-flow (S)	Inter-ception (I)
Eucalyptus hybrid,1658 trees/ha	India		80.8	7.7	11
Shorea robusta,1678 trees/ha	India		66.4	8.30	25.3
Tropical rainforest	Janlappa, Java	4480	-	-	20
Pinus roxburghii,1156 trees/ha	India		74.3	3.6	22.1
Alstonia scholaris,1675 trees/ha	India		57.0	17.0	26.0
China-fir plantation	Taiwan	1165	91.7	0.8	7.5
Zelkora plantation	Taiwan	868	90.7	1.4	7.9
Natural hardwoods	Banco,Cote d'Ivoire	-	86.9	-	-
Natural tropical rainforest	Yapo,Cote d'Ivoire	-	77.0	-	-
	Garamba, Zaire	-	74.0	-	-
	Puerto Rico	-	-	18.0	12.2
	Kenya	-	-	-	18.2

Table 10.6

STEMFLOW AND INTERCEPTION IN A TROPICAL FOREST IN TANZANIA
(after Jackson 1971)

(i) Average stemflow for storm classes

Storm size (mm)	0-5	5-10	10-15	15-20	20-30	30-40	40-50
Stemflow (mm)	0	0	0.1	0.2	0.4	0.5	0.9

(ii) Rainfall and interception

Rainfall P (mm)	1	2.5	5	7.5	10	15	20	30	40
Interception I (mm)	0.7	0.9	1.2	1.5	1.8	2.4	3.0	4.2	5.4
I/P (%)	70	36	24	20	18	16	15	14	13

to be significant factors of interception.

It has been claimed that secondary tropical forests characterized by a dense undergrowth are probably a more effective screen against rain than a virgin forest which is more open at the ground level.

Evapotranspiration. Tropical forests are an important source of water to the atmosphere because they transpire water in larger quantities than most other forms of vegetation. Evaporation from bare ground and shallow-rooted plants declines rapidly in the absence of rain, while a dense tropical forest continues to transpire water from deeper soil horizons. Evaporation also takes place, both from the canopy surface and the litter.

Singele (1981) reported that the evapotranspiration from the canopy of a forested basin may exceed that of pan evaporation by as much as 20 to 30%. Wilson and Henderson-Sellers (1983) also confirm that fluxes of vapor from such dense canopies of tropical forests are often higher than from tropical oceans. Lomme (1961) found that evapotranspiration reaches $2000\text{-}2300\,mm\,y^{-1}$ with a total rainfall of $4200\,mm\,y^{-1}$ in Java. It also attains $2500\,mm$ on the tropical coast of Queensland with $4175\,mm$ of rainfall; and in the Congo basin, $1230\text{-}1510\,mm\,y^{-1}$ under forest and only $950\text{-}1100\,mm\,y^{-1}$ under the derived savanna in the same climatic region. The value reported by ORSTOM for Guyanese catchments (in the Amazon region) is $1600\,mm\,y^{-1}$. In another study in western Java, during the 18 months from 1979 interception losses were 20% of the $4880\,mm$ rainfall input, mean transpiration was

2.6*mm d* $^{-1}$ and total annual evapotranspiration for 1980-1981 was 1500*mm*.

The quantity of rain water that evaporates from the canopy normally represents a loss to the water resource of the particular area so far as water yield is concerned; so is the water evaporated from the forest floor litter. Actual evaporation and transpiration from tropical forest continues throughout the year at near potential rates in many cases. For instance, in two small experimental catchments near Manaus in the Amazon basin, transpiration and interception together accounted for 80.7% and 74.1% out of the average annual rainfall of 2100*mm*. The corresponding values for transpiration alone were 62% and 48% (Salati and Voge 1983). Similarly, results from the Loweo catchment in the Yangambi forest reserve in Zaire show that 63% of the catchment rainfall of 1500*mm* was returned via evapotranspiration (Sengele 1981).

The role of litter. The tropical forest, and indeed any forest cover, produces litter (Table 10.7) which protects the soil from rainfall impact and disintegration of structure and filters out the fine particles that may clog the larger pore spaces. In addition, the forest furnishes food and protection to insects and animals which

Table 10.7

ANNUAL LITTER PRODUCTION IN SOME TROPICAL RAINFORESTS (*t ha*$^{-1}$)
(after Orimoyegun 1986)

Forest type and location	Rainfall (*mm*)	Leaves	Other	Total
Lowland rainforest				
Brazil	2100	5.6	1.8	7.4
Mora excelsa, Trinidad	-	6.9	-	-
Omo, Nigeria	1787	7.2	-	-
Olokemeji, Nigeria	1523	4.7	-	-
Lower montane rainforest				
Puerto Rico (1500 *m*)	-	4.8	-	-
New Guinea (2500 *m*)	-	6.4	1.2	7.6
Upper montane rainforest				
Mor Ridge forest, Jamaica	-	4.9	1.7	6.6
Mull Ridge forest	-	5.3	0.2	5.5
Wet slope forest	-	4.4	1.2	1.5

permeate the soil and increase soil permeability. Infiltration rates are therefore usually high under forest cover where the forest floor layer is well developed. Where it is disturbed by logging or removed by fire, protection may be decreased sufficiently to cause overland flow. The forest floor is therefore the part of the forest that affects infiltration.

It can also be argued that the balmy air (high humidity and slight wind plus continuous moisture fluxes) prevents the forest soils from drying, and hence from cracking as in the temperate zone; they do not even harden, so that they keep a certain permeability which allows infiltration.

Finally, the presence of a certain amount of humus in the top soil assures a soil structure favorable to infiltration. In the Yagambi forest in Zaire, with an annual rainfall of 1850*mm* , D'Hoore (1961) reports that yearly decomposition of organic matter in the litter produced by a secondary forest reaches 50%. In the Olokemeji and Omo forest reserves of Nigeria, the proportions are 52 and 58% respectively (Orimoyegun 1986). The coefficient of decomposition of humus reaches 68-76% under these forests as against 6 to 12% under beech forest in California.

Infiltration and overland flow. Infiltration and overland flow are processes representing the two sides of the same coin since what does not infiltrate the soil almost invariably runs off the surface as overland flow. Infiltration rates are influenced by diverse factors which include pedological and slope conditions as well as the nature and type of the forest cover. The resulting spatial variations together with the temporal variations of infiltration make it difficult to compare results obtained in different forested tropical basins. Kling et al (1981) found that high vertical drainage of soils in one of the three watersheds studied by them resulted in high infiltration rates and hence low overland flow. On the other hand, slowly permeable sub-soil characteristics of the other catchments, in the same area of the Amazon basin in French Guyana, are responsible for low infiltration and transformation of 60 to 70% of the incident rainfall into surface runoff for intense storms, and for the high erosion of 1 $t\,y^{-1}ha^{-1}$ (Roche 1981).

Cailleux (1959) also concluded that overland flow is insignificant in the primary tropical forest of French Guyana.

Overland flow can be classified as either saturation overland flow from the wet riparian zones of flood plains, or Horton overland flow as a result of rainfall intensity exceeding the infiltration rate.

In a small experimental watershed (the Kali Mando basin in south central Java) with humid andosols on quaternary volcanic ashes, Bruijnzeel (1983) observed that

237

Horton overland flow is not common. Even rainfall intensities of 200mm h^{-1} were not sufficient to yield such flow on non-compacted surfaces of the watershed.

Northcliff and Thornes (1981) observed that in large storms and under rainfall of very high intensity, the hydraulic conductivities of the latosols, near saturation level, are quite high in the upper horizons, so that Horton overland flow is quite rare in the Barro Branco experimental forested watershed in Reserva Ducke N.E. of Manaus in Brazil. Rapid response comes essentially from saturated overland flow on the flood plain.

Measurements by ORSTOM (Dabin 1957) on experimental plots at Adiopodioume, near Abidjan in Cote d'Ivoire, also reveal very little overland flow: from 1 to 3% of the total rainfall under forest, with a maximum of 7.8% during a downpour of 193mm. The terrain in this case is composed of Tertiary sands, but the slope is only a few degrees. The overland flow process under forest is essentially discontinuous and the water infiltrates the soil every few meters. This makes its measurement complex and the interpretation difficult. For instance, it has been observed that under the tropical forest overland flow often originates in concentrations of throughfall dropping from trees; it frequently moves the litter, accumulating it against obstacles further down, and its waters spread in different directions. This discontinuity increases for less intense storms.

Rougerie (1960) found that discontinuous overland flow in the forest of southern Cote d'Ivoire was more frequent than in the forest of Bahia in Brazil. He notes that in Cote d'Ivoire, the litter is never abundant and water runs underneath it. In contrast, the litter in Bahia is thick and the top soil is composed of a spongy humid horizon, both of which retard overland flow.

The explanation seems to lie in the manner of decomposition of the organic matter, but the roles of the type of litter, microclimate and organisms are not fully understood. It is known, however, that the impact of man has been more important in West Africa, and the earlier periods of shifting cultivation could have been responsible for some modification of the land that now facilitates overland flow, and a certain amount of soil degradation that diminishes the soil permeability. Accurate comparative measurements are necessary to resolve the unknown variables.

Bonell et al (1983) however found that widespread overland flow is common during the summer monsoon in the undisturbed rainforest in northeast Queensland, in Australia. They explain the process by reference to the ralationship between rainfall intensity and soil hydraulic properties. They find the surface 100mm of the soil to be so highly permeable (with a saturated hydraulic conductivity of about 32m d^{-1}) as to absorb even the normally high peak monsoonal intensities. But the hydraulic conductivity decreases sharply with depth so that the 100-200mm zone has a value of only 1.5m d^{-1} and the 200-1000mm zone a value of only 0.3m d^{-1}. The subsurface layer is quickly saturated during prolonged rainfall events, which results in saturated overland flow as defined by Kirkby and Chorley (1967) and rapid subsurface flow (Fig.10.4). Not surprisingly, therefore, some 47% of the total annual streamflow appears as quick flow (overland flow plus shallow subsurface flow) while frequently more than 45% of the rainfall from individual storms appears as quick flow in the catchment.

Forest lands have been generally regarded by hydrologists as areas that have optimum infiltration and negligible overland flow. The above evidence from Australia on the influence of soil permeability shows that this is an over-simplification.

There seems to be no doubt however that fire, trampling and compaction (by humans, cattle or vehicles) and logging disrupt the natural forest and usually cause marked changes in normal infiltration, often initiating overland flow. On the other hand, it is sometimes asserted that when the forest floor is maintained in an undisturbed state, although the forest cover is removed, infiltration remains unchanged and overland flow does not occur (Gilmour et al 1982).

Runoff regimes. A variety of river regimes can be recognised in warm humid regions. In a simple approach there can be recognised:

    (i)   equatorial rivers of the humid tropics;
          (a) with one peak,
          (b) with two peaks,

    (ii) rivers of wet/dry regions

Equatorial rivers with one peak such as the Brantes river in Indonesia (Fig.10.5) are produced by heavy annual precipitation of over 1750mm, without a pronounced dry season. The precipitation pattern may contain two periods of increased precipitation, but only one flood peak is produced.

Equatorial rivers with two peaks are produced by precipitation regimes with monthly totals over 100mm and an annual total over 1750mm. The basins are dominantly covered by tropical forest. The river Essequibo in Guyana is given as an example (Fig.10.5), though the secondary peak is much smaller.

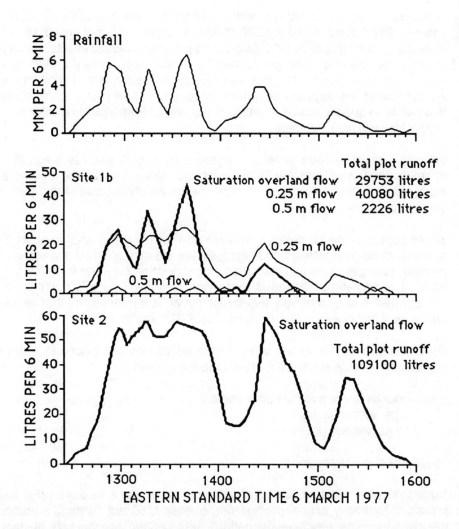

Fig.10.4 Continuous record of rainfall, saturation overland flow and
subsurface flow for two rainforest sites in North Queensland,
Australia (from Bonell et al 1983).

The equatorial regime is typical of the continuously wet sub-climate. A relatively
large flow is continuously present in the river channels and both intra-annual and
inter-annual variations are very limited. Sometimes, however, the seasonal variation
is more marked in the case of streamflow than of precipitation, because the drier
months also experience higher evapotranspiration. A major distinctive hydrological

240

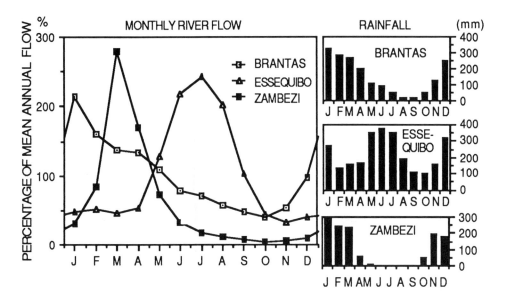

Fig.10.5 Three types of rainfall and runoff regimes in warm humid regions.

feature of this area is the existence of a dense network of rivers with stable flow and occasionally unstable channels. Two of the largest rivers of the world, the Amazon and the Congo, collect most of their runoff from the equatorial zone.

Rivers of tropical wet/dry regions (e.g. the Zambezi, see Fig.10.5) are found in regions covered by a variety of savanna and forest, and the regime is influenced by the length of the dry season. The seasonal effect of the rainfall distribution is well pronounced in the hydrological regime of the rivers.

In the non-continuously wet areas not affected by monsoons or trade winds, runoff follows rainfall but with a lag of two to six weeks. The lag derives from the fact that the initial rains are used to replenish the soil water deficit, and also from the storage in channel reaches especially in medium to large rivers. In smaller rivers, the response is faster. The base flow contribution to streamflow may be significant during the dry season under favorable geological conditions (Fig.10.6).

Areas with a pronounced seasonal break also have an important, though not very dense, network of permanent rivers which exhibit significant variation in both seasonal and inter-annual flow. This is complemented by a network of ephemeral streams. Erosion action and sediment discharges are larger than in the

241

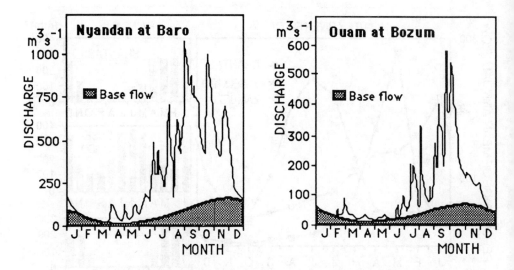

Fig.10.6 Stream flow and base flow for two rivers in wet/dry regions
not affected by monsoons or trade winds.

continuously wet equatorial areas, as a result of variations in the drying and wetting of soils and changes in vegetation cover. More rapid concentration of runoff also occurs in river channels, and higher specific maximum discharge results in particular catchments. As depicted by a number of rivers in West Africa, inter-annual variation of flow is significantly larger in this zone than for the equatorial zone.

The rivers of the monsoon region present large variations in flow, both seasonally and annually. The variations derive from changes in the intensity, area of origin or areal extent of the monsoons or trade winds. Indeed, Volker (1983) asserts that, in general, the regime of all rivers in southeast Asia is governed by the monsoons (and the trade winds). Typical hydrographs (Fig.10.7) are those of the Mekong, and the Huai Bang Sai in Thailand. The flood season results from the rains brought by the monsoon in May-October, and the low flow from the dry season. The regime of the Mekong river is simple in type while that of the Huai is flashy. In the latter, the flood and dry seasons are clearly seen, but owing to the differences in catchment characteristics, the distribution of runoff is different. The Red river in Vietnam also has flash floods in spite of its large size (120 000km^2).

The variability of flow increases with a decrease in vegetation cover, and these variations together with large rates of runoff produce large quantities of sediment discharge, specially where farming activities are widespread.

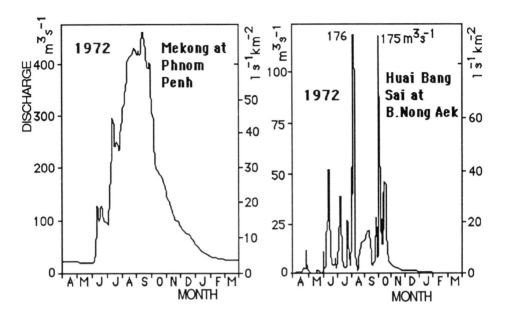

Fig.10.7 Contrasting hydrographs of two rivers
in southeast Asia (from Volker 1983).

River basins in areas affected by tropical cyclones experience high flood peaks. In Thailand for instance, in the Mae Klong river basin with an area of some 32000km^2, the heaviest rainfall occurs during August and September as a result of heavy monsoon rains superimposed upon torrential rains brought by tropical cyclones. As a consequence, flash flooding usually takes place towards the end of the rainy season when the absorption capacity of the catchment is low. The hydrograph of the Mae Klong (Fig.10.8) exhibits the flash flood components superimposed on the basic regime. Similarly in Taiwan, where all the five rain-generating mechanisms operate from May to October, most of the runoff occurs during this season.

Differences between the tropical river regimes can be also traced in the shape of their flow duration curves (Fig.10.9) which show that the river regimes of the equatorial rivers are more stable than those of the wet/dry regions.

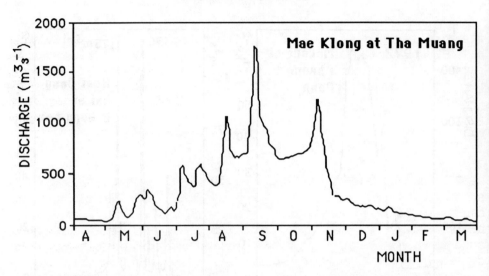

Fig.10.8 Flow of the Mae Klong river (from Champa and Boonpirugsa 1979)

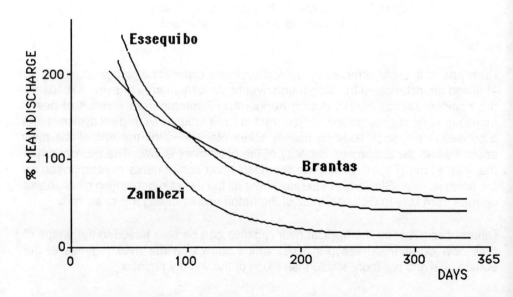

Fig.10.9 Flow duration curves for three rivers in humid warm areas.

244

Effect of swamps. In addition to the river basins, intermittent headwater swamps, known in parts of Africa as dambos, can be considered as belonging to the areas with a catchment response (Balek 1983). In contrast with ordinary swamps they have a distinct boundary. According to Hindson (1955) some of the intermittent headwater swamps remain wet late in the dry season largely due to the seepage which occurs along their margins and which results from slow subsurface drainage from the upland areas between the dambos.

They are found in the upper parts of the stream network; there the erosion cuts valleys where the rock is less weathered and covered by small deposits of soil. The shallow location of the impermeable surface is another typical feature. The typical catchment response is shown in a hydrograph (Fig.10.10) where three components can be traced: groundwater outflow from the forest and the swamp, and a typical surface runoff prolonged by the thick vegetation till the middle of the dry season.

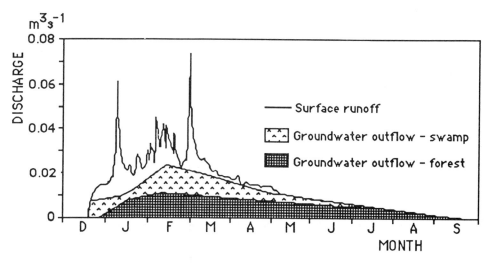

Fig.10.10 Typical hydrograph components of a headwater swamp.

The headwater swamps are rather small in size, but the total number of them in Africa alone has been estimated as 10^4-10^5. Table 10.8 shows the water balance of a typical dambo in Central Africa. In general, they have a stabilizing effect on the regime of the river headwaters, delaying the surface runoff process.

Table 10.8

ANNUAL WATER BALANCE OF A HEADWATER SWAMP (DAMBO) IN ZAMBIA

Drainage area	1.43 km^2
Area of dambo	0.15 km^2
Rainfall on drainage area	1330 *mm*
Evaporation from free water surface	1710 *mm*
Evaporation outside dambo	1320 *mm*
Evaporation in dambo	1075 *mm*
Outflow	255 *mm*

Storm Runoff. In general, the rivers of the humid tropics exhibit rapid responses to precipitation inputs as shown by the hydrographs in Fig.10.11 which represent conditions in part of the West African coast, Martinique, New Caledonia and Brazil. Hydrographs from small and some medium basins are typically single-peaked unless reflecting more complex patterns in rainfall intensity. Bruijnzeel (1983) found that surface runoff volumes Q_s (*mm*) of 42 storms recorded during 1975/76 in the Kali Mando basin in south Central Java are related to storm rainfall P (*mm*) by:

$$Q_s = 0.009 \, P^{1.415} \tag{10.1}$$

with $r^2 = 0.90$.

He found that surface runoff (quick flow) makes up 5-7% of monthly runoff in the wet season. Dagg and Pratt (1962) earlier reported a figure of 8.9% for a Kenyan basin of similar geology. In a forested basin in the Dominican Republic, the quick-flow contribution was 20% (Walsh 1980). However, for the area of northeast Queensland already mentioned, 45% of the total annual streamflow from individual storms appears as quick flow in the wet season. In the above examples, it is possible that the Javan and Kenyan basins have smaller contributing areas than the other basins cited, and hence the smaller proportions of storm runoff. Non-uniform methods of hydrograph separation might also account for part of the differences.

The minimum contributing area (MCA)(minimum area which contributing 100% of the effective rainfall would yield the measured direct runoff) was 1.3% of the total basin area for the Kali Mando basin, where MCA's of more than 5% of basin area are rarely attained during storms.

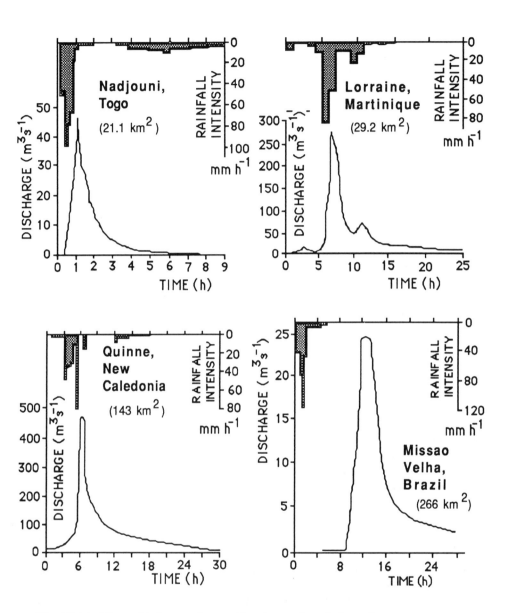

Fig.10.11 Storm hydrographs and hyetographs of rivers in
the humid tropics (from WMO 1987)

In collaboration with the government of Indonesia, the Institute of Hydrology (1985) assembled and analyzed 1001 station-years of flood data from 110 stations in Java and Sumatra. A prediction for the mean annual flood was developed in terms of catchment area, mean annual maximum one day rainfall, channel slope, and proportion of lakes and reservoirs in the catchment. Conversion factors for different return perods were also estimated.

Groundwater. Groundwater is an important component of the water balance in the humid warm areas, particularly in the humid tropics. The large volumes of infiltration available throughout the year make groundwater accumulation more continuous in the region than elsewhere. This is a result of adequate rainfall which ensures the opportunity for essentially constant percolation of water below the root zone, specially in the rain forest. Even in more seasonal areas of the region, where the continuity is interrupted, a large fraction of the total rainfall infiltrates to the saturated zone.

In addition to groundwater in alluvial lowlands, coastal plains and deltas, which will be discussed elsewhere, aquifers in the interior region may extend over vast areas (Mink 1983).

Alluvial sediments also form important interior aquifers. Much of the interiors of Asia and Africa consist of crystalline basement complex rocks which contain meagre groundwater in the weathered mantle or in faults and cracks. Such sources may nevertheless be very important for rural water supplies. Extensive sedimentary basins containing exploitable aquifers occur in Africa and South America.

Few studies of groundwater recharge seem to have been undertaken in the humid warm areas. In the Tapei Basin (Taiwan), some $240km^2$ in area, streamflow infiltration is a major component of recharge of the principal aquifers in the area of outcrops, but direct infiltration of rainfall is believed to account for most of the recharge to the shallow alluvium and upper water bearing zone. Water balance studies estimate annual recharge of the principal aquifer as $100 \times 10^6 m^3$, of which 9% originates in the major creek areas (Hsueh 1979).

Summary of water balance components. It is difficult to present typical water balance values for humid warm areas. Nevertheless, a few examples can be given indicating at least how the water balance is complicated and dependent on the climate and altitude. As an example for the equatorial regions, Fig.10.12a shows the water balance for Kribi in the Cameroons. There are two rainfall peaks, and the figure also shows how the soil water is used during a brief period in December and January. A small deficiency is recharged shortly after the beginning of the rainy season.

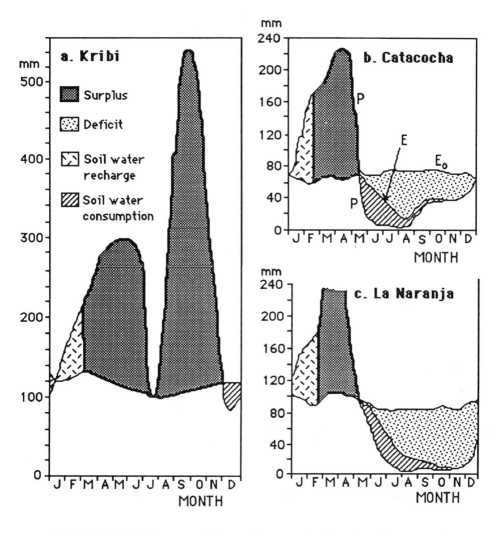

Fig.10.12 Monthly water balance diagrams for three locations
in the humid warm regions.

For the rivers of wet/dry regions the period of soil water deficit is more pronounced. Fig.10.12c is a water balance diagram for La Naranja, Ecuador. The altitude of the station is 528m and the rainy season continues for 4.5 months. The soil water deficiency rises steadily from May till November and then the soil water is recharged within 1.5 months.

In higher altitudes under otherwise similar conditions the soil water deficit is less pronounced, as can be seen in Fig.10.12b. The station at Catacocha, Ecuador, is 1860m above sea level. Owing to the altitude the potential evaporation is lower than at La Naranja.

The two further examples in Fig.10.13, for Yagambi in Zaire and Narajangan in Bangladesh, bring out the importance of interannual variability, since they show the rainfall distribution for a particular year as well as the average values. This highlights the difficulty of showing a "typical" water balance.

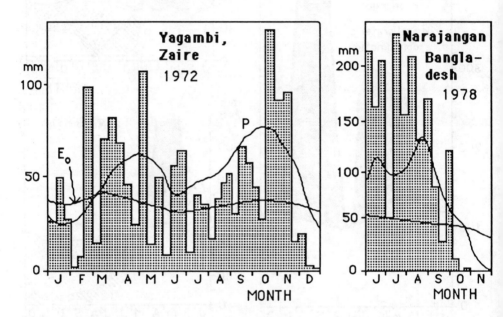

Fig.10.13 Decad (10 day) rainfalls in a particular year, with average values of rainfall and potential evaporation.

The following general points about water balances in the humid warm areas can be made:

(i)   There is high rainfall variability and occurrence of relatively dry periods all the year round even in the Congo basin.

(ii)   Evapotranspiration is comparatively stable.

(iii)   The water retention capacity of the soil (a function of structure and land use)   helps to dampen part of the observed fluctuations in soil water and streamflow patterns.

(iv)   In a typical monsoon regime, where rain-bearing cloud formations may be kept away for several days or weeks even in the middle of the monsoon season, farmers in southeast Asia adapt their agricultural practices by puddling and bunding their rice fields in order to collect water during the heavy rains for use by plants during the extended dry spells which may occur.

The hydrological cycle and water balance components for storm rainfall conditions are illustrated by ECEREX's (MAB) work on three small watersheds (areas 1.5, 1.45 and 1.45*ha*) with slopes ranging from 15 to 25%, monitored for 90-116 storm events in the Amazonia region of French Guyana (Roche 1981). Fig.10.14 shows the distribution of the various balance components. For example, on A basin, rainfall was 3470*mm*, of which about 46% was disposed of through evapotranspiration, 19% by surface runoff, 8% by subsurface runoff and 27% by groundwater recharge not drained by the stream. Soil erosion was 0.7*t ha-1 y-1*. For the range of rainfall intensity prevalent in the study area one would expect interception by forest canopy to account for 18-20% of the gross rainfall (see Table 10.5). This means that some 41% of the observed evapotranspiration was caused by interception.

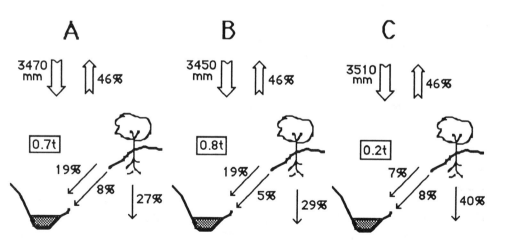

Fig.10.14 Water balance components for three small watersheds in French Guyana (from Roche 1981).

Erosion and sediment transport

In warm humid regions with catchment response the erosional processes are in many ways more serious than in regions with a moderate climate. For instance, it has been assessed that the rivers of Java carry more silt in one day than rivers of temperate regions of the same drainage area in two centuries.

The following factors are decisive in the erosional processes of warm humid areas:
- the type of climate,
- the degree of destruction of vegetation,
- the ability of the vegetation to regenerate,
- the leaf area index of the vegetation,
- the state of soil protection against violent temperature changes,
- the age of the river basins,
- the rainfall pattern.

Table 10.9 shows the erosion rates for West African streams (Grove 1971); they may be far below possible maxima. Dunne (1979) observed erosion ranging between 20 and 200 $t\,y^{-1}\,km^{-2}$ in various parts of the tropics. Various formulas for the assessment of tropical soil loss were summarized by Balek (1983). A comparison of soil loss for tropical and temperate regions made by Kirkby and Morgan (1980) is in Table 10.10.

According to Water Resources Planning Commission (1977), owing to deeply weathered fractured rocks, erosive geologic characteristics, steep topography and intense rainfall, the rivers of Taiwan have "the highest gradient in the world, great flow variation and serious soil erosion". Hwang (1985) reports that the existing 27 reservoirs in Taiwan have severe silting problems, specially those in the southwest. The rate of reservoir sedimentation in the south-west is reported to be 3 times the rate in the Yellow River and 10 times higher than in the U.S.A. It leads to 9-22mm of soil loss per year, and average depletion of 0.54 to 3.58% in the initial storage capacity of 12 of the reservoirs.

As Oyebande (1981) noted, sheet erosion, which affects much larger areas and which wreaks more severe damage in the area of soil loss, is no less significant than gully erosion which produces more spectacular and sometimes monumental features, as in the Agulu-Nanka and Okwudolor-Amucha areas of southeastern Nigeria. Soil erosion in forest-protected areas is not important in spite of the high rainfall intensity, under-developed ground flora and relatively thick humus layer. River bed erosion is however common in the lower reaches with flood plains and braided channels, and results in the washing away of large amounts of

Table 10.9

EROSION RATES FOR SOME RIVERS IN WEST AFRICA
(after Grove 1971)

River	Niger	Benue	Niger-Benue	Congo
Discharge $(10^{12} m^3 y^{-1})$	0.1	0.11	0.21	1.2
Erosion rates $(10^6\ t\ y^{-1})$				
in solution	5.5	4.5	10.0	98.5
suspended	9	22	31	31.2
combined	14.5	26.5	41	129.7
Specific erosion				
$(t\,y^{-1}\,km^{-2})$	19	77	37	37
$(mm\,y^{-1})$	0.008	0.036	0.015	0.0148

Table 10.10

SOIL LOSS IN THE TROPICS AND TEMPERATE REGIONS
(after Kirkby and Morgan 1980)

Vegetation	Slope (%)	Soil loss $(t\,y^{-1}km^{-2})$ Tropics	Temperate
Forest	-	9 (rain forest)	3
Grass	4-9	5 - 20	7
Bare soil	4-19	10000-17000 (humid tropics) 1800-3000 (savanna)	1000
Maize	-	30	12800
Coffee	-	2200	
Banana	0-7	1500	
Manioc	0-7	9000	
Sorghum	2	300 - 1200	
Groundnuts	2	300 - 1200	
Rice	-	17 - 289	

unconsolidated soil material. If the forest cover is removed, soil erosion is accelerated because of the impact of rain and resulting runoff. One example of a young gully in an active gully zone is at Okwudolor, near Owerri in Imo state of southeastern Nigeria. The gully, formed in 1984 as a result of the construction of a road, was 670m long, 70m wide and 26m deep in September 1986.

The magnitude of the erosion caused by removal of protective vegetation cover depends on soil structure, land form, rainfall characteristics and the land management systems adopted. Examples from Indonesia, Trinidad, Papua New Guinea, Ivory Coast, and Cameroon show that the annual rate of new soil formation (via weathering) for ultisols is 0.011-0.045mm (Owen & Watson 1979) and for alfisols 0.07mm (Boulad et al 1977) in the humid tropics. Thus it may take hundreds of years to develop 1mm of fertile top soil, and hence the need to prevent or check its loss.

The data in Table 10.11 on suspended sediment transport and soil loss data from Nigeria and peninsular Malaysia (Oyebande 1981; Fatt 1985), for basins of comparable size, illustrate the magnitude of problems associated with erosion and sediment transport in the humid tropics.

Table 10.11

SUSPENDED SEDIMENT CHARACTERISTICS OF SELECTED RIVERS
IN THE HUMID TROPICS
(Sources: Oyebande (1981); Fatt (1985))

River	Country	Basin area (km^2)	Water discharge (m^3s^{-1})	Sediment conc'n (ppm)	Sediment yield $(t\,y^{-1}km^{-2})$	Soil loss $(t\,y^{-1})$
Katsina-Ala	Nigeria	22,000	672	60	64	1,410
Cross	Nigeria	16,900	1,076	36	72	1,230
Su-men	Taiwan	763	-	-	5,701	4,350
Sg. Pahana	Malaysia	25,600	363	73	33	845
Sg. Langat	Malaysia	1,240	32	1,900	1,550	1,922
Sg. Perak	Malaysia	7,770	170	215	145	1,127
Sg. Kelantan	Malaysia	11,900	525	66	92	1,095

There are wide spatial and temporal variations in sediment concentrations in the rivers of the humid warm areas. Suspended sediment discharges are thus better calculated from relationships between point and average sediment concentrations when point daily sediment sampling is undertaken. It is usually more practical to use the relationships between sediment concentration and river flow. Such relationships are however only approximate and vary according to the season and hydrological conditions. In the humid tropics, with two or more seasons, sediment concentrations and discharges are higher during the transition from the drier to the wet season and vice versa. In practice, it is advisable to develop separate sediment rating curves for each season, as may be indicated by plots of the data.

Bordas and Canali (1980) reported the sedimentological behavior of four experimental basins located within the Forqueta River catchment (3000km^2) of South Brazil. Seven storms between December 1978 and June 1979 were used to characterize the hydrological and sediment responses of the basins (Table 10.12).

Table 10.12

STORM RAINFALL AND SEDIMENT PRODUCTION IN FOUR PILOT BASINS OF THE REPRESENTATIVE BASIN OF THE RIO FORQUETA, SOUTH BRAZIL

(a) Catchment characteristics

Catchment	A1	B1	A3	C3
Slope	40	20	40	50
Cover	Crops	Crops	Forest	Forest
Area (*ha*)	922	532	678	334

(b) Storm rainfall

Rainfall	No	Date	Max intensity (mmh^{-1})	Mean intensity (mmh^{-1})	Duration (*h*)	Total rainfall (*mm*)
Group 1						
Simple isolated	24	12.02.79	23	7.3	4	29
rainfall events	37	18.05.79	16	3.8	9	34
	38	22.05.79	26	4.9	11	54
Group 2						
Two simple events	27	9/10.05.79	20/21	2.5	20/17	50/41
separated by	29	4/5.04.79	12/12	2.8/4.6	21/5	60/23
24 hour interval						
Group 3						
Complex storm	30	15.04.79	10	1.7	21	35
pattern	39	04.05.79	10	2.1	19	40

(c) Sediment production   (C=bed load, S=suspended load, T=total load)

Rainfall No.	Load type	Sediment load ($kg\,km^{-2}$)			
		A1	B1	A3	C3
Group 1					
24	C	3680	3513	0	0
	S	1020	654	X	X*
	T	4700	4167	X	X
37	C	0	0	0	0
	S	754	2068	0	X
	T	754	2068	0	X
38	C	3880	14676	1032+	0
	S	1802	12173	1708+	X
	T	5682	26849	2740+	X
Group 2					
27	C	4149	4020	0	0
	S	2900	616	1.3	X
	T	7049	4636	1.3	X
29	C	x	x	0	0
	S	1334	8946	1	X
	T	X	X	1	X
Group 3					
30	C	X	X	0	0
	S	208	7040	>1	X
	T	X	X	>1	X
39	C	0	0	0	0
	S	150	81	2.3	X
	T	150	81	2.3	X

*Non-zero, but not determined.

It was concluded that slopes have practically no influence on overland flow and sediment production under forest cover; and on more erodible terrain, erosion is not significant except when the rainfall intensity exceeds $10mm\,h^{-1}$ with a duration of more than 4 hours. The threshold was observed to depend on the soil water at the beginning of the storm rainfall. It was particularly notable that the highest sediment discharge had a concentration of $39g\,l^{-1}$ and that during October 1969 the reservoirs for trapping sediment were completely buried three times.

The magnitude and proportion of the bed load is however surprising when compared to the assumption of less than 10% of the total for Nigerian rivers of comparable and larger sizes. The particle-size distribution and slope factors may be responsible for the apparently large difference.

Exogenous and endogenous interactions

Seven of the ten largest rivers of the world originate in the tropics, namely the Amazon, La Plata, Congo, Madeira, Orinoco, Tocantins and Rio Negro. The powerful streams of the Madeira, Tocantins and Rio Negro are tributaries of the Amazon and therefore the list of the main tropical rivers can be extended as indicated in Table 10.13 . Here the Mekong, Irrawaddy, Niger, Zambezi, Nile and Sao Francisco join the family of the ten largest tropical rivers. The Irrawaddy has its headwaters outside the humid warm region and a great part of the Nile flows through dry areas, but both rivers are to a great extent products of a humid warm climate. The water balances of these basins are shown in Fig. 10.15.

Table 10.13

GREAT TROPICAL RIVERS

No	River	Mean annual discharge (m^3s^{-1})	Drainage area (km^2)	Specific discharge $(l\,s^{-1}km^{-2})$	Length (km)
1	Amazon	212 000	6 140 000	34.5	7025
2	La Plata	42 400	3 104 000	13.7	4580
3	Congo	38 800	3 607 450	10.7	3400
4	Orinoco	28 000	1 050 000	28.0	2400
5	Mekong	13 500	810 000	16.6	4200
6	Irrawaddy	13 400	431 000	31.0	2150
7	Niger	7 000	1 091 000	6.4	4160
8	Zambezi	3 800	2 200 000#	1.7	3250
9	Sao Francisco	3 800	650 000	5.8	3720
10	Nile	2 950	2 881 000	0.9	6500

including part of Kalahari

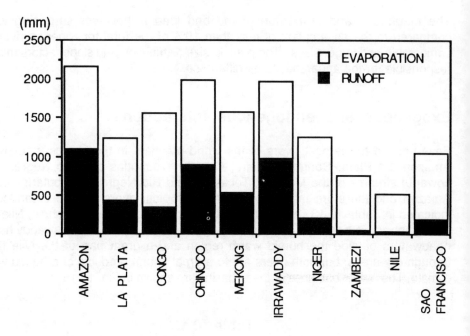

(mm)

Fig.10.15 Annual water balances of the ten largest tropical rivers.

Because these rivers and their basins are under the influence of very complicated conditions, the term "typical river regime" cannot be related to any of them. At certain sections along their lengths, each of these rivers tends to exhibit different regime types which reflect the climate and physiography of the catchment area immediately above the section. While the great rivers derive most of their flows from the humid warm areas, they also have substantial interactions with other climatic zones. The Niger is an interesting example of a river with significant exogenous interactions. Its headwaters are derived from a humid tropical region in Guinea, from where it traverses a sub-humid and semi-arid zone between Koulikoro (in Mali) and Niamey (in Niger) which represents 54% of its upper basin area, to enter Nigeria where it encounters progressively more humid areas on its descent to the Atlantic ocean through the equatorial zone of the Niger delta area. Thus the river system has water balance conditions (Table 10.14) which show a progressive decrease in specific discharge from Singuiri in Guinea to Niamey, after which it begins to increase. In the zone between Koulikoro and Niamey, 55-65% of the runoff is lost through evaporation and infiltration in an extensive swamp of the "interior Niger delta". In Nigeria, the Niger receives exogenous inputs from the north-west and central parts of the country, and the single peak of the simple regime becomes two. It later receives its major tributary, the Benue, which smoothes out the regime and

Table 10.14

WATER BALANCE OF THE NIGER RIVER SYSTEM SHOWING EXOGENOUS AND ENDOGENOUS INTERACTIONS

Station	Climate	Area $(10^3 km^2)$	P (mm)	E (mm)	Q (mm)	Q_{spec} $(l\,s^{-1}km^{-2})$
Singuiri, Guinea	Equatorial	70	1640	1220	420	13.3
Koulikoro, Mali	Sub-humid	120	1550	1155	395	12.5
Niamey, Niger	Semi-arid	371	1094	1013	81	2.6
Lokoja, Nigeria (largely)	Sub-humid	724	1221	1027	194	6.0
Mouth, Nigeria	Equatorial	1140	1250	1048	202	6.0

produces a single peak similar to the simple pluvio- tropical regime. This latter type is maintained to the Niger's outlet.

The humid warm areas are continouously interacting with the atmosphere immediately above them and through which they interact with other climatic regions through continuous transfer and exchange of energy and water.

There is a significant aspect of endogenous interaction which has often been discounted. This is the role of recycling, which has been de-emphasized (Unesco 1971). The accepted concept is that evapotranspiration from the land is transported away from an area (Linsley 1957). However, the works of Lettau et al (1979) and Salati et al (1979) have shown that the part of the precipitation in the Amazon basin from recycled (endogenous) water vapor is as significant as that formed from advected (exogenous) water vapor.

The Andes form an effective western barrier to the horse-shoe-shaped Amazon basin, so that the main influx of advected water is due to the trade winds from the east and the apparent main efflux is via the river system to the east. Thus, in the Amazon basin with an annual mean precipitation of 2000-2400 *mm*, major water recycling might be expected both from forest canopy evapotranspiration and locally formed cumulus, as well as from large-scale convectional activity within the basin. The mean turnover time for the water vapor was found to be 5.5 days in 1979 (Salati et al 1979). Measurements of the isotopes in the rain and river water have confirmed the importance of recycled water in the water balance of the whole basin. The studies conclude that only 52% of the precipitation in the basin was accounted for by the inflowing water vapor from the Atlantic Ocean; the remaining 48% was recycled (endogenous) vapor within the basin, which thus represents an important source of water.

Current pattern of water and land use

Constraints to optimum water utilization. In the humid warm areas, as in most other environments, the basic water resource problems are of three main types, namely: shortage, flooding and pollution (Oyebande 1975). Water shortage may be brought about by meteorological and hydrological conditions and/or by growing demand for water. The shortage may be permanent or seasonal. Flooding and pollution hazards are most severely felt in urban centres of developing nations, where large and ill-organized concentrations of people and economic activities both cause and aggravate the effects of the problems. In developing countries, rural communities tend to suffer more serious shortages of improved water supplies than their urban counterparts, who are closer to the seats of power and decision making.

There are aspects of availability and affordability of technology to harness the water resources, since water is a resource only in so far as there is a demand and there are technological and economic possibilities of using it. These possibilities include availability of funds to develop the water resources to provide a sound economic base for balanced growth in various sectors - domestic, industry, agriculture, electric power generation, transportation, recreation and tourism. An area where most developing nations of the humid warm areas encounter the most intractable problems in water and land use relates to provision of an effective management framework to administer water development and use efficiently. There is a need, for instance, to decide whether water is a social service to be provided free and if the governments can afford it; or whether a realistic, if not economic, price should be charged for water which is becoming increasingly expensive to produce. The existing pricing policies certainly encourage inefficiency and waste in water use. There is thus a problem of making water services more self-sufficient instead of constituting financial drains (Oyebande 1978).

Finally, instead of looking at the supply side alone, it has become inevitable in many developing nations which occupy the warm humid areas to pay attention also to the demand side of water use. There are rural communities and sectors of urban centers where per capita water use should be increased, but there are also others where wasteful use needs to be checked. In general, these countries need to check population growth which in many cases is up to 3% per annum. Given this rate of population growth, and the low annual rate of improvement in water supplies, it is difficult to match water needs, much less raise standards of living. This is one of the reasons why the objectives of the UN Drinking Water and Sanitation Decade will remain unattainable well after 1990 for many of the developing countries.

Land use and its relation to the water regime. A very high proportion of the warm humid regions is used for agriculture, and water availability dominates tropical agriculture.

A very common feature of tropical farming is shifting cultivation (known as "ladang" in Indonesia, "bewar" in India, "milpa" in Mexico and "roca" in Brazil). Under such a system a plot of old forest overlying fertile soil is cleared and then the brushwood is burned. Perfect timing of this work is essential, because the brushwood has to be dried out before being burned, but still no time should be left for fast growing plants to regenerate. With the uncertainty of the rainy season, work and crops can be easily spoiled. This practice therefore requires good natural knowledge of hydrological conditions. The annual harvesting of quickly growing crops begins early during the rainy period, while later crops are harvested during the dry season. Clearing proceeds steadily outwards from the village until another settlement has been reached. Some agriculturalists believe that because cultivation of this type is practised on soils of inferior quality, burning improves the infiltration, and the soil is enriched by phosphorus and potash.

"Wetland" (paddy) cultivation is widespread in southeast Asia, and at present it is a type of farming very much encouraged in the humid tropics because the abundance of water and not the quality of the soil is the decisive factor. The negative aspects of this type of cultivation are the requirements for grading, and in hilly regions very laborious preparation work on the soil surface, such as terracing, is required. As well as an abundant amount of water, dry sunny periods are needed for the ripening of wetland crops, and therefore the method has always been popular in hilly regions under the influence of monsoons.

Cash crop farming with a variety of crops for family consumption and also for the market requires soils of good quality and availability of water, possibly by irrigation. The selection of the cash crops reflects the particular environmental conditions and water availability.

Plantations are the final stage of the tropical land use pattern. While cash crop farming uses local domestic plants, introduced types of crops are grown on plantations and this in the long term may affect the hydrological cycle. Classical examples are the planting of rubber trees in Malaysia in the 19th century and the many attempts to change land use in wetlands. The latter attempts have resulted in crop improvement on deep soils underlying the swamps, but the acceleration of the water flow through formerly swampy areas has contributed to the destabilisation of

261

the flood regime in low-lying regions of the basins.

In contrast to shifting cultivation, "paramba" type cultivation has been practised on the west coasts of India and Sri Lanka. Its advantage is that it does not upset the natural ecological balance because a variety of useful plants are grown simultaneously all the year round. Each paramba is a self-supporting unit, and provided that the composition of the vegetation closely approaches natural conditions, this kind of cultivation can be very successful.

In many forested areas much of the forest has been replaced by plantations. Deforestation for agricultural and other commercial purposes has an impact through the degradation of the soil. The same can be said about bushfires, whether controlled or uncontrolled. Changes in the hydrological regime are even more drastic than those created by slow continuous felling, when open stands are covered by at least some vegetation.

Though the spread of irrigation has fostered a tremendous surge in crop production over the last generation, more than 80% of the world's cropland is still watered only by rain (Brown et al 1986). These lands produce two thirds of the global harvest. Much of the rainfed (dryland) agriculture of the world is practised in the humid warm areas where the crops are grown annually, though with varying risk of failure. Some of the land can be brought under irrigation, but with the high cost of irrigation projects ($10,000-15,000 per hectare in much of Africa) expansion of irrigation may not be a feasible solution to raising food production in many areas in the future.

Two water and land management approaches are being strongly recommended for reversing the economic and ecological decline of developing nations, many of which inhabit the region. The first is re-introduction of agro-forestry in its various forms, to rebuild traditions (specially in Africa) in which crops, livestock and trees are integrated as a matter of course. The second and more promising approach in the humid tropics is to grow crops without ploughing at all. Since 1974, researchers in the International Institute of Tropical Agriculture in Ibadan, Nigeria, have been studying low-cost minimum-tillage and non-till systems for humid and sub-humid regions (Lal 1981). Non-till methods involve planting directly into a stubble mulch and using herbicides for weed control. The approach can reduce soil losses to nearly zero, increase the capacity of crop land to absorb and store water, and reduce the energy and labor needed to produce a harvest. Minimum-tillage systems that maintain the productivity of the land offer an alternative to shifting cultivation which is still predominant in the humid tropical areas of Africa and Central America, and offer a chance to remove some marginal land from cultivation entirely.

Hydrological impacts of human activities

Types of impact. Wet humid regions today are faced with extensive and rapid changes. Unfortunately man's impact on the tropical hydrological regime is predominantly negative. The deforestation of vast areas has perhaps the most significant impact. Not only the quantity but the quality of water is influenced by the clearing of forests. The problem of cutting down tropical rain forest, sometimes called "forest mining", has become a matter of serious concern. High and Nations (1980) estimated that about $250,000 km^2$ of tropical forest is vanishing annually. The region of the Amazon tropical rain forest appears to be seriously endangered. Ranjitsinh (1979) studied forest destruction in Asia and the South Pacific and found that in this region about five million hectares are lost annually. In Thailand the proportion of the forested area in the country decreased from 58 to 33% within 25 years. In many parts of India, Pakistan and East Africa improper land use can be blamed for increased effects of flooding. Mooney (1976) believes that the resettlement of thousands of families along the arteries of the Transamazonia Highway, together with the replacement of the forests by roads, farms and ranches, threatens the tropical rain forest ecosystem.

Many problems are related to the construction of high dams in the warm humid areas; among the most serious are siltation, erosion and weed growth. The chemical and biological balance of rivers is also influenced by the construction of large reservoirs. A considerable amount of water is lost due to evaporation from the increased surface area of the water. Transformation of riverine ecological conditions to lacustrine ones, combined with the seasonal effect of the storage of floodwater, produce various types of thermal and chemical stratification, the effects of which are not easily predicted; fish are almost always influenced by a changed river regime.

Intensive cropping may result in soil degradation after several years of profitable production. This has been proved in schemes of intensive cropping of sugar cane and coffee in Brazil, peanuts in Senegal and cocoa in west Africa. In addition, the destruction of soil structure and wind and water erosion contribute towards the degradation. Sudden changes of the land surface by bush fires can increase the runoff volume tenfold and soil losses 100-3,000 fold.

The main problems resulting from industrial and mining activities and urbanization are microchemical pollution by hydrocarbons and nitrates. In addition to pollution by metals, water salinity increases, particularly in mining areas.

Impacts on erosion and sediment transport. Tin mining, timber logging, agricultural development and urbanization and infrastructural development have been identified as the human activities which caused severe erosion and substantial sediment generation in peninsular Malaysia (Fatt 1985). All the rivers with sediment yields exceeding $600t\,y^{-1}km^{-2}$ are said to be those with substantial mining activities. Sediment discharge from several tin mines in Kuala Lumpur produced concentrations that range from 5000 to 20 000 *ppm*. In 1955, tropical forests covered 75% of the area; the coverage was reduced to 55% in 1974 and 50% in 1980. It has been found that streams from logged areas carried 8 to 17 times higher sediment loads than before logging.

Soil erosion may increase from a low level of 0.2 $t\,y^{-1}ha^{-1}$ under virgin forest conditions to an alarming rate of 600 to 1200 $t\,y^{-1}ha^{-1}$ (Brunig 1975).

As in other parts of the humid warm areas, the rapid socio-economic developments of the past two to three decades have brought about accelerated infrastructural development such as construction of roads, houses, and other massive engineering works. An example was reported of erosion that produced $10^5\,t$ of sediment from an urban area (Kuala Lumpur) of $3km^2$ over a 2-year construction period. Cases of spectacular gullies in south-eastern Nigeria resulting from road construction have already been mentioned.

Impacts of deforestation on runoff. Information relating to the effects of man's activities (through deforestation) on hydrological processes is scanty in the humid tropics, but reviews by Hibbert (1967) and Oyebande (1987) have provided valuable summaries of available information. From Hibbert's summary, a practical limit of increased runoff of $4.5mm\,y^{-1}$ was obtained for each per cent reduction in forest cover. First year responses to complete deforestation vary from 34 to 457mm of increased runoff. There is however no strong correlation between the amount of increase in runoff and percentage reduction of forest cover. The highest amount of first year increase in water yield (among 39 studies reviewed by Hibbert) is reported for a 100% clear cutting of a watershed at Kimakia (in Kenya) under tropical forest. The amount was 457mm followed by 229 and 178mm during the second and third years, from the 35ha watershed.

The following are some of the highlights of the review by Oyebande (1987):

(i)   The increases in runoff for the first and second year after clear cutting of a tropical mountain forest in Central Taiwan during 1978-79 were 58% and 51% respectively of the annual flow. Low flows also increased by 108% and 293% in these years.

(ii)   The clearing of a 44*ha* watershed in Ibadan, Nigeria in early 1979 produced a significant increase in total runoff of 340 *mm* or 23% of the annual rainfall, from a negligible pre-clearing streamflow.

(iii)   Logging of North Creek watershed in N. Queensland, Australia led to an average increase in annual runoff of 10.2% or 293*mm*. Soil water levels in the experimental basin remained higher and soil water deficits were drastically reduced, perhaps because of reduction in transpiration demand.

(iv)   The effect of clear cutting in humid warm areas is shorter-lived than in less humid areas, due to rapid regrowth of vegetation.

(v)   A change in the forest cover of not more than 15% may not cause any measurable streamflow response.

(vi)   Maintenance of strong soil structure and hence high infiltration after clear cutting will enhance the contribution of base flow to the total runoff.

(vii)   The maximum first-year increase in runoff from 100% deforestation is of the order of 600*mm* in the humid tropics.

Studies of the effect of forest removal on peak flow produce more dramatic results, similar to changes in erosion and sediment yield.

Impact of over-exploitation of groundwater. In rural areas of the humid warm region, groundwater is often under-used, because of the abundance of rainfall and surface water. However, where conditions are favorable, small groundwater schemes should be encouraged, particularly where there may be temporary shortages of the surface supply or when the surface water becomes polluted.

In contrast, both renewable and non-renewable groundwater are often heavily exploited in and near urban areas, resulting in frequent reports of sharp falls in the water-table. As an example, the Taipei groundwater basin ($240 km^2$) in Taiwan has developed into a metropolitan area which includes Taipei city in the east. Accelerated growth in industry and rapid growth of population increased the demand for water for industrial, municipal and domestic supplies. Numerous bores were drilled and a maximum annual pumpage of $435 \times 10^6 m^3$ was withdrawn in 1970. Land subsidence caused by over-pumping was first noticed in 1961 and reached a maximum of 2.09*m* in Taipei city in 1976. From 1957-76 about 64% of the basin area subsided more than 0.5*m* and 21% sank more than 1.5*m*. The consequences

265

were that the flood levees around the city had to be raised from 1.0 to 1.9m and about 400ha of farmland became perennially inundated with saline tidal flows (Hsueh 1979).

Potential for future impacts.   Some of the hydrological effects of artificial changes in the physical conditions of catchments have been discussed above. This may mean that the hydrological data series are not homogeneous, so that as long-term records become available their significance for direct practical application is greatly reduced. Volker (1974) thus opined that methods must be found to predict in a more rational and systematic way the hydrological effects of artificial changes in a river basin.

It is also significant that, in the near future, a greater percentage of the total water resources of the earth will be exploited than at present (see Chapter 1). The hydrological effect of the control and use of water on a global scale has been negligible in the past, but if storage and diversion increase so as to reach a substantial percentage of the potential, the situation may change.

A general circulation model by Lettau et al (1979) predicted the effect of complete deforestation of the Amazon basin (6.3x10^6km^2): increase in albedo from the present 0.13 to 0.16 affecting an area of 2x10^6 km^2 ; increase in precipitation by an average of 75mm y^{-1} for the Amazon basin; and air temperature increase near the surface of 0.55 ^{o}C. However, Henderson-Sellers (1981) used a three-dimensional model to predict a decrease in rainfall of 600mm y^{-1} and a small temperature increase. It is to be hoped that the ongoing work on general circulation models will enable the prediction of the effect of further interaction on the weather and climate of the earth in general and of the humid tropics which represent an important energy source region for the rest of the world.

Implications for hydrological data collection and management

The problem.   The humid warm region exhibits peculiarities of climate, particularly the number and types of rainfall-generating mechanisms which directly affect the seasonal and annual variations in streamflow and the intensity of erosion phenomena. The large discharges and their rapid variation, the high humidity and temperature conditions, the problem of accessibility of certain locations together with the delicate ecological balance, and the underdevelopment and lack of funds affecting most of the region have serious implications for operational hydrology and the management of water and associated land resources.

These peculiarities not only influence the type of equipment to be used, but also the whole approach to collection, storage, and analysis of hydrological and related data. The region faces the dilemma of either applying a relatively low level of appropriate technology which is more within its grasp in terms of operation and maintenance and of using its relatively cheap manpower; or aiming for comparatively advanced equipment and technology, including satellites for data measurement, which makes it more dependent on the more developed economies for their supply, maintenance and modification.

Tschannerl (1979) recognized three types of hydrological data in developing countries:

- historical data which should be used whenever possible; in practice this means the full use of long rainfall records for the extension of stream flow records;

- data which have not previously been widely used and which can be processed without delay for the special purposes of water resource schemes; data obtained from tree rings, mud, pollen etc, and through the methods of isotope hydrology are in this category;

- data collected at the cost of postponing the initiation of water resource schemes, or in other words data from networks established during the preparatory stage of a project.

Network design and density.   The density of stations should be primarily dictated by the variability of the elements to be measured and the acceptable errors of estimation at an ungaged site, but financial constraints are often the over-riding factor.
In most of the countries in the humid tropics, the paucity of hydrological stations is a real problem. Some countries such as India and Malaysia may have reasonable densities in some areas, but lack stations in areas not easily accessible. In Central America and most of West Africa, the greatest number of stations are rain gages, and there are far fewer recording than standard gages. In Central America, a number of type A meteorological stations have been installed since 1967, but most of the network stations still cover only rainfall, temperature, relative humidity and evaporation by Piche evaporimeters (Garcia 1983). The Amazon basin (almost $7 \times 10^6 km^2$) is sparsely populated. In Brazil, within which lies $4.8 \times 10^6 km^2$ of the basin, the two government institutions responsible for hydrometeorological information have until recently paid more attention to the developed regions of higher population density. The prevailing conditions in the Amazon make it prohibitive for the economies of the Amazonian countries of Bolivia, Brazil,

Colombia, Ecuador and Peru to install and operate an adequate conventional network. Only a few humid warm countries have areas where the density of stations is high. These include parts of Australia, Cote d'Ivoire, northern Venezuela and Costa Rica (WMO 1983).

WMO (1983) makes recommendations for minimum network densities and length of operation of the stations in the humid tropical region. The climate of wet tropical zones with continuous seasonal precipitation not affected by monsoons or cyclones is fairly uniform. Such sub-regions are considered acceptable for location of stations at a density of one in 2500-2000km^2 for rain gages and one in 5000-4000km^2 for flow gages (Table 10.15). The humid tropical zone does not contain many areas with very high elevations (above 1000m), but where topography changes rapidly, a slight increase in density is necessary. Such densities can still yield satisfactory errors of 5-10% for mean annual precipitation and runoff.

Table 10.15

RECOMMENDED MINIMUM NETWORK FOR HYDROLOGICAL DATA
(STATIONARY REGIME) IN HUMID WARM AREAS
in km^2 per station (from WMO 1983)

Type of climate sub-region	Topo-graphy	Station density for precipitation network	Station density for streamflow network	Duration of operation for stationary regime (y)
1. Continuously wet	Flat	2500	5000	10-20
	Rugged	2000	4500	
1a. Wet, with seasonal variation (not affected by monsoons or tropical cyclones)	Flat	2000	4000	15-20
2. Affected by tropical monsoons	Flat	1000	2000	20-60
	Rugged and coastal	800	1600	20-60
3. Affected by tropical cyclones	Flat	600	1200	60-100
	Rugged and coastal	250	500	60-100

Garcia (1983), though perhaps giving a counsel of despair inspired by the gloomy economic horizon, was being realistic when he wrote: "the quest for long streamflow records, a cornerstone of many hydrological analyses, may be a futile effort in some areas of Central America, even if funds were available to substantially increase the present networks". He suggests that emphasis be placed on data quality rather the quantity or length of some of the records. He also recommends the adoption of a "maximum information" approach, which seeks to mobilize every bit of available information, including data acquired through periodic monitoring of catchment conditions, to validate the historical series.

The rain gage densities for the five sub-basins of the Amazon are only about one-tenth of the recommended levels in Table 10.15. Special attention should be given in the tropics to the establishment of experimental and representative basins. Here an intensive observation of a very dense network can replace gaps in the existing conventional network. Guidance for extrapolation of the data outside the experimental areas has been given by Balek (1983).

Table 10.15 shows that for the continuously wet and seasonal areas not affected by monsoons or tropical cyclones, the reduced time variation of hydrological and meteorological elements keeps the required duration of measurement low. Operation for a period of 10-20 years may be satisfactory, although longer periods are recommended where possible.

For areas affected by monsoons and trade winds, the density of the stations and period of observation should be increased to take care of the greater variability of the hydrological and meteorological elements (WMO 1987).

The large time variation in tropical cyclone conditions makes it imperative to increase the density of stations so as to obtain adequate data on the hydrological consequences of such phenomena. For the same reason, it is necessary to operate both flow and precipitation stations over a long period, probably between 60 and 100 years (Table 10.15).

The heavy budgetary constraints that affect the development of hydrological networks in most countries of the humid warm areas have compelled them to seek alternative approaches as in the case of the Central American countries and the Amazon. Paradoxically, it seems that the development and use of advanced hydrological equipment technology recommends itself highly for data measurement and transmission. Advanced equipment deserves serious consideration for providing solutions to some of the many operational problems of the region, provided adequate maintenance arrangements are available.

269

WMO (1983) has also made recommendations for the distribution of sediment gaging stations and the frequency of sampling. In addition, catchments in which the natural vegetation cover has been removed by man or natural events require sediment gauging.

Data storage and analysis. Without continuous air-conditioning, paper documents deteriorate in the hot and humid environment of this region. WMO has therefore recommended the storage of hydrological data on floppy disks for microcomputers, and on microfilm.

Two factors hinder the use of statistical analysis for the estimation of flood peaks and occurrence in the humid warm areas, particularly those affected by tropical cyclones. These are the relatively short records available for most of the stream gages and the heterogeneous population of rainfall and runoff events. It is now being recognised that the major events caused by tropical cyclones for instance require different methods, since mixing these events with others may lead to erroneous conclusions.

The various current approaches to the problem of non-homogeneity of storm rainfall events in the tropical cyclone sub-region include:

(i)   Treating the tropical cyclone data as outliers in a f itting obtained by using conventional statistical methods (Dubreuil and Vuillaume 1975; Guiscafre et al 1976)

(ii)   Use of a complex frequency distribution (Canterford et al 1981).

(iii) Calculation of separate probabilities for each rain-producing mechanism. The probability of the non-cyclonic rainfall or peak flow is calculated using conventional techniques. For the (purely or predominantly) tropical cyclone rainfalls or peak flows, all the events recorded in the particular region would be considered using regional studies (Garcia 1983; Coleman 1972).

In the case of the third approach, each event would receive a probability of occurrence in the given station or catchment, which is the product of the probability calculated by means of conventional techniques and the probability of the tropical cyclone occurring at a given station or catchment. Another advantage of this regionalization approach is that it could be applied to any point or catchment of the region, including the ungaged ones.

Meteorological and hydrological equipment. In order to eliminate record discontinuities in the humid warm areas, specially the humid tropics, it is necessary to take particular care in the choice of equipment. Sturdiness and simplicity of installation are desirable options.

Difficulties are often encountered in the operation of unmanned electronic equipment in the environment of a tropical forest. For example, of the 36 weather station months of operation during 1979-1981 by a U.K. research team working in West Java (Institute of Hydrology 1985), only 4 were of sufficient quality for use in detailed modelling studies. Good quality data were only obtained after two major "tropicalising" modifications were carried out, involving the incorporation of lightning protection circuitry on the pulse counting boards, and the addition of aspirated psychrometer sensors to the automatic weather stations. The latter additions were necessitated by the low wind speed conditions at the site,typical of most tropical situations.

Equipment failure is not always due to the tropical environment, but often to the bias of the equipment designs to temperate conditions. There is thus an urgent need for designers and manufacturers of equipment meant for use in the humid tropics to incorporate humid tropical conditions into their design studies. Such conditions include constantly high humidity; frequent occurrence of lightning; low wind speed; constantly high temperature; high rainfall intensities; and low pH in water. All these call for design of robust equipment together with appropriate materials and accessories that will stand the conditions of the region.

Implications for land and water management

Land management. High rainfall intensities and large volumes of surface runoff make it difficult under certain soil conditions to design economic mechanical soil conservation systems that will not fail at frequent intervals. Increasing emphasis should be placed on agronomic practices which simulate forest cover conditions.

Intensive agricultural systems in the humid tropics often rely on high levels of inputs such as agricultural chemicals, in order to maintain economic levels of productivity. It is evident that such inputs are soon lost from the production sites to the stream systems, and under certain soil conditions they may accumulate in the deeper horizons, thus leading to future problems.

Agro-forestry, non-till and minimum-till systems are a possible solution to both problems of soil conservation and loss of inputs.

271

Water management.   From the hydrological point of view the feasibility of agricultural and industrial development in developing countries should be based on an inventory of available water resources.   This includes surface water, groundwater, soil water, precipitation and waste water. The availability of water must be examined in the light of the technologies that are currently available. For further improvement of the inventory, the availability of water has to be studied in space and time. Here the hydrological approach becomes more complicated because the optimization of water use has to be analyzed in the light of pre-established economic goals that take into account the physical, political and sociological constraints pertinent to each region or country.   Transfer of appropriate technologies from temperate regions is not always feasible, because of the different economic background.

Management of water resources requires an understanding of both the nature and magnitude of hydrological processes at different scales. In the humid warm region, the intensity and continuity of hydrological processes assist the calibration of rainfall-runoff models, but the basic rainfall and streamflow data are still required. The physical difficulties of obtaining good quality data are exacerbated by the lack of funds provided to national data collection authorities.

About 80% of water in humid warm areas is used for agriculture. An improvement in the water supply to the fields is one of the most effective means of increasing food production. Some reserves can be found in the better use of monsoonal rains by increasing infiltration , which otherwise flows away as surface runoff.   Sometimes the situation can be improved by removing the phreatophytes and upland scrub and thus reducing the evapotranspiration. A considerable improvement can be achieved by effective irrigation.

For irrigation itself three aspects should be considered; the source of water, its transport, and its application. Ecological conditions have invariably been the decisive factor in the selection of the method of water transport. In the wet/dry regions, extensive schemes have not always been proved themselves to be the most efficient. Medium-sized irrigation schemes on 40-4000*ha*   are preferable.

At present, industry in tropical developing countries seldom accounts for as much as 10% of the total water use. It may therefore appear that wastewater treatment and recycling are minor problems which do not require an immediate solution. However, by the end of this century a high proportion of the world's population will be living in tropical developing countries, and increased industrialization and the consequent increasingly significant chemical contamination will be one of the main features accompanying this development.

272

Many developing countries in warm humid regions have considered economic growth and industrialization as key development priorities, and preservation of the environment has not been given the same weight. However, in the light of the experience of industrialized countries, it has become clear that urgent anti-pollution measures are necessary in advance to avoid an ecological upset, even though this may mean slower economic growth in certain industries.

The choice between small and large water management projects in humid warm areas is generally a debatable issue. Large scale water projects may not always be suitable and are, in some cases, not viable or feasible as a result of the large capital investment which is not easily available in developing countries, and the long gestation periods required. In addition, changes in the hydrological regime may result from large or regional water resource projects.

The current or prospective population growth of over 3% per year in many of the countries of the humid warm areas poses a formidable problem. Experience indicates that it will be difficult for these countries to keep pace with the water demands of the population. There is thus an urgent need for population control.

Finally, the developing nations of the humid warm areas need to adopt realistic social policy objectives in relation to water services. Efforts should be geared towards achievement of self-sufficiency in financing viable water projects, adopting appropriate technology and an optimum scale of production. This means that realistic water rates have to be charged users, while wasteful use is vigorously curbed.

11. Dry warm sloping land

Introduction

For the purpose of this book, dry warm areas have been defined as areas with an annual potential evaporation over $1000mm$, lying within the hyper-arid, arid, semi-arid and sub-humid classifications of the MAB map "World distribution of arid areas" (Unesco 1979). This encomposses a very large part of the earth's surface (Fig.11.1), including more than half of Africa, a large part of the Middle East, important parts of India and Pakistan, most of Australia, large regions in North and South America including the coastal zones of Peru and northern Chile, and parts of the Mediterranean coast of Europe.

On a global map it is impossible to exclude all the relatively small areas of higher land where potential evaporation is less than $1000mm\,y^{-1}$. It should also be noted that local aridity is influenced by geology and soils. When these conditions are very favorable, some areas in the sub-humid and even semi-arid zones may have hydrological features not too unlike those of humid areas. On the other hand, if conditions are unfavorable, arid features may be found in regions where the precipitation is above the upper limit of the sub-humid zone.

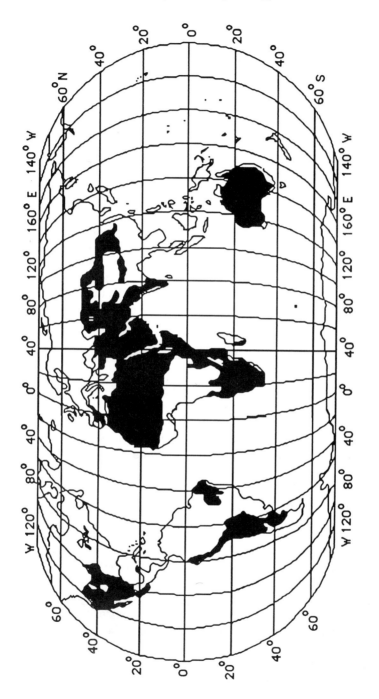

Fig.11.1 Generalized map of the warm dry region

Large parts of the dry warm area are occupied by flatlands, and there are also numerous small areas of flatlands more or less randomly distributed in the region. For this reason it is not practicable to show a small scale world map which delineates flatlands from areas with catchment response, which have a more or less well organized natural stream network. In this zone, areas with catchment response correspond to basins with a slope exceeding $0.1\,m\,km^{-1}$ for large rivers and 0.5-1.0 or even up to $3m\,km^{-1}$ for small watercourses, taking into account aridity and soil types.

In view of the size and global distribution of this hydrological region, it is only possible to indicate the main hydrological features and the main problems relating to water use and management.

A general characteristic is that trees and shrubs are sparsely or very sparsely distributed, and the hydrological effects of canopy and litter are negligible, with the exception of some sub-humid areas with shrub savanna in tropical areas and scrub (maquis) in sub-tropical areas. In the hyper-arid, arid and some semi-arid zones the plant cover is negligible, and before the first rains there is almost no plant- water interaction. Germination and growth of plants, soon after the first rain, change the situation: a slight interception by plants begins, and the evaporation of the soil-plant complex increases, the roots extending deeply into the soil. The depth of the plant root zone is more important than in temperate and humid areas, particularly for the few trees.

Where the soils are permeable, part of the rainfall enters the unsaturated plant root zone and infrequently penetrates to groundwater. The remainder becomes overland flow which is concentrated into the network of stream channels. Depression storage is negligible. Part of the soil water may also enter the drainage network, but most is transpired by plants or evaporated from the soil surface. Where the water-table is sufficiently shallow, there may also be a contribution from groundwater to stream flow.

For impermeable soils an important part of the precipitation becomes overland flow and joins the drainage network. There is little that infiltrates into the soil, with the exception of fractures or karstic zones through which water can reach the groundwater system. Some water returns to the atmosphere by evapotranspiration. There is some depression storage. Again there may be a contribution from groundwater to stream flow when the water-table is shallow, or the groundwater may be recharged from stream flow when the water-table is deeper.

The hydrologic characteristics, particularly near rivers, may be strongly influenced by exogenous inputs from regions with higher precipitation. There may also be endogenous inputs from up-catchment flatlands, which affect the surface and possibly the subsurface hydrology of downstream sloping areas.

Water balance components

Precipitation. By the definition of the region, the mean annual precipitation is less than the annual potential evaporation, and is generally low with, in most cases, a completely dry period of at least 2-3 months. Table 11.1 shows the range of annual precipitation for the four sub-classifications of aridity. The values given in the Table correspond to the data provided by meteorological services. It must be noted that, depending on the type of rain gage used, they can be lower than the actual rain depth at the soil level (Sevruk and Hamon 1984; Chevallier and Lapetite 1986). The overlapping of the limits results from variations of the potential evaporation, the lower limit being in relation to the chosen limit of 1000*mm*.

Table 11.1

ANNUAL PRECIPITATION (*mm*) IN DRY REGIONS

Classification	Lower limit	Upper limit
hyper-arid	10	75-100
arid	40	400-500
semi-arid	200	800
sub-humid	500	1100

The interannual variability of the annual precipitation is high or very high, with a statistical distribution evidencing positive skew. This characteristic is particularly important in subtropical areas and in the Sahel in Africa. For example, in Gabes (Tunisia) the lowest yearly precipitation is 36*mm* and the highest is 534*mm* with a median value of 158*mm*. At the same place the highest daily precipitation recorded is 124*mm* and it could probably be over 200*mm*. Fig.11.2 shows the percentage difference between the 10-year or 100-year rainfall and the mean annual rainfall P_y, over a range of 100-1100*mm* for P_y, obtained from 21 rain gages in the Lake Chad catchment area. The difference is more important for humid years than for dry years, corresponding to the positive skew of the statistical distribution. Such a pattern is not necessarily general for the warm arid zone, but it is observed over large areas.

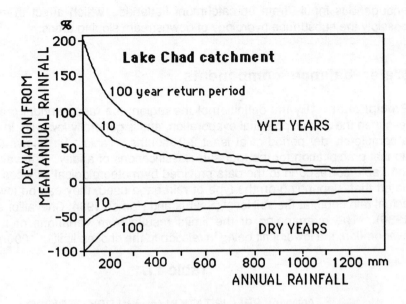

Fig.11.2 Difference between extreme and mean annual rainfall
for the Lake Chad catchment (from Colombani 1979).

Orographic influences on the mean annual precipitation are generally important,
with precipitation increasing with altitude. As a result, some high areas may be
outside the general limits set for the warm arid region, but it is difficult to map them
because of their small areal extent. Another orographic influence, the aspect of
land slope relative to wind direction, also has some influence on the depth of
precipitation.

The seasonal distribution of precipitation may vary widely from one place to another,
even over a relatively short distance. The main patterns observed are as follows:

(a)   A well defined rainy season in summer, with the second half of the dry
        season very hot. Most of the tropical areas, from the hyper-arid to the
        sub-humid zone, are in this class. The following situations may be
        observed: occurrence of one or several storms over a period of about six
        weeks (hyper-arid), or a rainy season over two to three months (arid), over t
        hree to four months (semi-arid), or over four to six months (sub-humid). For
        example, in the Lake Chad catchment area the July-September rainfall
        increases from 63 to 90% of the annual precipitation with a change in
        latitude from 8 to 14 °N (Fig.11.3). A variant of this category in very low

278

latitudes (e.g. Somalia, Kenya and Ethiopia) has two rainy seasons.

(b)   Rainfall concentrated near the end of autumn and the beginning of spring, with a warm or hot summer, or most rainfall concentrated in the winter. Most of the subtropical areas, from sub-humid to hyper-arid zones, have this pattern.

(c)   Rainfall occurring at any time of the year, generally in arid or hyper-arid regions (western coast of South America, the coast of Namibia and parts of the Sahara in Africa, and most of the Australian arid zone).

Some areas have transitional patterns between (b) and (c). All types of daily precipitation occur: low intensity rainfalls, high intensity convective storms in tropical areas, cyclonic storms (hurricanes principally in tropical areas) or continuous precipitation. High intensity storms occur in most of the areas under study. The maximum values of daily precipitation are lower than the world maximum records of the humid tropics, but not by a large margin.

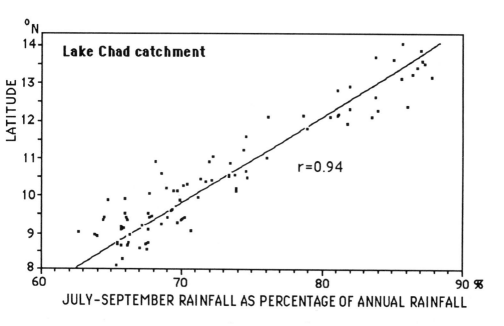

Fig.11.3 Relation between wet season rainfall and latitude in the Lake Chad catchment (from Colombani 1979).

Another important characteristic of the rainfall is the spatial distribution in a storm. Very often in arid zones the area covered by a storm is small, with a rapid decrease of rain depth around the point where the maximum is observed. A storm commonly covers an area from a few square kilometres to 100 - 200km^2, with more frequent values of 30 to 60km^2 in Africa. In mountainous zones the area covered by a storm is often smaller than elsewhere. There are also some regional differences: for example, in northeast Brazil a storm covers a smaller area than in west Central Africa.

This aspect of spatial distribution is important for studies on floods, erosion and sediment transport. It is rare for the area covered by a storm to be as large as the 1000km^2 observed in Boulsa (Burkina Faso) in 1962, though where typhoons occur the area covered (a strip centered on the axis of the cyclone track) may be very large, producing major floods. In some areas, particularly countries bordering the Mediterranean, frontal type rainfall on large areas can produce very large floods if it falls on saturated soils (Rodier 1985).

Rainfall intensity is also very important for generation of overland flow and floods, and for erosion. In tropical arid zones, rainfall ocurs generally as a convective storm of short duration, between 15 minutes and 2 or 3 hours. After some minutes of low intensity, high intensity rainfall occurs for the main part of the storm, which ends with a long low intensity tail. Depending on the area concerned, the intensity may be high (100 to 150$mm\ h^{-1}$ in five minutes). With cyclonic rains the intensity may be lower (15 to 50$mm\ h^{-1}$ in five minutes), as in the Mediterranean climate where however very high intensities can also be observed, as at Haffouz (Tunisia) in 1969, where the maximum intensity over five minutes was 212$mm\ h^{-1}$ (Colombani et al 1972).

Evapotranspiration. The theoretical maximum evaporation is given by the potential evaporation E_0. However in arid countries the actual evapotranspiration E is far below E_0 because of the lack of water availability during most of the year, and relatively low depression storage on sloping land. When a major storm occurs, an important part of the water runs along the slopes and joins the first order streams, another part is infiltrated, and a very small part is stored in depression storage which can evaporate before infiltration occurs.

As conditions of potential evaporation do not exist, the estimation of E is quite difficult, as it depends on many factors, including soil permeability, slopes, type of precipitation, and plant cover when it develops towards the end of the rainy season. Most of the evaporation comes from bare soils of various types and evapotranspiration of a soil-plant complex under a water stress not far from the wilting point.

In subtropical areas where the potential evaporation is a maximum during the summer, the actual evaporation at the same time is almost zero because there is almost no summer rainfall. In such a region, the aridity for a given rainfall depth is less severe than in tropical regions where there are higher temperatures during the rainy season.

The actual evaporation can be estimated from the water balance if groundwater recharge can be estimated, either from measurements of variations in water-table depth or soil water profiles, or by use of tracers in the soil water zone.

Potential evaporation can be estimated from Penman's equation (see Eqn.2.3, p.50). For warm arid countries it ranges from 1000 to over $3000 mm\ y^{-1}$. Fig.11.4 shows the variation of potential and actual evaporation with latitude in central Africa.

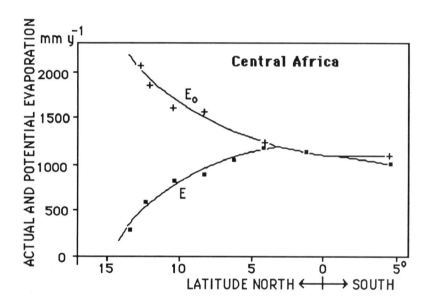

Fig.11.4 Variation of potential evaporation E_o and actual
evapotranspiration E with latitude, for longitudes
between 15° and 22° (from Riou 1975).

Infiltration. In this zone, infiltration generally does not penetrate very deep into the soil. For hydrological purposes, it can be described by two significant parameters:

(a)   The "initial loss", which is the depth of rainfall required before runoff begins from a small plot of very low slope; it includes microscale depression storage

(b)   The "long-term loss rate", which is the rate of infiltration on a plot when the runoff becomes constant during a period of continuing rainfall of uniform intensity. It corresponds to the infiltration rate as $t \rightarrow \infty$ in the Horton and Philip infiltration equations.

Table 11.2 is based on studies (Casenave 1982; Collinet 1980; Chevallier 1985) in the Sahel region of North Africa, from the hyper-arid to the sub-humid zone, using data from artificial rainfall on plots, from representative basins, and from neutron probe measurements of soil water content.

Table 11.2

TYPICAL VALUES OF INITIAL LOSS AND LONG-TERM LOSS RATE

Soil type	Initial loss (mm)	Long-term loss rate ($mm \, h^{-1}$)
very impermeable	3	2
impermeable	4 - 8	2·5 -16
permeable	12 - 25	20 - 50
very permeable with some vegetation	> 50	-

The figures in Table 11.2 are for conditions at the end of a long dry period. After several rains, both parameters decrease considerably, e.g. for impermeable soils the initial loss may be 2mm, while the long-term loss rate is 0-2$mm \, h^{-1}$.

The formation of a surface seal (Valentin 1981), due to the impact of rain drops on the soil surface, can make large areas impermeable, but this effect decreases and eventually disappears with increasing land slope. As a result, where the only variable is slope, there can be higher infiltration on sloping land than on flat land.

Except in the sub-humid zone, infiltration is generally low, particularly in impermeable soils, where it may be 10-40% of a heavy rainfall. Most of the infiltration is returned to the atmosphere by evapotranspiration, and any groundwater recharge from the soil water zone occurs only from extreme events, except possibly in karst or fissured rocks.

In the sub-humid zone, infiltration is likely to be higher, because of biological activity in the soils that makes them more permeable, and because there is a greater depth of precipitation. Groundwater recharge is then possible if permitted by the geological substratum, even with a higher consumption of soil water by the more dense plant cover. In texture contrast (duplex) soils, infiltrated water may move downslope through the relatively permeable surface soil, forming a temporary saturated zone over the impermeable subsoil. When this water-table meets the surface, the water runs overland to the first order streams. This occurs for example in the catchment of Oued Sidi ben Naceuv in the north of Tunisia.

Overland flow. Overland flow is a very important component of the water balance in warm dry areas with catchment response, both for water production and for erosion. Microscale depression storage can retain a significant part of small rainfall occurrences, but in larger storms a great part of the rainfall excess joins overland flow. The proportion of the precipitation which becomes overland flow varies widely, from 90% or more on impervious catchments with high slopes, to 5% or less on pervious catchments with low slopes.

Overland flow has been studied more by runoff measurements at the outlet of small catchments than by direct measurements on the slope. In Burkina Faso a comparison has been made (Albergel 1987) between low-slope plot runoff, using artificial rainfall, and catchment runoff. The plot rainfall excess L_i was measured for each soil type i, and the catchment rainfall excess L_{calc} calculated from

$$L_{calc} = \sum_{i=1}^{n} L_i \frac{A_i}{A} \qquad (11.1)$$

where A_i is the area of soil type i, and A is the area of the whole catchment. The ratio $K = L_{calc} / L_{obs}$ of this value to the rainfall excess measured at the outlet was 0.66 for Gagara east ($25km^2$) and 0.80 for Gagara west ($35km^2$), which can be considered an impermeable basin. The implication is that overland flow measured on a plot is partially infiltrated while running overland to the catchment outlet. An important part of this infiltration corresponds to transmission loss in the stream channel network, comprising storage in stream beds and banks and possibly recharge to groundwater. Depression storage may also be significant. On steeper basins the effect may be reversed, with values of K greater than 1, because the artificial rainfall measurements were made on plots which are almost horizontal.

Runoff.   As a result of the characteristics of precipitation described above, violent floods are typical in the warm dry zone. Slope also has an important effect on peak flood discharges, as shown by Fig.11.5 for a storm in the Sahel representing a 10 year flood in a region with a mean annual rainfall of 250*mm.*

The temporal pattern of runoff depends mainly on the precipitation pattern, evapotranspiration and slope. Intermittent flows are most frequent, though permanent flows may occur in subtropical and sub-humid areas or as a result of exogenous inputs from more humid zones. Three general patterns of runoff distribution can be distinguished:

(a)   Climates with a well defined rainy season in the summer. This includes most of the tropical areas from hyper-arid to sub-humid zones. For small basins, the runoff pattern is similar to the rainfall pattern, with some differences resulting from varying soil permeabilities and from plant cover growing after the first rains. On steep catchments with impervious soils, the runoff pattern is similar to the rainfall pattern, but with more pervious soils the first storms may give little or no runoff

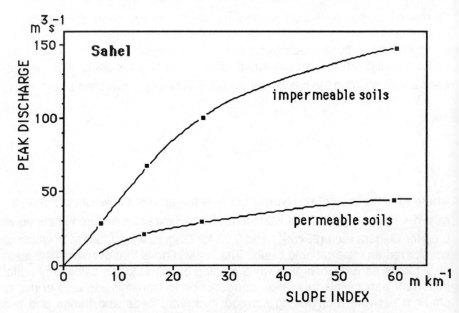

Fig.11.5 Relations between peak discharge and catchment slope
for a 70*mm* convective storm on a 10*km²* catchment
in the Sahel (from Rodier 1986).

compared with floods observed later in the season (Rodier1964). Between storm events, the runoff from small catchments usually stops, except in the sub-humid zone when the soils are permeable enough to store some water, or even in the arid or hyper-arid zones in an occasional year of very high precipitation. When the plant cover grows sufficiently during the rainy season, peak flood discharges at the end of the season are lower, even if they have a larger volume for the same precipitation.

With larger catchment areas, the flood hydrographs depend less on individual storms, but become more complex due to the various origins of runoff on the catchment. The drainage network causes important transformations of the runoff pattern. In the sub-humid zone, permanent flow becomes more frequent, but in arid and hyper-arid tropical zones it becomes less frequent as the catchment size increases. Even where there are springs, permanent flow does not extend far downstream.

(b) Climates with winter rainfall or with two rainy periods. This occurs mainly in the subtropical zone, with aridity ranging fron hyper-arid to sub-humid. Again the pattern of flood hydrographs on small catchments is similar to rainfall patterns, but in larger catchments is strongly influenced by the drainage network. Continuing flow between flood events is more frequent, because evapotranspiration is relatively less important when the rain falls during the cold or temperate season. In the sub-humid zone, permanent flow can be observed frequently even during the dry season.

In this zone the interannual variability of rainfall is higher than in zone (a), and the statistical distribution of maximum discharges has a strong positive skew, higher than for rainfall, particularly in the arid and semi-arid zones. For example, on Oued Zeroud in Tunisia the maximum discharge in 1969 on a $8500km^2$ catchment area was $2m^3 s^{-1}km^{-2}$ or $1700m^3 s^{-1}$ (Rodier et al 1970; Colombani et al 1972). The return period of this discharge is diffficult to evaluate, but the total discharge for the year was $2.5 \times 10^9 m^3$, compared with a median of $75 \times 10^6 m^3$ and a mean of $110 \times 10^6 m^3$.

(c) Climates with rain falling at any time of the year. This occurs in the arid and hyper-arid zones where floods are infrequent, and flow generally stops between two floods. There may be no flow for periods of a year or more.

285

To summarize, the annual hydrograph is closely related to the temporal distribution of the precipitation, and varies according to aridity as follows:

- Hyper-arid zone. One or a few short floods, with no runoff in some years and almost continuous floods for two or three weeks in exceptionally wet years.
- Arid zone. One or two series of floods each year, with floods normally separated by dry spells; but with pervious soils and/or close spacing between floods, there may be continuous flow for at least part of the rainy season.
- Semi-arid zone. A longer series of floods, with generally continuous flow in the rainy season.
- Sub-humid zone. Hydrographs generally similar to those in the neighbouring humid zone, with a succession of peaks and recession curves, but the period of low flow is more accentuated, and the river can be dry for a variable length of time.

The length of period without runoff is an important parameter, and varies from more than one year in the hyper-arid zone to zero in the sub-humid zone.

The volume of runoff varies considerably with slope, soil type, rainfall pattern and catchment size. As an example, for a tropical dry area with a median annual rainfall of $400mm$, the annual runoff on a $25km^2$ catchment ranged from $1.6mm$ for sandy permeable soils to $140mm$ for colluvium, clay and marls (Rodier 1975). For slopes greater than $3m\ km^{-1}$, slope did not have a great influence. The runoff decreased with increasing catchment size , so that the corresponding runoff figures for a $1000km^2$ catchment were 0 and $80mm$.

In the semi-arid and sub-humid zone, direct recharge of aquifers is more frequent and more important, but losses by evaporation along the river beds decrease sharply.

The interannual variability of runoff depth is very high. The value of the coefficient of variation is high and the skew coefficient is positive (McMahon 1979; Rodier 1985). Both increase with increasing aridity, with some exceptions (Nordeste Brazil for instance). Cyclonic precipitation (hurricanes) may sometimes affect the distribution by occurrence of very large runoffs (and peak floods), which increases the irregularity of the runoff distribution. It is more suitable to use median values to describe distributions than mean values, which are too much influenced by the high extreme values.

In subtropical areas the irregularity of the rainfall pattern can produce a great irregularity of runoff. For instance on the Oued Zita basin, a small $3.2km^2$ catchment area in south Tunisia, comparable annual rainfalls have given very different annual runoffs: with $P_y = 378mm$ in 1973-74, the runoff was $234mm$, and with $P_y = 320mm$ in 1975-76, the runoff was only $34mm$. The reason for this difference is that in 1973-74 a single rainfall in 17 hours represented almost $300mm$ of the yearly total of $378mm$ (the median yearly precipitation is about $170mm$ (Camus and Bourges 1983).

It is important to consider also that when there is a multi-annual drought as has been observed recently in the African Sahel, the consequences for runoff are very severe. When the yearly precipitation is far below the median value for several years in succession, runoff becomes very low.

The distribution of the main yearly flood is similar to that of the annual runoff, but flood peak discharges are increased by high slopes, and this influence is even greater than that of the climate. However for the same basin area, the same yearly precipitation and the same slope, floods in the tropical zone, although also violent, are not as large as those in the subtropical zone and in regions affected by tropical cyclones (hurricanes).

Groundwater. In catchment response areas, groundwater has not the same importance in the water balance as in flatlands, because infiltration towards deep groundwater is much more limited, unless there are many fissures or karstic formations. As has been said above, water can infiltrate into the upper part of the soil when it is pervious enough and may reappear further downstream. Shallow aquifers may permit a permanent flow, specially in sub-humid zones. Generally recharge and discharge of such aquifers is rather rapid. Most of the infiltration takes place along the river beds, and some in overland flow. The water thus infiltrated generally returns quickly to the rivers in the sub-humid zone.

Erosion and sediment transport

Erosion and sediment transport are very important in catchment response areas of the warm dry zone because of the sparse plant cover, and the high velocity of the overland and stream flow, combined with the effect of the slope. Erosion can be observed as sheet erosion in which the whole surface of the soil is lowered, or linear erosion in which soils are cut by rills and deep gullies, or bank and bed erosion in the channel network. The solids transported can be divided into fine suspended matter (with a sedimentation diameter less than $60\mu m$), coarse suspended matter (sedimentation diameter more than 60 μm) and bed load. Data are not very numerous, and not accurate because study of these processes is expensive and time consuming.

Sheet erosion has been studied by means of measurements on plots. Studies on such small areas (generally between 20 and 200m^2) give information for field erosion which is useful for agronomic purposes. Statistical studies have led to formulas such as that of Wischmeier and Smith (1982), established with 10 000 plot-years of data and known as Universal Soil Loss Equation. Unhappily it is not universal, but it can give useful results after local adjustment with new measurements, but only for sheet erosion. As soon as rills and gullies appear, the Wischmeier formula cannot be used, and it is therefore inapplicable for most of the catchment response areas in the warm arid zone. Many authors have found relations between sediment delivery at the outlet of a catchment area, and different parameters such as slope, catchment area, precipitation etc. Generally the results are not very good, and extrapolation to other regions is not possible. Some facts seem to be reliable: the amount of fine suspended matter carried in a river is not determined by the hydraulic conditions in the stream flow, but by the amount which can be supplied from the catchment upstream. In many cases it seems possible to determine an upper limit of concentration of suspended solids for a flood on a given basin, when the flood volume exceeds a critical value. This limit on the concentration seems to be a characteristic of the basin (Demmark 1980; Rodier et al 1981).

Bed load transport is not well known because measurement is difficult, except for indirect measurements in natural or artificial lakes. For more than one hundred years many formulas have been developed to calculate bed load transport, but the results are not at all coherent. The bed load transport at a given time in a given place on a river depends not only on the hydraulic characteristics of the river but also on the past chronology of river discharges. Numerical modeling may be the only possibility for calculation, and this requires much further study (Abdalla Sharfi 1986).

Langbein and Schumm (1958) published a graph illustrating the variation of erosion with precipitation. Maximum erosion is observed for 250 to 350 *mm* of annual precipitation when there is enough rain to produce significant runoff, but not enough to permit the growth of plant cover to protect the soil.

The concentrations of suspended solids vary greatly in dry warm areas, from less than 1$g\,l^{-1}$ to more than 500$g\,l^{-1}$. The highest values are observed in the arid and semi-arid zones of the subtropical region.

The annual sediment delivery expressed as specific erosion also varies greatly, from $50t\ km^{-2}$ to $50\ 000t\ km^{-2}$ as on Zeroud (Tunisia) in 1969 (Colombani et al 1984). Sometimes it has been possible to evaluate the part of the sediment due to sheet erosion and the part due to bed and bank erosion. During a single flood in 1973-74 on Oued Zita in south Tunisia, it was estimated that sheet erosion was $2mm$ and bed erosion $5mm$, very high values of erosion typical of these regions, while the mean concentration of suspended matter was at least $45g\ l^{-1}$ and the maximum concentration at least $84g\ l^{-1}$ (it probably exceeded $100g\ l^{-1}$) (Camus and Bourges 1983; Colombani et al 1984).

On the Kountkouzout basins in the sahelian zone in Niger, with a semi-arid climate and a mean annual precipitation of $400mm$, concentrations of suspended matter varied from $1g\ l^{-1}$ to $20g\ l^{-1}$ for catchment areas varying from 2.6 to $10.6km^2$. The annual specific erosion varied from 65 to $215t\ km^{-2}$ on cultivated slopes of 1% and from 1450 to $1950t\ km^{-2}$ on cultivated slopes of 12% (Dubreuil and Vuillaume 1970).

In the Cameroons, with a sub-humid climate ($P_y=1000mm$), the yearly specific erosion on a $6.6ha$ catchment area on Mayo Kereng was $640t\ km^{-2}$ and the concentration of suspended matter varied between 0.12 and $2.8g\ l^{-1}$. A large part (about 60%) of the solid transport is bed load in this case. Plant cover is a good protection against sheet erosion in this region (Pelleray 1957).

In conclusion, erosion and sediment transport are generally more important in the arid and semi-arid zones than in the hyper-arid and sub-humid zones, and in the subtropical rather than in the tropical zone. However, erosion in the sub-humid zone may become important because of agricultural practices; when the natural plant cover disappears, overland flow and runoff are able to carry a high sediment load.

Water and land use

In this region the two main problems affecting water use are the lack of water for a significant time each year, and water excesses involving floods and erosion.

The possibility of water supply from aquifers, the depth of those aquifers, the possibilities of recharge, and the quality of the groundwater are also important, though aquifers have not the same importance as in flatlands.

Salinity and water quality are very important for management and operation. The nature and the volume of sediments may have a strong influence. All these characteristics are closely tied to the origin and development of water and land use.

Agriculture. Most human activities need a continuous water supply. Agriculture is a main exception, since many crops need water only for a limited number of months. The seasonal patterns of precipitation and runoff often do not fit the timing of these water needs. In subtropical areas the water supply stops in summer and autumn, which causes stress on trees and shrubs. In tropical regions the lack of water is an impediment to dry season agriculture.

Human activities must be adapted to the difficulties related to the hydrological regime, or possibly the water resources may be controlled in order to cope with human needs. A combination of both attitudes is often required. There are good examples of appropriate solutions that have been found in the past, which have tried to fit the agricultural techniques to the natural water regime. For example, in the Nabatean terraces in the Negev desert (Israel) and in south Tunisia, the slopes of the upper parts of small catchments are cleaned up to facilitate surface runoff which is concentrated into first order streams, sometimes by construction of small channels. The sediment from the catchment area is stopped by very small dams (made of stone blocks without mortar) and very low earth dikes. Consequently, series of terraces of silt and sand are created in the stream beds. Water accumulated in this deep soil in winter and spring allows crops of vegetables and cereals upstream and fruit downstream. In the Matmata hills in Tunisia, small rock dams with little spillways are built along the valleys. Deep terraces called "gessours" accumulate behind the dams, permitting valuable production of various crops. In the tropical Sahel the runoff downstream of small watercourses stops in ponds where sediment is deposited. When the ponds dry up, the peasants plant out the sorghum that has been previously seeded on small plots. Enough water remains in the sediments for the complete growth of the sorghum crop. In many developing countries of the dry warm areas, the main agricultural resources are produced by rain-fed agriculture.

Another possibility is the large scale management of water resources, but it must be stressed that management of water resources is expensive and must be supported by valuable crops. This is not possible in many areas like the African Sahel, where the soils are often too poor to support rich crops. In such situations the only possibility for management is the modification of the cultivated areas by the farmers themselves, using cheap processes mainly supported by manual work. When the crops are sufficiently valuable, more expensive techniques may be used, such as storage of water by dams, or the use of groundwater (if any) or of water coming from more humid regions upstream (exogenous inputs). There are however many obstacles to these techniques due to the hydrological characteristics described above. It is not easy to transform the soil surface by terracing or other methods

290

without increasing peak floods and erosion during exceptional rainfall (Heusch 1985). During the last century in Maghreb (in the subtropical zone of north Africa) the erosion due to human activities has probably been multiplied by five as a result of the destruction caused by roads and other forms of damage (Heusch 1985).

Drip irrigation can be used, but it requires some technology and it may be too expensive in most places.

Another technique is research to find new crop varieties which fit the rainfall pattern. For instance some varieties of millet can develop a crop in three months, while others need five, six or eight months. Some plants can grow even with high water salinity, which permits the use of very poor quality water such as occurs in oases in the arid and hyper-arid zones.

In many cases it is very difficult to find an acceptable solution to the drought problem in dry warm areas.

Stock grazing. Extensive stock grazing generally uses the driest land areas (except in the hyper-arid zone). The stock need food and water, if possible in the same place. Nomads know where the rain falls and drive their herds accordingly. However, too many oxen around a single water pond or well during a drought has dangerous consequences upon soils and plant cover, and it is necessary to increase the number of ponds, small dams, wells and bores. Possibly a better solution is to reduce the number of oxen. Goats, which are numerous in subtropical regions, may be a calamity for the scarce plant cover, and are an important factor leading to desertification.

Fuel wood. People need wood for cooking and also sometimes, in the subtropical zone, for warming themselves during cold days in winter. This is also a problem for desertification, with no evident solution unless a cheap fuel to replace wood can be found.

Urbanization. The main problems in urban areas are water supply, sewage disposal, and protection against floods. Water supply can be maintained by water storage either behind dams, or in large cisterns with management of their catchment area, or from groundwater if there is any. It is always a difficult problem in this region, except when there is an input of exogenous water coming from a humid zone upstream.

Sewage must be treated before it is returned to the river, where discharge may be zero or very low during the dry season. After treatment, it can be used to recharge the groundwater by infiltration from shallow ponding basins, or sometimes by means of injection wells.

Particularly on higher slopes, floods may be very dangerous due to the high velocity of the water. If sediment transport is important this is a supplementary danger. Dikes may be necessary to protect urban areas, or dams may be built upstream in order to reduce peak discharges.

Mining and tourist trade. These two activities use relatively little water (unless it is used for site processing of mine ores) and may support a relatively high cost for it, unlike irrigation water. The use of deep aquifers, if any, is possible or water may be piped over a long distance; even carriage by trains or trucks is possible in special cases. However there may sometimes be a conflict between the needs of agriculture and tourism in sea-side resorts like the coastal area of Tunisia or even in the south of Spain, where there was a severe drought for some years before 1986.

Human impacts

Past hydrological impacts of human activities. Human activities have had an increasing effect on natural hydrological regimes because of the exponential increase in human population, due largely to technical advancement and medical progress. The areas with catchment response in dry warm regions have difficulties in supporting this increase. The natural hydrological pattern may be transformed in three ways: by construction of large structures such as dams, large channels and so on, by agricultural and grazing practices (including small structures) and by urbanization. This influence may be very important when it has lasted for centuries on very large parts of the region. It is specially evident in the subtropical zone where there has been a high population for many centuries, as around the Mediterranean. It is principally the influence of agricultural and grazing practices which have modified the natural pattern. In the sub-humid zone the clearing of natural plant cover in order to obtain crops and grazing land have caused higher erosion, with the best part of the soils being removed, and another consequence has been an increase in flood peaks. Over-grazing in warm areas has often modified the floristic composition of these pastures, the interesting plants for stock being progressively replaced by shrubs of lower palatability, and in the more arid areas a desertification process has taken place, resulting in different hydrological processes at the microscale, typically more runoff and less infiltration.

More recently, large structures have greatly modified the hydrological regime in many places. Peak floods may be limited by special dams; for instance floods of $17000 m^3 s^{-1}$ occurred on the Oued Zeroud in Tunisia before the construction of the Sidi Saad dam which is expected to reduce peak floods downstream to a maximum of 10 000m^3s^{-1}.

Building of roads and railways may cause new gully erosion during the construction period. Mining activities also can increase erosion by a factor of 50-100 , or even more with open-cut operations.

Urbanization causes very specific problems, with the decrease in surface permeability producing increased overland and stream flow, and the need for disposal of a large volume of sewage which is too often returned to the local river, even if its natural discharge is low or zero. Pollution may spread far downstream in the river, with many consequences for fish, aquatic plants and the people who live there and use this polluted water for their own supply of drinking water.

Potential for future human impacts. As noted above, the exponential increase of the population, specially during the last century, has produced a need for food which has very important consequences. In the subtropical zone in particular, such high population densities result in a catastrophic situation, because of the necessary development of agriculture to produce enough food. Too often agricultural practices are at variance with the climatic conditions. This is true also in the tropical zone, for instance in Africa and in South America, where the population density is less critical but is increasing rapidly and where also the effects of droughts seem to be more intense. Almost everywhere in dry warm areas where agriculture is the main activity, countries may pass beyond the limits of sound management of water and soils, even involuntarily, because it is not easy to know the appropriate long term limits. Effects that will be significant only after ten, twenty or fifty years may be neglected by governments that have so many urgent problems to solve.

The main dangers that may be mentioned here are desertification, erosion and sediment transport, water pollution, lowering of water-tables, and modifications of the drainage network causing unknown effects on the hydrological regime. Desertification is a well known problem as a result of its recent increase. Many people think that the main factor in desertification is a change of the climate, with the occurrence of severe droughts for several years when the annual precipitation is far below the median value; but the main factor is the growth of the population, and the climatic hazards only reveal the dramatic effects of this growth. The first and the main action to consider is to persuade these communities that they must drastically control this increase. All the other technical practices are also necessary, but of no use if birth control is not efficient.

Desertification may lead to a diminution of infiltration into the soil and consequently to a higher overland flow and higher peak floods. Desertification induces also a change in the plant cover, mainly in sub-humid areas. Fifty years ago in Africa the

293

soil remained without cultivation for several consecutive years, and the plant cover was woodland with many small trees and shrubs. This cover is now changing rapidly, with the trees being cleared either for fuel wood or in order to obtain more land for crops. The new plant cover, if any, consists of cultivated plants or a poor new natural cover. Soils are no longer well protected, and erosion produced by a higher overland flow is more important, contributing to the degradation of the landscape. Another consequence is the loss of storage capacity in the dams which receive and trap a larger volume of sediment. For instance, in Algeria each year $20 \times 10^6 m^3$ of dam capacity is currently being lost. All these phenomena have been observed for centuries in the subtropical zone, and now they can also be seen in the tropical zone and will extend greatly in the future.

12. Small high islands

Introduction

An island is defined as a land mass surrounded by the sea and with an area which is small compared to the global land surface.   Islands have a widespread distribution but their greatest concentration and the most populated occur in the tropical regions of the Caribbean and the Pacific.

There is justification for distinguishing small islands in a hydrological context and the qualifying features are reviewed below.   A precise delimitation of a critical area is not feasible but an upper boundary within the range 5000-10 000km^2 seems appropriate.   The vast majority of islands have areas below these values and the relative few of larger size have logarithmically increased proportions (Fig.12.1), such that their hydrology is more akin to that of continental masses.

The discussion in this book treats small islands in two categories, high islands which show catchment response in this chapter, and low islands with flatland hydrology in Chapter 18.   Before detailed consideration of these two categories, it is appropriate to consider the more global controls of climate and geology, and some other common issues.

Small land area promotes rapid water loss into the surrounding seas, whether by

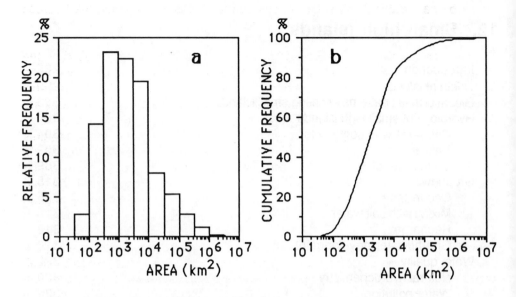

Fig.12.1 Distribution of areas of islands greater than 36km^2.
Data source: Showers (1979).

surface runoff or groundwater discharge.   Water source options are limited and critically dependent on the "island" occurrence of rainfall, runoff and recharge. Surface water storage schemes, other than those of very small size such as are associated with rainwater catchments, are rarely feasible, for both topographic and socio-economic reasons.   In the context of development, water supply problems are essentially single island problems.   Problems of technology, water rights, pollution and management are more regional in scope and can relate to groups of islands.

Small islands have a shared climate with the surrounding ocean but the maritime influence is widely pervasive beyond climatic correlations. Sea water is invasive to aquifers with which it is in contact and also exercises an important control on littoral ecology. Transfer of fresh water across sea barriers is constrained by cost but sea water is readily available for desalination, where other factors, chiefly economic, allow.

A fundamental distinction referred to earlier is between high islands and low islands. High islands are typically composed of volcanic or hard continental type rocks. Low islands are mainly composed of limestone. Resource options are greater in high islands, whereas low limestone islands are critically dependent on the resources of thin fresh water lenses overlying sea water within the bedrock aquifers.

The classical concepts relating to the provision of water supply are based on the planned availability of sufficient volume through storage reservoirs or aquifers and in places via perennial streams, aquifers, aqueducts or pipelines. These concepts must be modified to suit the situation in small islands where perennial streams are rare and surface storage generally not feasible. Groundwater commonly represents the most important and reliable supply source, the development of which requires the applicability of a range of techniques consistent with the variability of the associated aquifers. In particular a major consideration is the occurrence and controls to behavior of the freshwater lens systems which require a detailed assessment of water balance and of the seawater-freshwater interactions in often complex, fissured, karstified or layered aquifers.

Population distribution in the small islands throughout the world shows great variability, both regionally and locally. On high islands, populations tend to be small and concentrated in littoral areas, more particularly on the leeward drier sides. On low limestone islands, population densities can sometimes be very large, which puts a severe strain on water supply. Regional dispersion is most marked in the vast Pacific Ocean.

The economic level of the majority of island countries is generally low. Independence has come fairly recently to many small island countries and a desire to attain the life style practised by earlier colonial powers has tended to impose a strain on basic resource development, such as water. There are institutional, economic, political and legislative constraints and adverse geographical factors, all of which affect planning and development. The ethnological background of the populations, particularly in the Pacific region, is very diverse, often within a single country, and legislative practices derived in part from the colonial heritage and now being influenced by customary laws are rarely coherent and effective for more advanced development and controls. In most independent small island countries, little formal legislation on water resources exists and statements of policy are usually generalized and without a legal water code. Customary laws typically provide full ownership of water resources to the landowner on whose land the source occurs. With continuing development it is inevitable that conflicts will arise if land access is to be obtained or adequate protection of water sources is to be ensured. Although it is a common feature in traditional societies that private ownership is readily sacrificed for public benefits, the arrangements are difficult to formalise and at best are based on short-term lease arrangements. State ownership is not favored and there is merit therefore in placing most emphasis on establishing long-term user rights. In most island countries there is an urgent need for the enactment of water

297

legislation which will be both effective and acceptable.

Water supply development is mainly for domestic supply with occasional applications in irrigation or hydropower. The importance of groundwater for phreatophytic trees and plants which form important alternative sources of water or food, such as the coconut palm or the taro root, must also be stressed.

The overall importance of groundwater, which increases in significance in the small low limestone islands, requires a close concern for potential pollution. Although a wide range of pollution occurrences have been identified, it is generally held that the main threats to small island water resources are the inappropriate use and handling of pesticides and inadequate sewage disposal. Legislation is particularly critical to ensure protection of such highly vulnerable aquifers as fissured limestone or vesicular lavas. Groundwater resources may also be much affected by general land use practices. Urban development on low islands is a case in point. On high islands, deforestation may result in a reduction of stream low flows which can critically affect water supplies.

The proximity to the sea must inevitably invite consideration of desalination or other alternatives such as the importation of water by ship. Desalination has an obvious attraction for islands with water resource problems and particularly so for low limestone islands with dense populations and a freshwater lens at risk from over-abstraction, high drought incidence and pollution. The main constraint to desalination is cost.

The concept of a "hydrologic island" has been proposed by Matalas (1987) as a land unit whose geometry is delineated by coincident physical, biological, social and political boundaries. There are clearly some valid comparisons in respect of these features with the hydrology of geographic islands. The most evident difference occurs in relation to the role of the maritime environment which not only has major effects on current hydrological processes but is also related to fundamental earth processes which have had important consequences for morphology and geology.

Some 18 000 islands appear on the topographic maps prepared by the United States Geological Survey (Matalas 1987). The actual numbers must be several times larger and will fluctuate in accordance with the level of the sea. A list of the 400 larger islands between $36km^2$ and $2.2x10^6km^2$ (Greenland) has been assembled by Showers (1979) and the data have been used to prepare the summaries in Fig.12.1. Islands not included in this list are likely to be very small and probably associated with groups of larger islands. On this assumption, the data plots probably reflect the more fundamental geological processes which control the development of islands. The logarithmic distribution of island areas with a median value of 1725km^2 should be noted. The large majority of islands have areas less than 10 000km^2.

Origin of islands

The origin of islands merits consideration because of the detailed correlation which exists with morphology and geology. Three main classes are generally recognised:- continental islands, islands of the island arc system, and oceanic islands.

Table 12.1 shows the main structural and physiographic units of the earth's surface. The areas of islands are in general substantially less than the areas of the related global unit.

Table 12.1

AREAS OF WORLD STRUCTURAL UNITS AND PHYSIOGRAPHIC PROVINCES[*]

	$10^6 km^2$	Percent of world surface
Continent	149.	29.1
Submerged Continent	55.4	10.9
Ocean	306.5	60.0
(i) Abyssal ocean floor	151.4	29.7
(ii) Ridge and rise	118.6	23.2
(iii) Continental rise	19.2	2.8
(iv) Island arc and trench	6.1	1.2
(v) Volcanic ridges and volcanoes	5.7	1.1
(vi) Other ridges and elevations	5.4	1.1

* data from Wyllie (1971)

Continental islands are either detached parts of an adjacent continental land mass or emergent land areas on the continental shelf, slope or rise. The former are generally composed of hard resistant rocks of similar types and with similar cover of soil and vegetation as occur in the adjacent coastal mainland areas and with a similar climate. The islands off the west coast of Norway are an example of this type and

299

result from the submergence of a dissected coastline. The second type are generally composed of softer rocks and sediments of more recent formation. The Bahamans constitute a typical example. They are low flat islands composed of reef limestones and dolomites, overlying a considerable thickness of up to $400m$ of similar sediments deposited in a sinking basin formed on the continental shelf or possibly on the ocean floor beneath the continental slope.

Island arcs occur on the continental side of deep oceanic trenches and may have some continental affinities. They are associated with subduction zones. The best known occurrences of this group are the Antillean arc in the Caribbean, the Indonesian arc at the easten extremity of the Alpine-Himalayan tectonic belt, and the western side of the circum-Pacific tectonic belt which extends between New Zealand and Alaska, and has by far the greatest length. They include larger islands such as Java and Sumatra, which are excluded from the present hydrological context. Island arcs occur on sites with a history of high tectonic activity and vulcanism extending to recent times. Transgressive sedimentary sequences may be present on the flanks of uplifted masses. The volcanics are predominantly andesitic in composition and include lavas and true pyroclastics in addition to volcaniclastics deposited by erosion or tectonic instability.

Oceanic islands rise from the ocean floor and are predominantly volcanic but may include limestone either as fringing reefs or as platforms overlying a volcanic base which may be no longer exposed. The volcanic islands rise from the ocean floor either along the mid-oceanic ridge systems (Iceland, the Azores and St Helena in the Atlantic, and the Galapagos and Juan Fernandez in the Pacific are examples) or on more isolated and scattered volcanic ridges, such as Hawaii and French Polynesia and other chains in the central Pacific. Oceanic islands differ from island arcs in a number of important aspects. They are derived mainly by volcanic eruptions without strong associated orogenic activity and are composed predominantly of basaltic lavas. Pyroclastic rocks do occur but are less abundant.

Carbonate reefs are associated with islands throughout the tropical and to some extent subtropical ocean regions. Although other reef building organisms are associated, corals provide the main reef framework and can exist only in clear warm water with normal ocean salinity and less than $60m$ depth. Fringing reefs and raised reefs are common features on islands generally but atolls which are roughly circular reefs with no land mass in the central lagoon are predominantly associated with oceanic islands. The theory of atoll evolution propounded by Darwin in 1839 has received confirmation in more recent studies. The concept is of an initial development of fringing reefs around a rising volcanic mass. Subsequent sinking, probably as a result of isostatic adjustment, may eventually result in the formation of an atoll (Fig.12.2).

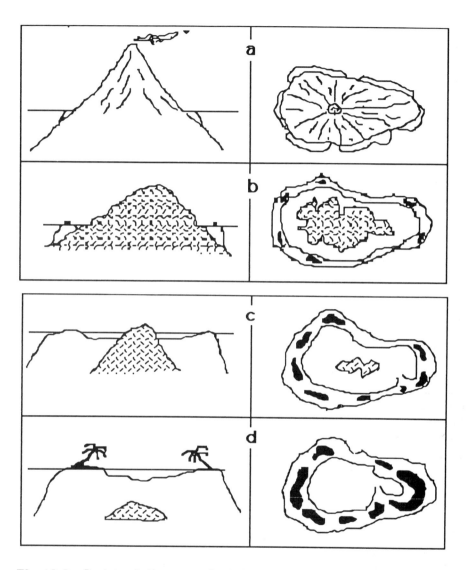

Fig.12.2 Series of diagrams depicting various stages in the geologic history of an atoll. An emergent high volcanic island (a) forms as the initial stage. After cessation of vulcanism, the volcanic core begins to subside (b) until little is above sea level (c). The final stage is the familiar ring-like structure that characterizes most atolls (d).
After Ayers (1984 a).

Sea-level changes can have important hydrogeological consequences in relation to primary lithology of developing carbonate sequences or in the processes of karstification. Basalt volcanic sequences can also be affected. Many basalt lava piles have built up initially below sea level and the consolidated rocks have low permeability. Sub-aerial basalt flows in contrast have high permeability. Subsequent land subsidence which is a common feature in the growing volcanic domes will carry the high permeability rocks to below sea level with significant effects on the occurrence of the freshwater lens.

The causes of sea-level changes may be tectonic, volcanic, or land subsidence caused by isostatic imbalance of the floating lithosphere. Such changes tend to be irregular and variable from place to place. Other more regular widespread changes are likely to be eustatic in origin associated with global changes in sea level such as those resulting from the periodic growth and melting of continental ice caps in the Pleistocene. In the Pacific region, some 30 or so ancient shorelines have been recorded with elevations up to $365m$ above present sea level and to $1800m$ below. The deepest levels occur on some flat topped sea mounts (guyots). There is general agreement that the various shorelines between $-90m$ and $+75m$ are of eustatic origin and many have been dated by radiocarbon content of associated sediments and show reasonable accordance with dated ice age and interglacial periods. Calculations of sea-level changes based on the maximum extent of Pleistocene ice caps accord with this interpretation. Shorelines outside this range are thought to have a tectonic or related origin but the cause is often obscure. In some cases, there apppears to be a progressive sequence in oceanic island chains in which early growth and emergence is followed by submergence and eventually atoll formation. In other cases, atolls have been subsequently uplifted. Various explanations have been proposed ranging from continental drifting away from a high volcanic ridge feature to elastic movements in the lithosphere caused by volcanic loading (McNutt and Menard 1978).

Some general water resource correlations can be recognised which relate to the age and origin of the various rock types occurring on islands. Older continental rocks tend to be more resistant and less permeable, other than karstic formations where the reverse is true. Surface infiltration rates will depend much on the nature of the soil profile which will vary with age, bedrock type, and the climate. Surface calcretisation will also reduce the permeability of surface carbonate rocks. The andesitic volcanic sequences of the island arc systems have a predominance of submarine flows and ill-sorted pyroclastics, both of which tend to have low permeability and porosity. Sub-aerial basaltic sequences in contrast often produce highly permeable and porous aquifers related to the abundant vesicular and fracture features in a multiple pile of thin flows.

302

Global climatic patterns affecting small islands

Small islands largely share the climate of the surrounding oceans. Climatic data for the ocean regions are understandably much less than for continental areas but in recent times the use of satellite imagery and radar techniques has greatly increased knowledge. The broad climatic patterns are determined by the general circulation of the atmosphere caused by the differing radiation balance of high and low latitudes and the seasonal fluctuations. Other regional controls include the position of large land masses and major ocean currents.

The basic climatic patterns occur in shifting latitudinal belts and may be summarized as follows:

(i)   A broad belt of strong circumpolar westerlies with polar front cyclonic rainfall extends over the middle and high latitudes. Rainfall in consequence is greater on western coasts and western offshore islands than on the eastern side of the continental land masses.

(ii)   The easterly trade wind belts. These winds in ocean regions are generally warm and moist and are the source of abundant rain except on the leeward side of major continental masses.

(iii) An equatorial belt of high pressure with light winds and low rainfall.

(iv)   The Inter-Tropical Convergence Zones (ITCZ) occur at the general junction between the trade wind belts and equatorial zone of high pressure, with consequent upward air motions which often result in copious rain.

Cold ocean currents reduce evaporation and rainfall, and the effects of such currents are apparent in oceanic regions west of Africa, Australia and South America. More irregular and transient atmospheric effects include low pressure systems associated with higher latitude disturbances which occasionally interrupt the trade wind and equatorial belts and bring heavy rain. The relatively wet and cool climate of the northern Bahamas is attributed mainly to such effects. Of greater impact are tropical cyclones which are intense storms consisting of spiralling hot moist air systems and capable of sustained wind speeds of up to 200$km\ h^{-1}$ with gusting up to 400$km\ h^{-1}$. They develop in low pressure areas over ocean regions close to the ITCZ and develop into cyclones in a belt between 50° and 30° north and south of the equator. The high winds are accompanied by storm surges which constitue the main cause of loss of life during these disturbances. Small islands and coastal regions are most vulnerable to cyclones which rapidly dissipate over large land areas. The source of a cyclone is commonly remote from the main disturbance and only a few such systems, between 80 and 100 each year, grow to this extreme degree, and the conditions which control their development are unknown. Attempts to prevent cyclone formation have so far proved fruitless.

Although the atmospheric circulation patterns provide the basic controls to rainfall occurrence, local features result in very significant spatial variations. It has been estimated for example (Takasaki 1978) that the mean annual rainfall of the Hawaiian Islands is 1778*mm* whereas that of the adjacent ocean regions is about 635*mm*. The higher rainfall is an orographic effect and low islands will have comparable rainfall to the local ocean area. These divergences must be borne in mind when examining regional rainfall maps. To illustrate the various features, maps and station data from the southwest Pacific region are shown in Figs.12.3 to 12.5 and Table 12.2. The dry to very dry zones are associated with divergent easterly wind systems. Annual rainfall is moderate to low with a high degree of variability (note Christmas Island and Tarawa with coefficients of variation of 42% and 64% respectively). A correlation of monthly rainfall with monthly potential evaporation demonstrates that water deficits can occur in any season. The wet to very wet areas are associated with the trade wind belts and the ITCZ. Variability also decreases as the mean annual rainfall increases. The orographic differences are expressed by Suva (F in Fig.12.4) on the windward side and Nadi (E in Fig.12.4) on the leeward side of Viti Levu, the main island of Fiji (see Fig.12.5).

Hydrology of small high islands

Contrasts with continents. The comparisons and contrasts that need to be made are between islands and continents and between large islands and small islands (less than 10 000*km²*). Table 12.3 lists physiographic and runoff data for continental and island collectives. Runoff data are from the Unesco (1978 b) compilation of world water balance and in the case of the island collectives must be regarded as approximations obtained by extrapolation since there is likely to be a paucity of observational information. The data suggest that island collectives, in comparison to continental, are wetter, have shorter streams and less potential for both surface and subsurface storage. To extend the comparisons further, the physiographic data have been separated for large and small islands and these are shown in Table 12.4. The greater slope of the small islands in combination with the smaller land area is likely to increase further the disparity on stream length and potential of storage. Because of the logarithmic size distribution, the few large islands will clearly be much closer in physical response to continental collectives than the small islands. A further important feature distinguishing small from larger

304

islands is also implicit. The constraints on surface storage and perennial surface runoff result in groundwater being the more reliable and more important water supply source. This is unusual for land areas with such relatively high rainfall and constitutes a significant hydrological feature which is reflected in an appropriate emphasis on development issues, informational requirements and technological problems.

Table 12.2

ANNUAL CLIMATIC DATA FOR SELECTED STATIONS[*]

| | Rainfall | | | | PE | Elev. | Record |
| | Mean | SD | CV | G+ | | | Period |
	(mm)	*(mm)*	%		*(mm)*	*(m)*	
Tarawa	1981	831	42	4.7	1890	1	1941-1980
Christmas Island	832	537	64	2.6	-	3	1951-1984
Rotuma	3526	418	12	76	1672	26	1951-1980
Nieu	2041	457	22	20	1577	21	1905-1972
Nadi	1873	441	23	18	1556	19	1942-1984
Suva	3117	532	17	33	1415	6	1951-1981
Raoul Island	1535	363	24	18	1337	38	1937-1978
Chatham Island	895	166	18	58	853	44	1951-1981

[*] From Prasad and Coulter (1984)   + Gamma shape parameter

Table 12.3

MEAN VALUES OF AREA, HIGHEST ELEVATION, SLOPE INDEX AND SPECIFIC ANNUAL RUNOFF FOR CONTINENTAL AND ISLAND COLLECTIVES (from Matalas 1987)

| | Arithmetic Means | | | Geometric Means | | | Specific |
| | Area | Elevation | Slope | Area | Elevation | Slope | Runoff |
	(km^2)	*(m)*	Index	(km^2)	*(m)*	Index	$mm/y/km^2$
Continental collectives	1 023 420	3 120	0.001	281 420	2 380	0.008	345
Islandic collectives	22 950	1 060	0.045	1 980	640	0.008	1 812

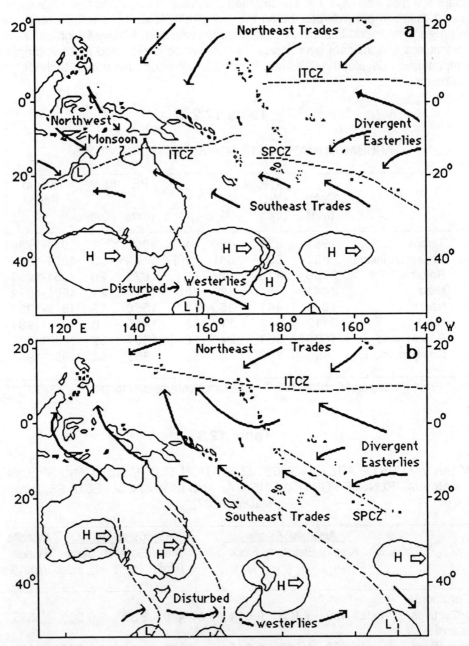

Fig.12.3 Typical circulation patterns in the southwest Pacific: a. Summer; b. Winter. SPCZ is South Pacific ConvergenceZone. (After Steiner 1980).

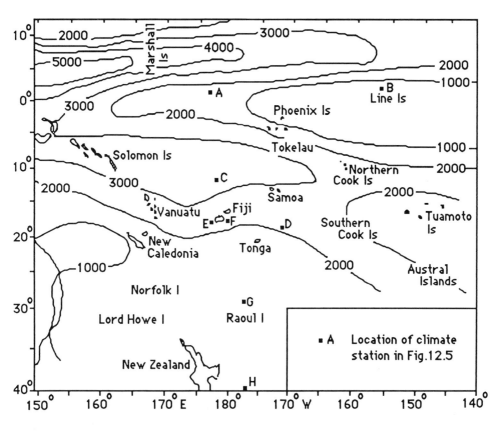

Fig.12.4 Mean annual rainfall (*mm*) in Central and
South Pacific (after Hessell 1981).

Table 12.4

ARITHMETIC MEANS OF AREA, HIGHEST ELEVATION AND SLOPE INDEX
FOR LARGE (>10 000km^2) AND SMALL ISLAND COLLECTIVES

Islands	Number (Shower 1979)	Area (km^2)	Elevation (m)	Slope Index
Large	71	124 129	1 996	0.0174
Small	340	2 180	867	0.0510

307

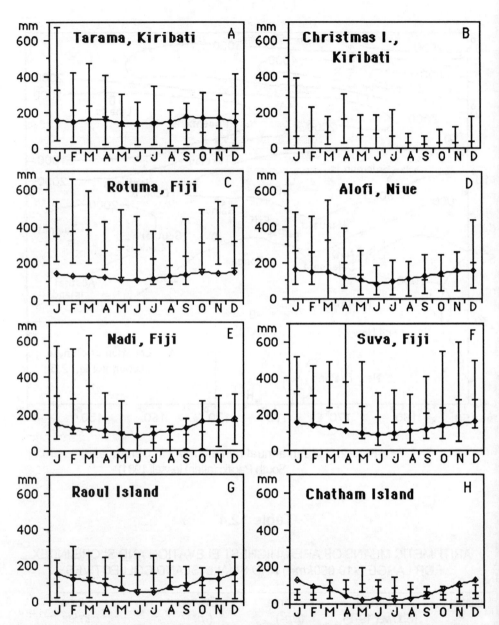

Fig.12.5 Monthly values of rainfall and average potential evaporation for some South Pacific islands. Rainfall values shown are 90 percentile, mean, and 10 percentile. Potential evaporation calculated from the Penman equation. (From Prasad and Coulter 1984).

Rainfall. Rainfall in the islands of higher latitudes is mainly associated with temperate and subpolar low pressure systems. Small islands in these latitudes are mostly of continental type with climate closely associated with and affected by adjacent continental masses. In the tropical regions, rainfall is predominantly related to the trade wind belts and movement of the ITCZ and less frequently to tropical cyclones and frontal disturbances from higher latitudes. The rainfall of high islands is strongly influenced by orographic effects with greater rainfall on the windward slopes. On islands of more moderate relief and elevation, orographic rainfall is better distributed than in islands of higher elevation. This is apparent in the isohyetal maps of the Hawaiian islands (Takasaki 1978). Rainfall is less on the leeward sides of high islands and in small islands in the rain shadows of larger islands. In these locations, there is also greater variability and it is common for much of the annual precipitation to occur in a few large storms associated with topical cyclones or frontal disturbances.

Surface runoff. High islands are composed largely of volcanic or hard continental rocks, sometimes with marginal areas floored by alluvial sediments or raised limestone platforms. Apart from the rainfall, runoff is strongly controlled by the morphology, the surface cover of soils and vegetation and the basin geology.

Drainage frequency is affected by the infiltration capacity of the surface materials, soils and exposed bedrock. It may be negligible on limestone areas or on recent and very permeable volcanic rocks which have thin soil cover. High islands are commonly of high relief and drainage basins are small. In the Hawaiian islands, for example, drainage basin areas average between 3 and 5km^2. Basin size will tend to increase with area. In the main island of Fiji (Viti Levu, 10 429km^2), 44 out of 54 gaged basins have areas less than 100km^2. The steep slopes of the watersheds and valley sides enclose little channel storage. Stream hydrographs have high peak flows but low total volume. Thus the time required to flow through a drainage basin is very short and flash floods are a common result of intense storms. Table 12.5 lists some maximum observed floods in small islands which are relatively among the highest in the world. Flood damage can be compounded by high sediment load amounting at times to mudflows, a feature particularly common in the tropical island arc groups with unstable landscapes associated with recent vulcanism and tectonic activity. Alluvial flats which have been formed by surface flooding are naturally most vulnerable to this hazard. Damage and human risk is greater on the larger islands many of which, such as the larger Indonesian islands, have dense populations on the alluvial flats because of their agricultural potential. Table 12.6 lists flood data from three large islands in the tropics, Java, Sumatra and Papua New Guinea. Although close comparisons cannot be made, it is clear that specific discharge is higher in basins of smaller area and this effect is likely to be emphasized in the even smaller basins of small islands (see Table 12.5).

309

Table 12.5

MAXIMUM OBSERVED FLOODS IN SMALL ISLANDS *

Country	River	Basin Area (km²)	Annual Rainfall (mm)	Maximum Discharge (m³s⁻¹)	Specific Maximum Discharge (m³s⁻¹km⁻²)
Atlantic Ocean					
Puerto Rico +	La Plata	520	-	2700	5.2
Martinique +	Riviera Blanc	4.31	5500	120	27.8
Cape Verde #	Ribeira Brava	6.7	350 253	37.8	
Indian Ocean					
Reunion #	Grand Bras	23.8	5000	750	31.5
Pacific Ocean					
Molokai •#	Halawa	12	-	762	63.5
Kauai •#	Wailua (South Fork)	58	-	247	42.6
Tahiti#	Papenoo	78	5000	2200	28.2

* Data from IAHS catalogue of maximum observed floods
\+ Island arc   # Oceanic islands   • Hawaiian group

On the basis of assembled runoff data on several small Pacific islands (Fig. 12.6 and Table 12.7), there is some evidence to indicate that total yields (proportion of runoff to rainfall) are generally and at times substantially less than those of basins in the larger Japanese islands. Comparable yields occur only on Babelthuap and Kauai, both islands which are relatively old with a well-developed soil cover. Part of the difference could relate to higher evaporation but a further reason for the apparent disparity could relate to a significant component of deep infiltration to a basal aquifer which does not reappear as base flow but discharges as marginal springs and/or dispersed groundwater. The effect is most marked in the oceanic islands composed of permable basalts and less so in the island arc islands such as Fiji.

Streams on small islands are ephemeral except where they occur in areas with good reliable rainfall such as on the windward slopes in the trade wind belts or where they are largely supplied by adequate groundwater reserves. Little information is available on baseflow runoff for streams on small islands. Studies by Kinoshita et al (1986) on streams for larger islands should be essentially comparable for small

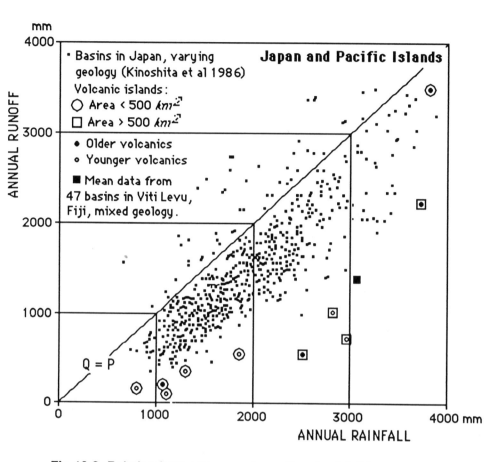

Fig.12.6 Relation between annual runoff and rainfall in some
Pacific islands and catchments in Japan.

islands, except perhaps in respect to total yield, and demonstrate a clear correlation
of the low flow indicators with geological factors. The indicators show a progressive
decrease in the following order: - Quaternary volcanics, Tertiary volcanics and
granite, hard Mesozoic-Palaeozoic sedimentary rocks. The factors which control
low flow are the storage and transmissivity of the shallow aquifers in contact with the
stream channel, the infiltration capacity of the surface soils, and the excess of the
rainfall over the evapotranspiration. The Japanese studies would suggest that age
is the main control of the volcanic sequences which presumably relates to the
infiltration capacity of the overlying soil cover. Some equivalent studies on a
number of basins in the island of Viti Levu, Fiji (unpublished report of the Institute of

Table 12.6 FLOOD DATA FROM THREE LARGE TROPICAL ISLANDS
(Data source: Institute of Hydrology 1985)

Country/basin	River	Years record	Mean annual rainfall *mm*	Basin area km^{-2}	Mean annual flood m^3s^{-1}	Specific discharge $m^3s^{-1}km^{-2}$	Max. observed flood m^3s^{-1}	Specific discharge $m^3s^{-1}km^{-2}$
Group A: basins < 600 km^{-2}								
Sumatra								
431	Bt Agam	16	2340	346	108	0.3	434	12.5
316	Bt Anai	9	3450	110	116	1.1	365	3.3
Java								
043	Kali Serayu	48	3424	50	60	1.2	205	4.1
Papua New Guinea								
PNG044	–	16	2000	208	295	1.4	1056	5.1
PNG035	–	13	3000	218	239	1.1	652	3.0
Group B: basins > 600 km^{-2}								
Sumatra								
818	W. Besai	9	2430	662	294	0.4	926	1.4
511	Bt Wari	6	3290	4579	2941	0.6	5764	1.3
Java								
044	Bengawan Solo	12	2122	1442	909	0.6	2336	1.6
003	Citarum	31	2479	4232	1447	0.3	2540	0.6
046	Bengawan Solo	11	2189	12429	2072	0.2	4745	0.4
Papua New Guinea								
PNG111	–	7	2250	746	1200	1.6	4252	5.7

Table 12.7

RUNOFF DATA FROM SOME ISLANDS IN PACIFIC REGION

Island	Age (My)	Area (km²)	Max.El. (m)	Mean annual values in mm and percentage of total			
				Rainfall	Runoff	Evap.	G'water recharge
(A) <u>Hawaiian Group: Oceanic Basalts</u> [1]							
Nihau	7.5	118	384	1029	226	741	62
				100	*22*	*72*	*6*
Kauai	3.5-5.7	890	1573	3768	2261	904	602
				100	*60*	*24*	*16*
Oahu	0.03-3.65	979	1208	2541	610	1016	915
				100	*24*	*40*	*36*
Molokai	1.2-1.8	420	1491	1856	557	1002	297
				100	*30*	*54*	*16*
Lanai	1.2-1.5	225	1011	1104	243	729	132
				100	*22*	*66*	*12*
Kahoolawe	1.03	72	443	767	153	537	77
				100	*20*	*70*	*10*
Maui	0.41-0.86	1174	3007	2841	1241	733	846
				100	*40*	*26*	*34*
Hawaii	0.00-0.45	6501	4139	2983	746	1313	915
				100	*25*	*44*	*31* (B)
<u>Palau Islands: Oceanic Basalts</u> [2]							
Babelthuap	Tertiary	c.350	242	3759	3571		
				100	*22*		
(C) <u>Norfolk Island: Island Arc</u> [3]							
Norfolk Is.		35	300	1318	210		
				100	*22*		
Individual catchment, grasses, 8 month period					10		
Individual catchment, forested, 8 month period					3		
(D) <u>Fiji: Island Arc</u>							
Viti Levu (33 basins)		1684[4]		3050	1347[5]		
				100	*44*		

[1] Source: Macdonald et al (1983) [2] Source: Grimmelman (1984)
[3] Source: Falkland (1984 b) [4] Range 3.5 - 1472km²
[5] Base flow component is 538*mm*, equivalent to 18% of rainfall

Hydrology, Wallingford, UK) have indicated similar trends in the same rock groups but with much lower multiple correlation coefficients. It is tentatively considered that more detailed subdivision of the volcanics in relation to origin in addition to age is needed to differentiate more accurately the baseflow indicators. Basalt and andesitic sequences would be expected to respond differently as aquifers, and variable uplift or submergence of submarine of subaerial sequences could also exert major influence.

Vegetational cover should have an effect with forested catchments expected to show lower runoff ratios than grassed catchments (Table 12.7, Norfolk Island). The effect of replacing forest by grass, if a consequence of shifting cultivation, is also likely to increase runoff on eroded or compacted soils. Vegetational variations will also control soil water deficits and hence overall recharge with possible effects on baseflow and/or deep infiltration.

A knowledge of rainfall and runoff is needed for resource assessment, including rainwater storage systems, for consideration of flood hazard or for the calculation of groundwater recharge. Data on small islands are rarely of adequate duration. Records of less than 10 years are useful but can mislead on drought probability. The standard methods for network density, design, evaluation and processing apply for small as for large islands (Hall 1984) but for extrapolation to ungaged catchments, detailed considerations of slope, surface cover and basin geology must be taken fully into account. Unfortunately the statistical studies which would assist such correlations are rare ond the small amounts of assembled data demonstrate considerable variability.

Where a significant component of baseflow exists, groundwater and surface water development will need to be closely integrated since abstractions of the former must reduce base flow. Although surface storage is not generally feasible, multipurpose flood control structures can be used to assist groundwater recharge. Fog drip inducement is also a possibility worthy of consideration in high islands.

Rainwater catchments may satisfy basic water needs for domestic supply at best but are utilized when surface or groundwater supplies are inadequate in quantity or quality. They are most economic when large and sited in areas with a reliable and well distributed rainfall such as in the ITCZ. Various methods to optimize rainwater catchment systems have been developed using either computer techniques (Schiller and Latham 1982; Guillen 1984) or manual calculation (Pompe 1982). The basic principle is to calculate the storage to provide enough water for the period of the year where there is no rainfall or when it is insufficient to meet water needs. Rainfall records for some 20-30 years are desirable to provide an adequate basis for calculations, and optimization improves with the availability of shorter term records.

Groundwater

Occurrence. Because of the typically ephemeral nature of surface runoff on small islands, groundwater, whether as spring discharge or in aquifers, generally represents the most important water supply source.

Groundwater occurrence on high islands is complex and particularly so in the volcanic islands of the andesitic province due to their greater structural and lithological variability and the occurrence of thicker and more massive lava flows. Oceanic sequences (Fig.12.7) commonly consist of a series of thin flows with high vertical as well as high horizontal permeability, thus promoting deep downward infiltration. Perched aquifers overlie localized aquicludes such as ash beds, and high level aquifers may also occur, contained within vertical dikes. Springs which issue from high level aquifers may form the source of perennial streams. In the islands of Vanuatu, 37% of the present service cover is from springs and 21% from streams, many of which are spring derived. Springs in andesitic volcanics tend to be more frequent in occurrence but of lower volume than those in the basaltic rocks, a feature which correlates with the greater geological complexity and lower permeability and porosity in the former sequences.

On many high islands, there are developments of coastal plain sediments of alluvial or marine origin. Depending on relative permeabilities, the coastal plain sediments may act as a cap rock resulting in an artesian build-up of hydraulic head in the

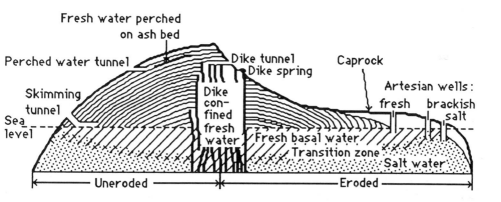

Fig.12.7 Cross-section of volcanic dome of oceanic basalt type, showing occurrence and recovery of groundwater (from Peterson 1984 b).

deeper bedrock aquifer or as an uppermost phreatic aquifer in a multiple sequence. Both circumstances have important effects on resource occurrence and development issues.

Mixing with salt water. In addition to high level or perched aquifers, groundwater in a basal aquifer floats upon and displaces more dense sea water with a shape which is consequent upon the areal distribution of recharge and the flux towards the coast. The fresh groundwater and sea water are separated by a transition zone the thickness of which is determined by geological and hydraulic factors. Flow movement through a branching, irregular porous medium promotes mixing of the two adjacent fluids (dispersion) and a consequential flux occurs of both fresh and brackish water towards the coast. Dispersive mixing is proportional to flow velocity and therefore the higher the flux, the greater the dispersion. The mixing process across a freshwater to seawater transition zone may be affected by the asymmetric velocity distribution since Peterson (1984 a) states that the greater the freshwater flux, the closer the transition zone will approach a sharp interface. Mixing is also promoted by natural disturbances of the interface which are occasioned by tidal fluctuations, wave action and variations in recharge whether seasonal or longer term. The size constraint and isolated maritime locations of small islands are reflected by highly dynamic groundwater responses producing a relatively wide transition zone.

The Ghyben-Herzberg approximation is the basis for most quantitative investigations but with limitations due to its basic assumptions that the saltwater is at rest and the freshwater flow is horizontal. Using the notation in Fig. 12.8,

$$h_s = \frac{\rho_f}{\rho_s - \rho_f} h_f \qquad (12.1)$$

If ρ_f, the density of fresh water, is taken as $1000 kg\ m^{-3}$ and ρ_s, the density of sea water, as $1025 kg\ m^{-3}$, h_s will be approximately equal to 40 times h_f. The 1:40 ratio relates to a sharp interface or the midpoint of a transitional zone.

An important consideration is the movement of the interface in response to changes in abstraction or recharge. The position of the interface is a measure of the storage and also controls potential water quality changes during abstraction due to upconing. Referring to Fig.12.9, Schmorak and Mercado (1969) give an expression for the critical pumping rate Q_c at which upconing in a confined aquifer will be incipient:

316

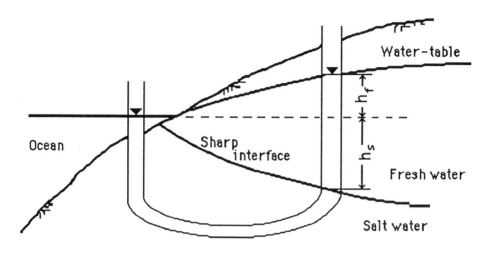

Fig.12.8 Definition sketch for the Ghyben-Herzberg approximation (from Barker 1984 b).

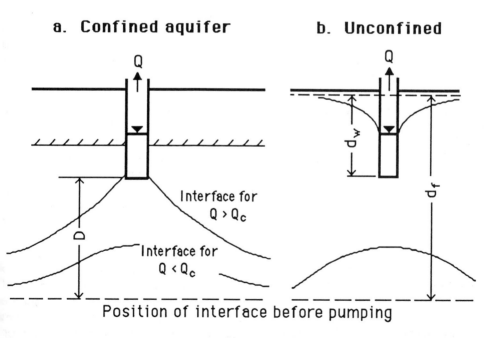

Fig.12.9 Definition sketch for upconing at a skimming well (from Barker 1984 b).

$$Q_c = \pi \, KD^2 / 40 \qquad (12.2)$$

while Chandler and McWhorter (1975) state that upconing will not occur in an unconfined system if

$$d_w < d_f / 3 \qquad (12.3)$$

Information on interface response is therefore important for resource prediction and development.

Recharge. Some information on groundwater recharge can be obtained by stream hydrograph analysis as described in Chapter 2. Evaluation of the deep infiltration to the basal aquifer in contact with sea water requires different techniques and a fuller discussion of these is given in the section on low islands where precise quantification of recharge is more critical. The freshwater component of the basal aquifer system on high islands may extend to considerable depths below sea level in accordance with static water elevations, and values of about $300\,m$ have been recorded in the Hawaiian group of islands. The aquifer may become artesian below a marginal cap rock of low permeability (Fig.12.7). Estimates of deep infiltration have been made for the islands of Hawaii (Takasaki 1978) and selected physiographic, hydrologic and other data are shown in Table 12.7. On the assumption that the runoff figures also incorporate any base flow component, deep recharge can show a range from 6 to 31% of mean annual rainfall. Low values all occur in the three small islands (Kahoolawe, Lanai, Nihau) which are in the rain shadow of larger islands. As is to be expected, evaporation shows larger proportions although it is numerically little different from the amounts in the other islands. The only other island with low recharge is Kauai in which the runoff component is high. The substantial component of deep recharge in several cases could well account for an apparent lessening of runoff yields, a feature referred to earlier.

Groundwater development. Development of dike-impounded groundwater is an attractive proposition because of the high hydraulic head and the relative isolation from sea water contamination. If tunnelling as opposed to drilling techniques are utilized, the massive loss of storage which commonly ensues during construction has disadvantages in lowering the head and decreasing lateral leakage into the basal aquifer. There are obvious advantages in retaining and developing high level storage to avoid excessive overdraft on the basal aquifer, particularly on

318

the leeward side of high islands where populations are high, aquifer recharge is low and the basal aquifer has often poorer quality water due to admixture with sea water. High level dike aquifers occur and are developed in the oceanic islands of Hawaii and Tahiti.

Development of a basal aquifer in contact with sea water must be such as to avoid excessive intrusion by upconing or the sea water wedge. Analyses of aquifer behavior and critical abstraction rates mainly make use of the Ghyben-Herzberg approximation which assumes a sharp interface (Barker 1984 b). On this basis, the intrusion distance L of the salt water wedge (see Fig.12.10) is given by:

$$L = KB^2 / 80\ Q \qquad\qquad (12.4)$$

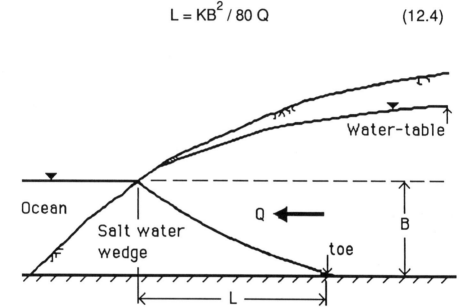

Fig.12.10 Definition sketch for a salt water wedge in an unconfined aquifer (from Barker 1984 b).

To reduce saline intrusion, shallow skimming wells, infiltration galleries or relatively large numbers of dispersed boreholes with low pumping abstraction are advantageous. Recharge barriers utilising surplus runoff and double pumping techniques may also be applied. Other factors may require compromise developments. To reduce upward leakage through an artesian cap rock and losses by evaporation, hydraulic heads need to be reduced, which will also cause saline encroachment. Knowledge of the hydraulic parameters of the basal aquifer within the transitional zone or interface is essential if reliable predictions on the results of development are to be made.

Water quality

Natural hydrochemistry. The chemistry of natural surface water and groundwater is controlled by the composition of the source rainfall and by various chemical and physical processes. The chemical processes mostly relate to interactions between water and rock material. The physical processes include fluid mixing which in the small island context mainly refers to sea water.

The chemical composition of island rainfall is dominated by ocean spray and in consequence typical ionic ratios such as Na:Cl and Na:Mg are comparable to those in seawater (Table 12.8 and Table 12.9). The rainfall is in equilibrium with atmospheric gases, particularly CO_2, and at normal atmospheric pressure the dissolution of CO_2 gives a pH of about 5.7. The islands in the tropics are remote from major sources of industrial gaseous oxide discharges which are responsible for the occurrence of acid rain. High levels of radioactivity did occur in the South Pacific following atmospheric testing of nuclear weapons but these have now dissipated. Some variations in the chemical composition of rainfall occur, both seasonally and in single storm events during which dilution increases with time. The feature could affect recharge calculations by the chloride balance since intense rainfall generating surface runoff is likely to have a different composition from rainfall at other times.

Table 12.8

WORLD MEAN RAINFALL AND SEA WATER COMPOSITION $(mg\,l^{-1})^*$

	pH	Na+	Ca2+	Mg2+	K+	NH4+	NO3-	SO42-	Cl-
Rainfall	5.7	1.9	0.08	0.28	0.03	<0.1	<0.1	0.58	3.9
Sea water	8.2	10 556	400	1272	380	0.1-2	<0.1-2	2649	18 980

* From Morrison et al (1984)

320

Table 12.9

RAINFALL COMPOSITION DATA FOR THE SOUTH PACIFIC *(mg l⁻¹)**

Source	pH	Na^+	Ca^{2+}	Mg^{2+}	K^+	NH_4^+	NO_3^+	SO_4^{2-}	Cl^-
World average	5.7	1.9	0.08	0.28	0.3	<0.1	<0.1	0.58	3.7
Niue	6.7	3.0	15	0.05	0.05	<0.01	<0.01	9.9	7.2
Tokelau	6.4	2.0	<0.1	0.4	0.1	0.01	<0.01	<1	2.0
Tarawa	6.2	2.3	2.0	<0.1	<0.1	n.d.	n.d.	<1	3.5
Queensland	5.5	2.5	1.2	0.5	0.4	<0.01	<0.01	<1	4.4

*From Morrison et al(1984)

During infiltration, the content of carbon dioxide increases in the soil because of its organic content and the chemical composition continues to change in accordance with chemical interactions with the host rock. These changes are accelerated with higher temperature and account for the increased rate of weathering of rock material in tropical regions. The breakdown of the main component minerals of basaltic and andesitic volcanic rocks, such as olivine, pyroxene and plagioclase feldspar, releases magnesium bicarbonate and silica which are dominant ions in associated groundwaters (Table 12.10). If sulphides such as pyrite are present in the host rock, groundwater may exhibit high sulfate levels in excess of potability limits. The total ion content of carbonate groundwater is generally higher than volcanic rock groundwater because of the ready solubility of limestone in weakly acidic circulating groundwater. The presence of other cations, notably magnesium, in the carbonate rock also increases solubility.

Table 12.10

ANALYSES OF SOME PACIFIC ISLAND GROUNDWATER
FROM BASALT ROCKS *(mg l⁻¹)*

Source	pH	Na^+	Ca^{2+}	Mg^{2+}	K^+	HCO_3^-	SO_4^{2-}	SiO_2	Cl^-	Fe
Oahu, Hawaii	7.7	12	24	15	5.3	156	1.6	50	15	0.43
Taveuni, Fiji	7.1	9.5	17	10	1.7	134	2.9	n.d.	2.9	0.28
Opolu, W.Samoa	7.6	18	21	n.d.	n.d.	120	n.d.	n.d.	31	0.16

* From Morrison et al (1984)

Water pollution. The chemical compositions of naturally occurring water sources of small islands are generally within acceptable limits of potability. Other than mixing with sea water, most observed occurrences of pollution or contamination can be traced to human influences. High metal levels have sometimes been observed in rainwater storage containers where runoff is from galvanised iron roofs or those covered in lead based paint. Elevated levels of organic matter have also been noted (Downes 1981) in water collected from roofs made of traditional materials. Biological pollution of pathogenic bacteria is a common occurrence in both surface water and groundwater and mainly relates to inadequate sewage disposal. Aquifers in fissured limestone or volcanic rocks and underlying thin soils are very vulnerable to this and other forms of waterborne pollution, particularly so when the aquifer is phreatic and at shallow levels, as is common in coastal locations. Water from unprotected wells or streams in inhabited areas commonly has coliform levels in excess, often grossly, of WHO limits. The second most clearly perceived risk relates to the inappropriate use and handling of agrochemicals, notably pesticides, some of which are very toxic and extremely long-lived.

Conclusions

Close comparisons of the hydrology of small high islands with large islands and continents are constrained by generally limited data on the former areas. Runoff characteristics are more akin to those of small mountainous catchments in the larger land masses. Baseflow runoff correlations with basin geology will be comparable but the hydrogeology of the basal aquifer and the effect of deep infiltration in the more permeable sequences combine to distinguish small islands from the larger land masses.

Section C

Flatlands

13. Flatlands with snow and ice

Extent of the region

Flatlands with an annual potential evaporation less than 500 *mm* extend across the northern periphery of the northern hemisphere, between Murmansk and Kamchatka in Europe and Asia, and between the lowland coast of the Bering Sea and northeastern Labrador in North America (Fig.13.1). Mapping the bioclimatic zones of the northern subarctic caused Ermakov and Ignatiev (1971) to conclude that these zones differ little between the continents. The similarities between types of climate and soil categories result in corresponding similarities of ecological conditions within landscape zones. There is no significant area of this region in the southern hemisphere: the part of southern Chile with potential evaporation less than 500 *mm* consists mainly of sloping land.

General climate and water/energy balances

The tundra environment of these flatlands, which are very common in the USSR, has the following broad characteristics:

(a)   A limited amount of incoming heat energy throughout the year, associated with the pronounced negative radiation balance at the earth's

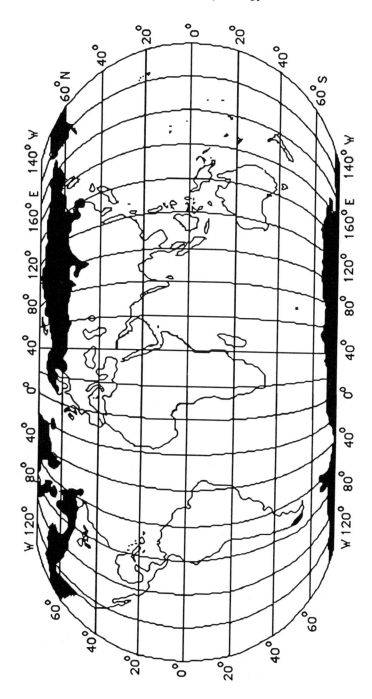

Fig. 13.1 Generalized map of areas with annual potential evaporation less than 500 *mm*

surface and the circulation in summer of cold air masses from the Arctic;

(b)   Development of cyclone activity and the resultant significance of frontal precipitation;

(c)   A great excess of precipitation over the energy available to evaporate it, resulting in very low values (0.2-0.4) of the RIA (see Chapter 1) $R_n/\lambda P$ (Budyko 1971);

(d)   Low average air temperature during the warmer months, and low air temperatures at the soil surface;

(e)   Resulting from the above factors, very wet soil conditions to a degree exceptional in non-tropical latitudes (Grigoriev 1970).

Distinctive hydrological features

Snow and snow cover. In winter a large part of the region is covered with snow, and in general there are over 200 days per year with snow cover on the ground. However the cover is not deep. While the annual precipitation ranges from 100 mm in the northernmost tundra to 700 mm at the southern boundary of the forest tundra, the proportion of winter precipitation does not exceed 25% in the north of Canada (Dreyer 1978), rising to 50% in the continental areas of the forest tundra.

Permafrost. Permanently frozen ground occupies 14% of the earth's land surface, and a considerable part of this is located in the northern hemisphere (northeastern Europe, north and northwestern Asia, and a large part of Canada), the remainder being in Antarctica. The thickness of the permafrost layer varies from 1-2 m to several hundred meters; in north Canada, the regional thickness can reach 500 m (Brown 1978). The permanently frozen ground is topped by a layer, a few centimeters to 1-2 m thick, which melts every summer and freezes completely in the winter.

River icing. This term is given to water which freezes on river banks and beds at their contact with frozen ground. These river icings are multiannual formations often of large size (areas up to 100 km^2 and volumes up to 0.5-0.6 km^3). Over 7500 icings, with a total volume of about 30 km^3, have been recorded in the northeastern USSR. The total volume in all river icings of the northern hemisphere has been estimated as 180-200 km^3 (Unesco 1978 b).

326

Marshes.   The very flat nature of the region, the continuous water surplus, and the presence of permafrost combine to result in poor surface drainage and hence a large area of marshes. On the average, marshes in the tundra zone occupy 50% of the land area, but the figure can be as high as 80-90%, e.g. in the Kanin peninsula in the USSR. The proportion of marshes is less where the relief is more dissected, as in the Kola pensinsula, where the value is 20% in the tundra and 30% in the forest tundra.

The marshy nature of the land is not particularly important for the accumulation of water or for regulating the runoff (Bay 1969). This is particularly evident in the permafrost areas, where there is a minimal probability of an increase in base flow in the spring, because the soil and underlying rock are frozen through. Water progressively accumulates in the soil during thawing, but the melted water fills the soil water zone; part is blocked at the lower horizons, and gets frozen (Roulet and Woo 1986).

Groundwater.   The contribution of groundwater to river runoff is limited by low hydraulic gradients, the thin layer of seasonally thawed soil, and the freezing of the rivers in winter. It varies from less than 1% to a maximum of 20% of the total annual river runoff.

The river regime

Endogenous runoff.   Rivers in the permafrost area have low winter flows, often 1-2%, or less, of the annual total. Many large rivers freeze to their beds and their flow ceases. The sharp decrease in winter flows is accompanied by the formation of an ice cover up to $2m$ thick, or more in some places.

River flows are at their highest during the melting of snow, and fall gradually at the end of summer. The importance of rain in causing high flows is low relative to the runoff due to snow melt. During the autumn rains, high river water levels are no more than 1.5-2 times the low levels.

The low contribution of groundwater to river flow results in a regime in which sharp rises in river levels are caused by increases in surface runoff. Catchments respond instantaneously to precipitation (Brown et al 1968), and runoff coefficients are high (0.7 or more) compared with other regions.

Most of the runoff occurs during the warm period of the year. In the Porcupine river (a tributary of the Yukon near the Polar Circle), 94% of the runoff occurs during May

-September. In the Ban river, slightly further south, 91% runs off during June-October.

As shown in Table 13.1, years of low and high water content (in the snow, the marshes and the soil water) result in different proportions of seasonal runoff. In average years, the seasonal distribution varies by less than 5% for catchments of different size.

Table 13.1

DISTRIBUTION OF SEASONAL RUNOFF IN THE WEST SIBERIA LOWLAND
(after Stezhenskaya 1966)

Percentage of the annual volume

Zone	Catchment area (km^2)	Latitude (°N.)	Spring	Summer/ autumn	Winter	Spring	Summer/ autumn	Winter	Spring	Summer/ autumn	Winter
			Average			Water storage High			Low		
Tundra	1000 - 10 000	68	75	20	5	68	31	1	64	31	5
	10 000 - 100 000	68	67	23	10	55	37	8	60	23	17
Forest tundra	1000 - 10 000	66	57	32	11	52	39	9	60	31	9
	10 000 - 100 000	66	56	28	16	58	35	12	57	23	20

The effect of lakes.   The above description of the runoff regime is typical of the region as a whole. However, a more even seasonal distribution of river flow is found in catchments containing lakes. The gradual release of lake water prevents many rivers in the Canadian northwest from freezing through to their bed, even in relatively small catchments (Dreyer 1978). From the data given in Table 13.2 for a small catchment in north Canada, it can be seen that the runoff comprises lake water, snowmelt and melting soil water; the direct role of rainfall is insignificant.

Exogenous rivers. There are many large rivers with their origin in other regions, which cross the tundra flatlands and discharge large volumes of water into the Arctic ocean. The Ob river carries 399 $km^3 y^{-1}$, the Lena 535 $km^3 y^{-1}$, and the Yenisei 564 $km^3 y^{-1}$ (L'vovich 1971). While flows in the small rivers (catchment areas up to 1000 km^2) formed entirely in the tundra and forest tundra vary between a few and some tens of $m^3 s^{-1}$, the flows in the large exogenous rivers range from several hundreds to tens of thusands of $m^3 s^{-1}$ (Fig.13.2). During the main runoff period, the ratio between high and low flows is 10-15 in the exogenous rivers, and 40-50 in endogenous rivers.

Table 13.2

WATER BALANCE OF A SMALL CATCHMENT
(after Roulet and Woo 1986)

Components of water balance (*mm*) - May 15 to August 5, 1983

	Inflow			Discharge			
Snow melt	Rain	Lake runoff	Ice and water in soil	Surface runoff	Base flow	Evapo-ration	Ice and water in soil
146	34	430	251[a]	331	1	223	306[a]
			251[b]				265[b]

a- assuming 100% saturation of peat with ice
b- assuming 70% saturation of peat with ice

Rainfall - runoff relations

It can be concluded that river flow is mostly due to surface runoff which is mainly related to water storage on the catchment. Bay (1969) found a linear relationship between runoff and water storage for north Canada, while Dreyer (1978) found close relationships between precipitation and both surface and total runoff for the tundra and forest tundra of the USSR (Fig.13.3); the linear correlation coefficients were 0.86 and 0.84 respectively.

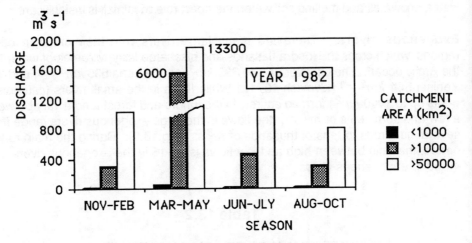

Fig.13.2 Seasonal variation of average monthly flows of rivers in the tundra zone of the European USSR.

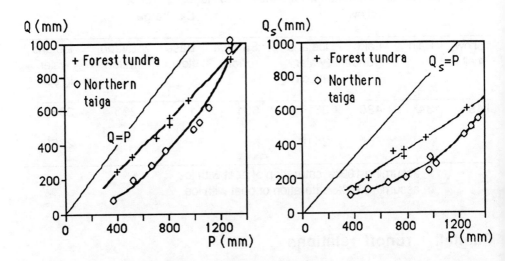

Fig.13.3 Relationships between total runoff Q and surface runoff Q_s, and annual precipitation P, for forest tundra and northern taiga in North America (after Dreyer 1978).

In the European part of the USSR, the average annual river runoff from the tundra is 300 *mm,* from an average precipitation of 665 *mm* (L'vovich and Chernogaeva 1974). Similar values are found in the tundra and forest tundra landscape of Siberia (Nikolaeva and Chernogaeva 1977) and of north Canada (Dreyer 1978).

In view of the close correlation between water storage and runoff, it is reasonable to assume that variations in these quantities would be directly related to variations in precipitation, which ranges between 100 $mm\,y^{-1}$ in the extreme north to 700 $mm\,y^{-1}$ on the boundary between the forest tundra and the northern taiga (sub-arctic evergreen forests).

14. Humid temperate flatlands

Extent and general characteristics

Humid flatlands with potential evaporation betwen 500 and 1000mm occur in the northern hemisphere in a belt roughly centred on latitude 50°, running through east Europe, Asia and North America (see Fig.8.1 in the chapter on humid temperate areas with catchment response). The natural land cover is taiga (subarctic evergreen forest) and mixed and broad-leaved forests.

Although the broad characteristics of the region are similar, the part in the American continent has higher values of precipitation and net radiation. Consequently, the annual evaporation rates in mixed and broad-leaved forests in North America are higher than the values in Europe by 100mm and 150-200mm respectively (Ermakov and Ignatiev 1971). Table 14.1 shows the average regional water balance.

Runoff formation processes

Infiltration. Surface infiltration depends on soil genesis, its grain size distribution, and the type of plant cover (forests, grasslands or cultivated areas). In meadow areas of the forest zone of the European USSR, the infiltration of spring precipitation is 80% in sandy soil, 67% in loamy sand and 47% in silty clays.

Base flow. An analytical model of the water balance (Chernogaeva 1974) shows that the total runoff within this region may range from 30% to 60% of the precipitation. The proportion of base flow within this total may vary from zero to almost 10%, and is determined by the precipitation regime, the geomorphology

Table 14.1

AVERAGE ANNUAL WATER BALANCES (*mm*) FOR EUROPE AND NORTH
AMERICA
(from L'vovich and Chernogaeva 1974; Dreyer 1978)

Area and zone	Precip- itation P	River runoff Q	Base flow Q_u	Surface runoff Q_s	Evapora- tion E	Potential evapora- tion E_o
Europe						
taiga	750	290	95	195	460	500
mixed forest	715	170	60	100	545	700
North America						
taiga	800	380	160	220	420	500
mixed forest	900	320	120	200	580	700
leaved forest	950	320	90	230	630	1000

and lithology of the area considered, and the presence of marshes and permafrost.
One example will be given of the effect on the water balance of the soil hydraulic
characteristics, as determined by lithology, climate and basin morphology.

Table 14.2 shows the characteristics of two taiga catchments with podzol soils and a
grass cover, in the European USSR. The precipitation and total runoff are close to
the average values for the region. However, the soils of the Ukhta are sandy, while
those in the Voloshka catchment are sandy loams and silty clays. As a result, the
proportion of base flow in the Ukhta exceeds the regional value (0.29) by 47%,
while the value for the Voloshka is 45% below the regional value.

Effect of catchment area. To characterize the runoff components in different
parts of the region, it is necessary to select rivers with a stable base flow and
catchments mainly in a uniform climate and landscape, which restricts the catchment
areas to the range 1000 - 50 000km^2. The base flow in small catchments is always
relatively less than in large catchments, because the river bed is shallower. In
catchments of 1000-2000km^2 in the middle of the European USSR, the base flow
reaches a stable value of 0.15-0.20$mm\,d^{-1}$.

Table 14.2

ANNUAL WATER BALANCE OF TAIGA BASINS IN THE EUROPEAN USSR

	Voloshka river at Toropovskaya	Ukhta river at Ukhta
Area (km^2)	7000	4280
Precipitation P (mm)	780	770
Evaporation E (mm)	449	423
Total runoff Q (mm)	331	347
Surface runoff Q_s (mm)	278	201
Base flow Q_u (mm)	53	146
Runoff coefficient Q/P	0.42	0.45
Base flow index Q_u/Q	0.16	0.42

As a result of the instability of the base flow and the low regulating capacity of the catchment, small rivers in the flatlands have very uneven hydrological regimes. In the small rivers of the European USSR, it takes only 2-3 weeks in spring for up to 70-80% of the annual runoff to be discharged. The wide range of soil and plant types, and of geomorphology, has a great effect on the formation of runoff in small catchments, so that it may vary from the regional value by an order of magnitude.

Frozen ground

In the taiga zone, the soil is frozen for about half of the year, and there are significant areas of permafrost. As stated in Chapter 13, the effect of permafrost is to reduce infiltration and base flow, and so increase the proportion of surface runoff in river flows, leading to a less stable flow regime. In the flatland catchments of the Lena and Vilui rivers in east Siberia, where the permafrost is deep, the base flow is 5% or less of the total river flow, and its absolute value is about $5 mm\ y^{-1}$.

As in the case of the tundra zone, a close relationship has been observed between total runoff Q and its surface component Q_s. The correlation coefficient between these variables for the forest zone of North America is over 0.94 (Dreyer 1978).

Marshes

Extent and origin. A feature of the region is an abundance of lakes and marshes, which is common in the latitudes of the mixed and broad-leaved forests. In the central plains of Canada and West Siberia, the combination of low relief with an excess of water results in marshes covering up to 70% of the area of some catchments. The area of marshes in Estonia, Latvia and Lithuania is 20, 30 and 40% respectively of the total area of flatlands. The largest single area of marshes in the world is the lowlands taiga of West Siberia, where marshes cover over 50% of the land surface.

The extent of these marshes is explained by the excess of precipitation over evaporation, the low relief, and the genesis of the area. When the glaciers retreated at the end of the Ice Age, they left behind a plateau with lakes which formed the central point for the development of marshes. The high surface runoff caused sphagnum sods with a high water capacity to accumulate at the margins of these lakes, and the marshes subsequently developed from these margins, from the early Holocene to the present time.

Effect on the water balance and runoff regime. Many scientists have estimated that the total runoff from marshlands is less than that from adjacent areas. However, the water balance research of Kulikov (1976) in the Vasjugan area of West Siberia indicates that the transpiration from marshes does not exceed that of nearby dryland forests. It can therefore be assumed that the earlier assessments of runoff from marshlands have been under-estimated, due to the difficulties of taking experimental measurements. The Vasjugan area, in the north of the catchments of the Ob and Irtush rivers, is a key area for the analysis of the environmental features of heavy oligotrophic marshes.

Fig.14.1 shows hypothetical diagrams of the components of seasonal runoff from marshes, well-drained forests and agricultural land. It shows that there is virtually no base flow resulting from highland marshes. These marshes contribute less than forest complexes to the local energy and water balances. In the case of large catchments occupied by more than 20% of marshes or permafrost, there is not so much a variation in the absolute value of the annual runoff as in the runoff regime. In the heavily marsh-covered catchments of the taiga rivers, the slower melting of forest snow and the delayed movement of snowmelt under low slopes, together with the storage capacity of the marshes, prolong the spring flood to an extreme extent, continuing through the beginning of floods from summer rainfall.

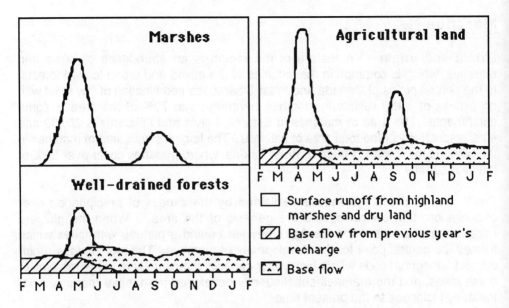

Fig.14.1 Seasonal runoff from three types of catchment.

Effect of drainage. The construction of drainage in marshes creates improved conditions for the runoff of marsh water. However, the capacity of the unsaturated soil zone increases considerably at the same time, and this tends to slow down or reduce the runoff. Again, lowering the marsh water level by drainage reduces the evaporation, but the use of these drained areas for intensive agriculture causes higher evaporation. Hence the overall changes in the runoff due to marsh drainage will depend on the relative influence of factors such as the density and type of the drainage network, the standard of the drains, the type and quality of land use, and the extent and location of the drained area within the catchment (Novikov 1981).

For smaller rivers an additional factor is the very significant role of the regional hydrogeological conditions, since changes to the runoff due to drainage may be caused by a considerable change in the process of groundwater recharge, and in the supply of snowmelt and rain water to the river network, following the installation of drains at depths comparable with river levels. The direction and extent of changes in the runoff regime will in such cases be determined by the specific hydrogeological features of the area.

When there are impermeable layers near the soil surface, the installation of a drainage system which cuts through these layers will help to transfer part of the surface runoff to the base flow which feeds larger rivers. As a result, the total runoff of medium and large rivers will not change, but it will diminish in small rivers.

When there is a considerable depth of impermeable soil in the catchment, the effect of deepening small rivers is to increase their drainage capacity and so lead to higher runoff.

It follows that the impact of marsh drainage on the annual runoff of small rivers may be positive or negative, depending on the physical and geographical conditions of the catchment.

15. Dry temperate flatlands

Extent of the area

The hydrological classification adopted for this book shows the main area of dry temperate flatlands extending in a latitudinal belt between about 40 and 50 °N, from Mongolia across the Caspian plain and around the north margin of the Black sea into parts of Hungary (see Fig.9.1 in the Chapter on dry temperate areas with catchment response). There are also significant areas in Spain and in the Great Plains of Canada. There are no areas of appreciable extent in the southern hemisphere, but the large flatland areas of Buenos Aires province in Argentina (Fig.15.1) are marginal to the sub-humid classification and also to the boundary between the temperate and warm zones. As this area and the Great Hungarian Plain are the best documented cases of marginally dry temperate flatlands, they will be used as contrasting case studies in this chapter. The average monthly rainfall and potential evaporation for two locations in Buenos Aires province are shown in Fig.15.2.

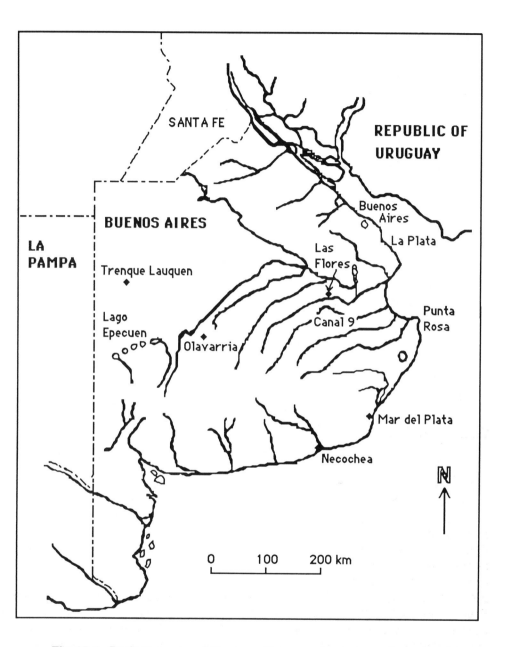

Fig.15.1 Drainage map of Buenos Aires province, Argentina, showing some artificial canals along natural stream beds.

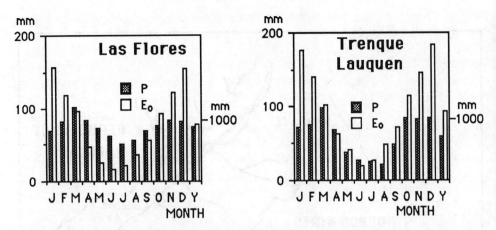

Fig.15.2 Monthly and annual rainfall P and Penman potential evaporation E_0, for two locations in Buenos Aires Province.

Hydrological system and processes

Vertical processes. The characteristic feature of large plains is that the surface has no slope at all, or the slopes are negligible (less than 0.1 $m\ km^{-1}$) and interrupted by local depressions. The runoff is extremely small compared with evaporation and infiltration, and the latter two almost completely balance the precipitation. This indicates that the models of conventional hydrology, based on the catchment response to precipitation, cannot be used to characterize hydrological processes in flatlands. Since the differences in elevation between different parts of the flatland are relatively small, there is an almost negligible amount of potential energy available to maintain horizontal water transport on the surface or in the groundwater system. It follows that the dominant hydrological processes are vertical water exchanges between layers at different levels in the soil water zone; these movements initiate the migration of salts, which may accumulate in the topsoil if the predominating direction of motion is upwards. Thus soil development and consequent agricultural productivity are greatly influenced by local hydrological processes.

As shown in Chapter 6, the average depth of the water-table in flatlands depends to a large extent on the relative values of average precipitation and potential evaporation. In the dry temperate zone there is not too much disparity between these quantities, and therefore the water-table is usually shallow and there is significant capillary rise from it to plant roots or the surface, for at least part of the year. This contrasts with the situation in dry warm flatlands without external inputs,

340

where the water-table is deeper and the groundwater is almost isolated from surface hydrological processes.

Dominant storages. The hydrological system illustrated in Fig.2.2 (p.42) classifies the storages into those in which water movement is essentially vertical and those in which it is essentially horizontal. As a result of the predominance of vertical exchanges in flatlands, it follows that the first group will have major importance. These storages comprise water held within and on the surfaces of plants, surface water stored locally in the micro-relief, and water in the unsaturated zone, both within and beyond the influence of plant roots. Groundwater should also be included, since in this environment its lateral movement is small compared with its vertical exchanges in the form of recharge and evaporation. Shallow unconfined groundwater systems, with long turnover times, typically extend continuously over large areas.

Drainage network. Instead of the sloping beds of rivers, the basis of the surface drainage network in flatlands is the local depression which stores the precipitation temporarily in shallow pools. Water in excess of the storage capacity of the pools moves on the surface as sheet flow or erodes shallow channels. Cascades composed of a series of depressions with negligible differences in elevation develop in this way. The small runoff channels are unstable, and may shift or disappear as a result of natural influences (extreme dry or wet periods) or changes in land use.

Since there is no pronounced surface slope, even the direction of overland and shallow channel flow may change in response to randomly occurring events, such as the spatial distribution of precipitation and wind direction. As a result, a single flatland region may on different occasions supply either of two rivers which flow into the sea hundreds of kilometers apart (Fuschini Mejia 1985). In such circumstances, the concept of catchment area loses its definition.

Morphology. At the macro scale, flatlands often have an average gradient (see Fig.15.3) which influences the direction of surface flow in extreme events. However the local water balance is largely determined by land forms at a meso scale of 0.1-1*km*. These land forms take the form of very shallow depressions and very mild slopes which, without altering the general slope, alter the direction and value of the local gradient. This results in a landscape consisting of unstable or permanent shallow lakes, with discharges which are usually unstable but in a few cases may be perennial.

The land forms and soils are the result of a succession of climates and events which developed mainly in the Holocene period. In the case of the Argentine pampa, the

341

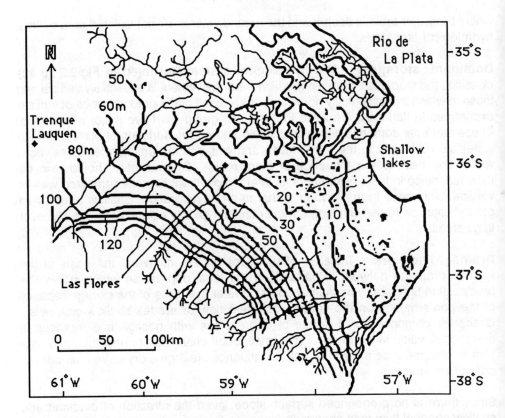

Fig.15.3 Contour map of the Pampa Deprimida in
Buenos Aires province, Argentina.

uncoordinated landscape corresponds to a geological period very much drier than
the present (Tricart 1973), and the present drainage difficulties exist because of
this. Paleo-drainage channels have some influence on the present hydrology.

In these circumstances, soil genesis produces a mosaic distribution of soil types
which is closely related to the local land forms. As will be developed later, this
mosaic pattern becomes the building block of the hydrological characterization and
modeling of flatlands, in contrast with the basin structure of areas with catchment
response.

Floods and droughts. Because of the low potential energy available for
horizontal water transport, both floods and droughts have longer durations and
affect larger areas of flatlands than is the case in sloping land.

342

When there is an excess of precipitation on flatlands, the shallow lakes become interconnected, and then large areas are inundated with water which moves very slowly. These are the only occasions on which surface hydrological processes occur at a macro scale, rather than in isolated systems. Not all floods cover the whole of the flatland area; this depends on the depth and spatial distribution of the precipitation. The inundation and waterlogging of the soil may cause deterioration or death of vegetation and sometimes animals.

As an example of a particular flood event, in 1980 the rainfall in one month in the province of Buenos Aires had a volume of $60km^3$, and the runoff to the sea was $5.5km^3$, leaving $54.5km^3$ to remain in storage or be evaporated. By means of satellite images, Dominguez and Carballo (1984) measured the flooded area as $286km^2$. During the following 5 months, most of the water surplus was evaporated, some moved into the groundwater system, and only a very small part became river runoff.

In the eastern part of the Argentine plain, the climate is more humid, and the result of a flood (often from artificial channels which break and overflow) is to completely saturate the soil for periods of years. This area forms marshes and swamps because there is insufficient evapotranspiration to remove the surplus, except in very dry years.

When there is a lack of precipitation, the immediate consequence of a drought is ameliorated by the shallow lakes which allow the survival of animals and the maintenance of pastures around their margins. Short droughts, usually due to a seasonal lack of water, are frequent in some areas, and do not cover the whole of the Argentine plain. Their impact is on the development of agriculture, as the stored soil water in these conditions may be insufficient for crop growth. Widespread and continuous drought affects the entire plain, and prevents crop growth without irrigation. Duran (1982) has tabulated the history of alternating widespread floods and droughts in the province of Buenos Aires, going back to the year 1574. He has also (Duran 1986) given examples of simultaneous occurrence of smaller floods and droughts in the same region.

Water balances

Local water balance. For a land area A with an arbitrary boundary (Fig.15.4), the water balance equation for a period Δt can be written:

$$(P - E) A + (Q_{si} + Q_{gi} - Q_{so} - Q_{go}) \Delta t$$

$$= \Delta(V_{int} + V_s + V_u + V_g) A \quad (15.1)$$

where
P = rainfall in period $\Delta\tau$

E = transpiration and evaporation from soil

Q_{si}, Q_{gi} = surface and groundwater inflows across boundary

Q_{so}, Q_{go} = surface and groundwater outflows across boundary

V_{int} = average volume of water in interception storage

V_s = average volume of water on surface

V_u = average volume of water in unsaturated soil zone

V_g = average volume of water in groundwater (relative to an arbitrary datum)

Δ indicates the change during the period Δt.

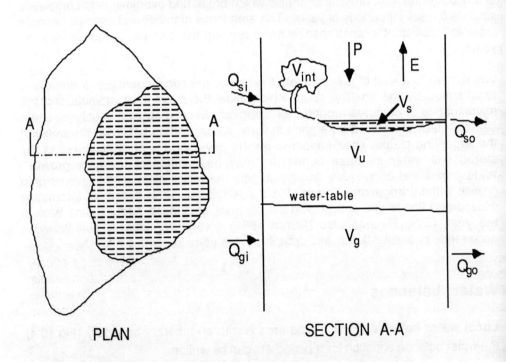

PLAN   SECTION A-A

Fig.15.4 Definition diagram for local water balance equation.

Although measurement of all the terms in (15.1) poses a formidable task, some simplifications can be made by choice of an appropriate balance period Δt. For example, the term V_{int} will normally be negligible relative to the other terms, for balance periods of a day or more. The term $(Q_{gi} - Q_{go})$ will also be negligible in the usual case of a groundwater system with a long turnover time and hence essentially steady flow. Again, if the balance period is a year, the term V_u will usually be negligible relative to the other terms in the equation. Thus the equation for the annual water balance can be written:

$$(P - E) A + (Q_{si} - Q_{go}) \Delta t = \Delta(V_s + V_g) A \qquad (15.2)$$

The terms on the right can be readily measured from the depth of surface water and the depth to the water-table, while the precipitation P is also easily determined. Difficulties then remain with the evapotranspiration and the surface flow terms.

The basic problem then in design of a water balance network is the measurement or estimation of the surface flows, which as we have seen may occur as shallow sheet flows or flows in shallow unstable channels.

For local water balances, the problem is minimized by selecting the boundaries of the area A in such a way that surface flow across them occurs only in extreme events. This concept has been developed in Australia by attempting to define areas of repetitive micro-hydrology (Australian Water Resources Council 1972; Chapman 1984) which meet these criteria in different landscapes. Definition and mapping of these areas, which may range in extent from a few to some hundreds of meters, can be achieved by aerial photo-interpretation or automated processing of multi- spectral data from satellites, aided by ground truth in the form of maps of soils and vegetation (Asmussen et al 1984; Marlenko et al 1984). Where surface storage is significant, an accurate surface elevation map is also essential.

Large-scale water balances. Following the concept described above, a large flatland can be viewed as a mosaic of similar areas which have a self-contained system of surface hydrology in all but extreme events. When these large events occur, water flows between these areas and across the whole flatland, and usually finds its outlet in a river, an artificial channel, or a lake. In each case, there is some concentration of the surface runoff, so that conventional hydrometric techniques can be used at this macro scale. Thus the problem of estimating water balances again becomes tractable at this very much larger scale. In these extreme events, flows between the individual mosaic elements of the landscape become internal re-distributions of water within the macro system.

Exogenous effects. The simple classification of a flatland into micro and macro systems is complicated by inputs from adjacent sloping land or more humid areas with greater runoff. Rivers from these areas interact with the flatland areas in two ways: by flooding the adjacent plain and by acting as drains for flatland runoff. Fig.15.3 shows the streams entering the Pampa Deprimida in Argentina from the southwest. Some of these streams are lost in the flatlands, some reach a well defined channel (Rio Salado) in the north, and others which previously reached a natural stream are now artificialy channelized and reach the sea (e.g. Canal 9, Fig.15.1).

In spite of its importance, the external water does not determine the hydrological characteristics of flatland areas, although there is usually a need for engineering works to control the interactions. The external water may have a significant role in the behavior of deep confined groundwater in flatlands, since the recharge zones are often in the external sloping area.

Implications for hydrological data collection. As has already been demonstrated, there are major differences between the appropriate and feasible techniques for hydrological data collection in flatland areas, from those which have been developed for areas with catchment response. In view of the predominance of vertical water transport there is a strong case for a more dense network of pluviographs (Kovacs 1984 a), and particular attention must be given to methods of spatial interpolation of both rainfall and evaporation data (Major 1984; Wales-Smith and Arnott 1985).

A basic requirement in flatlands is a network of stations to measure the depth of water in shallow lakes and the depth of the water-table. Measurements of the water stored in the unsaturated zone (V_u) should be made at selected sites, so that weekly or monthly water balances can be calculated.

An essential adjunct to hydrological data collection in flatlands is good maps of topography and the landform-soil-vegetation complex. Remote sensing from satellites can provide invaluable data in this regard, and also in the monitoring of the surface expression of major hydrological events.

Hydrological impacts of human activities

A consequence of the limited availability of potential energy in flatlands is that the hydrological regime may be modified considerably by human activities. As land use modifies the vertical processes of infiltration and evapotranspiration, it has a major input on the local hydrology.

346

Unsuitable agricultural practices , such as excessive tilling or over-grazing, can lead to a change in the soil, creating impervious horizons, decreasing the storage capacity and reducing soil productivity. Depending on the soil type and the position of the water-table, such practices can also lead to salinization of the topsoil.

In the Argentine plains, many shallow lakes have been filled and even cultivated. This, combined with reduction in infiltration, leads to an increase in the volume of runoff occurring immediately after heavy rainfall, and to an overall increase in the severity of both floods and droughts.

Depending on the areal extent of these influences on local hydrology, there may be more or less severe impacts on the hydrology of the whole flatland region. Other forms of land use may also influence the large-scale hydrology. For example, the construction of roads and railways (generally parallel to the sea) in the Argentine pampa impedes the continuity of slow overland runoff during widespread flooding and sometimes concentrates discharges into the road and railway drainage systems, which are never adequate. Another effect is created by canals which drain water from neighboring hilly areas across the flatlands to the sea. At times the water level in these canals is above the surrounding plain, and overflows or breaks in the embankments produce flooding in the adjoining area; this water often remains on the plain for prolonged periods.

Land and water management

Influence of vegetation. The selection of the plants to be grown in a given area of flatlands must take into account their impact on the local hydrology as well as their productivity and immediate economic benefit.

Natural vegetation in flatlands has been studied by methods that take into consideration the associations of vegetation groupings (Soriano et al 1984). The same method has been used to study cultivated vegetation, and it was found that this increased evapotanspiration, prevented waterlogging and produced more biomass.

Current research is using genetic techniques for the selection of plants or plant communities that can be used for pastures, which are resistant to droughts, waterlogging and disease, and do not impoverish the soil (Cahuepe et al 1982; Fioriti and Fuschini Mejia 1985; INTA 1977; Leon 1980; Parodi 1947; Sala et al 1981; Soriano et al 1977).

347

Effect of scale.   Agricultural practices have a determining influence on the storage and evaporation terms of the water balance. Water management in flatlands is therefore a compromise between micro-hydraulics in local areas, which optimizes the availability of water for vegetation, and macro-hydraulics for larger areas, which addresses the problem of excesses and shortages of water.

Micro-hydraulics. In Argentina, Barbagello (1984) has developed a micro-hydraulics system which has been successfully implemented locally in the Pampa Deprimida. In other parts of the world, polders with drainage systems and pumping stations have been developed (Volker 1984; Godz et al 1984; Kovacs 1984 a).

The elements of a micro-hydraulics scheme are an improved, eventually fertilized, soil, a pumping station which carries excess water to artificial storages through discharge canals, and an irrigation installation. By simultaneous use of all these elements, a net downward flow of water in the unsaturated zone can be maintained, preventing the rise of salts. The natural shallow pools in the landscape can be used or improved as the basis for artificial storages. This was recommended (though without foreseeing the pumping) by Ameghino (1884), who showed that the hydraulic-agricultural problem of the Argentine plains is a problem of shortage and seasonal distribution, rather than one of excess water. Tricart (1973) also recommended the use of the shallow lakes for artificial storage.

Macro-hydraulics.   The object of these engineering works is to remove large-scale water excesses from the flatlands as quickly as possible. The source of this water may be extreme precipitation in the flatlands themselves and/or floods which occur in adjacent sloping areas and must cross the flatlands to reach the sea. The design of these works therefore calls for an understanding of the hydrology of the whole region which has often been lacking in the past. Although several large canals have been constructed in the pampa of Buenos Aires Province, major flooding occurred in 1980 and 1985. One of the reasons was the suppression of natural storage at the micro scale, by filling in the shallow lakes and cultivating them.

Two general principles can be given for the design of macro-hydraulic schemes:

(i)   The provision of temporary storage for the floodwater from the sloping land by the construction of reservoirs which are normally kept empty (Godz et al 1984).

(ii)   Selection of canal routes that take into account not only topography but local soils and land forms, and paleo drainage paths where these exist.

348

By their nature, these macro-hydraulics works are used infrequently, but it is important to provide for continual maintenance to prevent deterioration from plant growth, trampling by stock, or vandalism. Pumping stations must be provided along the canals, to transfer water transversely from the adjacent plain. Levels and flows in the canals must be controlled by means of gates, and gaging stations must be used to develop a rating curve relating flow to both stage and hydraulic gradient.

Other engineering works which cross the flatlands (such as roads, railways and pipelines) must be designed so as to interfere as little as possible with the overland runoff and the operation of the canal scheme.

The Great Hungarian Plain. This section will conclude with a comparison of the land and water management systems which have developed over the last few centuries in two large flatland areas, the Great Hungarian Plain lying in the centre of the Carpathian basin, and the pampa of Buenos Aires province. The temperature is slightly higher in Argentina due to the difference in latitude, but the higher potential evaporation is partly compensated by higher precipitation, so that the water deficit E_0 - P is almost the same in the two regions.

Water control and land reclamation in the Great Hungarian Plain started more than 200 years ago, when the extensive animal husbandry was not able to provide enough food for the increasing population (then about 3 - 4 $p\ km^{-2}$). The initial works involved collecting the exogenous water from the sloping land east of the plain, and carrying it in a deep channel to the Tisza river. Both the new channel and the river were diked to protect low lying areas from flooding, and the slope of the Tisza was increased by several cut-offs which decreased the length of its meandering bed, and accelerated the propagation of floods. Large networks of canals were constructed to collect the excess endogenous water and convey it in a relatively short time to the main channels, where large pumping stations were constructed to ensure the discharge ot the excess water into the rivers even in periods of river floods. The capacities of the canals and the pumping stations were determined by fixing the longest acceptable period of inundation. As the level of cultivation was improved, the overall system was not modified, but the drainage period was shortened by increasing the flow capacity of the system. By the 1970's the length of canals was more than 30 000 km, the pumping capacity was about 1200 $m^3\ s^{-1}$, and the systems were able to drain the area in 14-15 days even in the case of extreme inundation caused by the simultaneous effects of snow melt and high precipitation (Kovacs 1978 b).

These systems met the agricultural demand until the development of intensive agriculture, which modified the requirements for water control. With the use of heavy machinery on large farming units, average yields can be achieved of

349

6-7$t.ha^{-1}$ for wheat and 8-10 $t\,ha^{-1}$ for maize. However, this requires not just reducing the time of inundation below a given limit during the critical period in the spring, but complete control of soil water during the whole growing season. This involves lowering the water level in the collecting canals and supplementing the systems with tile drains. As it is impracticable to deepen the main canals, booster pumps are provided between the collector canals (serving a unit of several thousands of hectares) and the main canals.

This more intensive drainage may lead to water shortages in the later part of the growing season. Part of the drained water must therefore be stored for possible later use, but shallow natural reservoirs are not suitable for this purpose because of high evaporation losses and the deterioration of water quality. The booster pumps however make it feasible to construct deep reservoirs, the water level in which is raised well above the plain (Kovacs 1978 b).

The province of Buenos Aires . The original inhabitants of this plain were a few nomads who developed a closed economy which did not interfere greatly with the natural habitat. The first white settlers brought with them foreign animals, such as the horse and hare, and foreign plants, such as the thistle, which had some effect on the landscape. The first exploitive use of the plain, by both white settlers and incoming indigenous tribes, was the export of cattle and horses, and their hides.

In the early stages, there were no fences, but the fields were divided by ditches and permanent dwellings were built. Subsequent evolution (Randle 1981) involved the general use of fencing, troughs and windmills, the improvement of pasture, and the planting of wood lots for timber and wind breaks. This formed a complex which was vulnerable to the effects of natural events such as floods, which therefore took on a greater importance than they had during the first stage of settlement. As agriculture was not a means of subsistence, but a means of exporting, and the immigrants were not the landowners, there was a tendency to excesses in cultivation and over-grazing, leading to deterioration of the soil and wind erosion, particularly in the drier western region.

The construction of the railways had a major effect on population distribution, and eliminated the long cattle drives which had previously been necessary. They also had a substantial effect on floods, not due to the bridges (which were built to generous designs, as shown by the fact that they are still in use) but to the embankments, which stopped the natural flow of large floods across the plain. When the water cannot follow its natural path, it either flows along the embankment until it reaches the next drain, or it stays in place, flooding the land until it evaporates. This artificial longer period of flooding damages the soil structure and the associated vegetation, which can change in floristic composition.

This situation was exacerbated by the increased use of trucks for transport of cattle and goods, which required the construction of a network of trunk and minor roads, again with embankments which interfered with the natural drainage. The trunk roads in particular are designed to be above flood levels and this can have serious impacts; in the 1980 floods, part of the recently opened Route No. 11 had to be blown up because of its interference with the flow of flood water.

The first large-scale approach to water management was the construction of the canals which were intended to solve the problem of removing flood water from the sloping land adjacent to the plain. These canals have been partially successful in this objective, but because they are also bordered by embankments, they too interfere with the natural flow of floods of local origin.

More recently, as a result of studies of the geomorphology (Tricart 1973) and the soils (INTA 1973) of the region, new agricultural techniques have been developed for soil conservation and fertilization. These techniques have been very beneficial in stopping abuses of the environment at the local scale, but they have been applied without taking the overall problem into account. As was shown at the 1983 Olavarria symposium on the hydrology of large flatlands (Fuschini Mejia 1984), there must be coordination between soil conservation works at the local level and civil engineering works at the larger scale.

Socio-economic aspects

Droughts and floods have an immediate economic impact, not only because of the loss of crops and animals, but also because of the destruction of the productive and residential infrastructure. Depending on the degree of destruction of this infrastructure, the population migrates with all the effects on social structures that this implies.

There is a graduated scale of damages which depends on the intensity, areal extent and duration of floods and droughts. It is not necessary for a flood to be widespread for it to have a damaging effect economically and socially, given that the interruption in transport produces a loss of wealth even in areas not directly affected. Serious damage is caused in floods by crop losses, the abandonment of homes, the dispersion of families, and the interruption of social and educational activities; but these may be only temporary, and the population will return and recover its losses in future years. However, if the flood is of such a magnitude that it destroys the physical structure of houses and workplaces, even in cities, the serious damage can become permanent and the population abandon the area. In some cases, there may be severe soil erosion due to the flood; in the 1985 floods in Argentina, the city of Epecuen and its neighboring areas had to be abandoned for this reason.

Local droughts produce impoverishment, but this can usually be included in the normal risks of agriculture, and does not lead to movements of the population. However, widespread and prolonged drought may endanger human life, and produce an increase in soil erosion which can be permanently damaging to productivity. This can lead to the abandonment of farms and the movement of the population, usually to the cities.

Flatlands are not only affected by the hydrology of adjacent areas, but also by their socio-economic level. For example, the province of Buenos Aires is bordered on the north and south by populations with a high socio-economic level, and this provides economic justification for many projects which cross the plain. As has been described, such projects can have a negative impact on the hydrology of the plain and hence also on its economy.

The contrasting development of water management in Buenos Aires province and Hungary, described in the last section, can be ascribed to the different socio-economic conditions. In the pampa the pressure is mainly for improvement of pastures and grazing conditions, rather than the development of intensive agriculture that would justify the high capital and running cost of water control on the Hungarian model. The main requirements are therefore the exclusion of flood water from the sloping areas, and the development of improved drainage within the flatlands, without increasing the hazard of seasonal droughts. These are complex problems which require the full understanding and cooperation of the local population for their solution.

16. Humid warm flatlands

Occurrence and general characteristics

A very typical pattern of tropical relief consists of shallow sediment-filled broad basins separated by the divides formed by fault blocks, mountain ranges and plateaus. A striking feature of these basins is their less developed or absent drainage system. For instance, it is believed that the drainage network in Africa was formed during the period of existence of Gondwanaland, and after its disintegration the rivers had to find their way to the sea.

The general climatological features of the humid warm tropics (see Fig.10.1) have been described in Chapter 10. Swamps and lakes dominate in the flatlands, but some morphologically different lakes have been formed in regions with increased tectonic and/or volcanic activity. Another common feature is the effect of backtilting, when the backswamps of the rivers have formed irregular marshy lakes and altered the drainage flow pattern and water balance. An example is given in Fig.16.1 of the area between Lake Victoria and Lakes George and Edward.

Boundaries between the flatlands and the areas with catchment response are not very sharp. Inflows and outflows often originate in hilly or at least rolling areas and thus a combination of the hydrological regimes under the influence of local water resources and transported ones is common.

A considerable part of the savanna belts belongs to the flatlands. Beard (1964) defines savanna as the natural vegetation on the highly mature soils of senile landforms which are subject to unfavorable drainage conditions, and which have

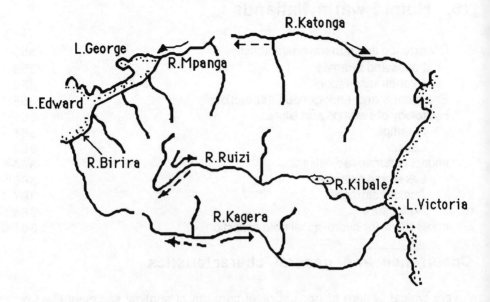

Fig.16.1 Effect of backtilting on the drainage network.

intermittent perched water-tables with alternating periods of desiccation and waterlogging with stagnant water. A vast belt of savanna, forest savanna and woodland savanna surrounds the rainforest and reflects various ratios of the length of the dry and wet seasons.

Savanna lands cover approximately 13% of the earth's surface. In Africa, savanna extends across the northern part of the continent and on the Central African Plateau. In South America vast areas of llanos exist in the Orinoco basin and campos in the central and southwestern basins, and in Australia they are found to the south of the tropical rain forests.

For tropical soils in the flatlands the following types of prevailing vertical water flux can be recognised (Mohr and Van Baven 1972):

-       a continuous downward movement occurs in regions with no dry seasons; the permanent inflow of water into a wet soil profile results in continuous leaching;

-       alternating downward movement and cessation is typical in wet/dry regions where the upper soil layers dry out seasonally and the water-table is at such a depth that capillary action cannot transport groundwater to the surface, and leaching occurs only during the wet season;

- ascending water movement occurs when there is very low rainfall and high evaporation from continuous transport of capillary water enriched by salts;

- alternating upward and downward movement occurs in wet/dry regions when groundwater can reach the surface layers and evaporate (see Fig.6.3).

Lakes and swamps. In the absence of the river network, conditions become favorable for the formation of standing water series in which "the water motion is not that of a continuous flow in a definite direction, although a certain amount of water movement may occur, such as internal currents in the vicinity of the inlets and outlets" (Welch 1952). Two types of standing water series in the humid tropics can be described by the following simple schemes:

(i)   Lake/pond - perennial swamps - intermittent swamp - land;

(ii)   Headwater area/lake - perennial headwater swamp - intermittent headwater swamp - land.

Some headwater swamps can be considered as belonging to the area with catchment response (see Chapter 10) because they have an inclined surface. In fact, contrary to the concept of standing water series, some authors have suggested that all swamps must have an inclined surface because the aquatic vegetation in them requires water that is flowing (Debenham 1952). The distinction is therefore one of degree.

In addition to the climatological characteristics described in Chapter 10, the formation of lakes and swamps in the tropics is related to the following features:

- flat relief and an impermeable soil or rock layer close to the surface;

- surface relief which can absorb the water excess from a larger basin during the wet season;

- increased density of the vegetation in slowly flowing or standing water;

- a runoff regulating system acting as a reservoir with balanced rates of inflow and evapotranspiration;

- a high ratio of surface area to water depth;

- an environment that allows the size of swamps to fluctuate from season to season and from year to year.

Thus both swamps and lakes, seasonal and perennial, occur whenever the morphological and climatological conditions are favorable to the water of the rainy season congregating in a locality faster than it can disperse. Thus a variety of lakes and marsh-ridden areas is found in the tropical flatlands, and they differ in size, composition of the vegetation, and hydrological regime.

Evapotranspiration. Evaporation from the free water surface and evapotranspiration from areas with aquatic vegetation is a dominant feature of the water balance in flatlands. In contrast with lakes, the water balance is clearly related to the vegetation. However the lakeshore vegetation may play an important role in determining the regime of lake inflows. In addition, the regime of outflows can be influenced by the formation and destruction of vegetation barriers (Fig.16.2).

Water balance calculations from different sources indicate that the evapotranspiration from tropical swamps can be much higher than the evaporation from a free surface of water. Hurst (1954) concluded that the evapotranspiration from the Nile papyrus can exceed the evaporation from a free water surface, and Van den Wert and Kamerling (1974) reported values of E/Eo in the range 3.0-4.0, for standing water series covered by water hyacinth.

High evapotranspiration is also produced by various types of phreatophytes and riparian vegetation.

A high variability of the evaporation and evapotranspiration from month to month and year to year has a significant influence on the variability of the hydrological regime of the flatlands. Fig.16.3 shows the fluctuation of potential evaporation observed in the Bangweulu lake/swamp system, sketched in Fig.16.4. As the area covered by vegetation varies from year to year, we can expect the ratio: E/E_0 to be highly variable in time and space.

Swamps occupy much larger tropical areas than lakes. Table 16.1 gives an estimate of the area of swamps in the tropics. The figures include swamps which are not in the warm humid tropics such as the South American intermittent swamps known as pantanal which originate from about 1200mm of rainfall falling during a short period on less permeable soil.

356

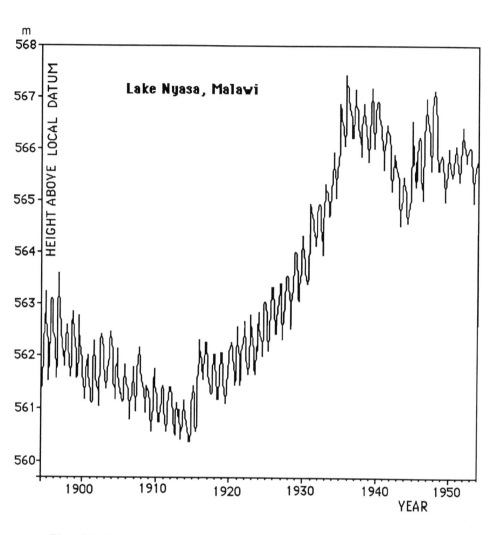

Fig.16.2 Long-term water level fluctuation in Lake Nyasa, showing short period oscillations caused by vegetation barriers.

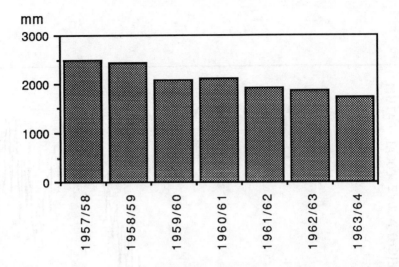

Fig.16.3 Annual potential evaporation in Bangweulu lake/swamp system.

Table 16.1

AREA OF SWAMPS IN TROPICAL PARTS OF THE CONTINENTS

Continent	Swamp area (km^2)
South America	1 200 000
Asia	350 000
Africa	340 000
Australia	2 000

Exogenous and endogenous interactions

Very few swamps and lakes have their hydrological regime under the sole influence of local water sources. Transported water contributes significantly towards the formation of swamps and lakes. For instance, in the coastal area of Asia the mangrove swamps are formed through a contribution of silt deposits, as a result of the erosional activity of the exogenous water source. The density of the mangrove population is closely related to the intensity of silt accumulation and to its origin.

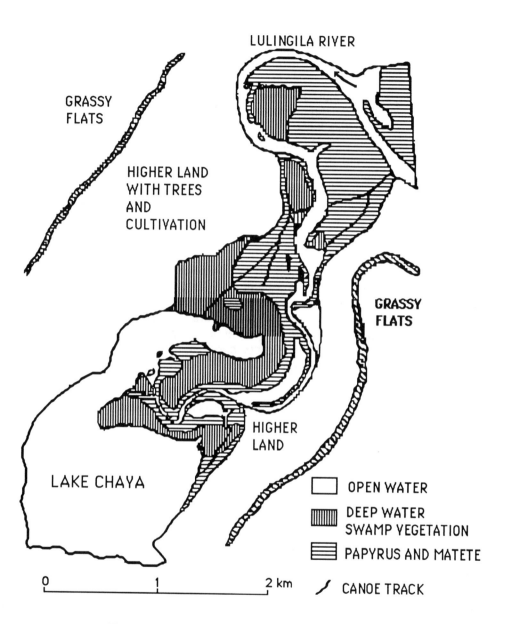

Fig.16.4 Sketch of part of Bangweulu swamp.

Deposits of granitic origin result in a dense population while those of sandstone origin are less favorable for mangroves. It can be concluded that a significant difference in swamp morphology is related to the source of swamp recharge.

Ordinary swamps are recharged by the precipitation falling on their surface, but contributing basins with catchment response serve as an equally significant source in many regions. Extensive flatlands (flood plains) are found in the vicinity of large rivers when the morphological conditions and inflow regime become favorable for temporary or permanent flooding. In a study of permanent tropical swamps in Papua and New Guinea, Taylor and Stewart (1958) found that a distinctive feature was the occurrence of herbaceous communities at a distance up to 1*km* from the river, while in seasonal swamps the same communities were found only in the surrounding pools of water. Obviously the formation of the herbaceous communities in swamps is related to the prevailing hydrological regime.

A symbiosis of lakes and swamps with micro-exogenous interactions is found in many parts of the humid tropics. Often the ratio of lakes and marshy areas fluctuates year by year and long term changes are also common. For instance, in Lake Valencia (Venezuela) a century ago there was enough water to form a permanent outlet. Since then the water level has dropped by more than 5*m* and the lake has become bordered by swamps and marshy plants. A sudden increase of the level of Lake Victoria in the 1960s flooded a great part of the coastal swamps (Fig.16.5).

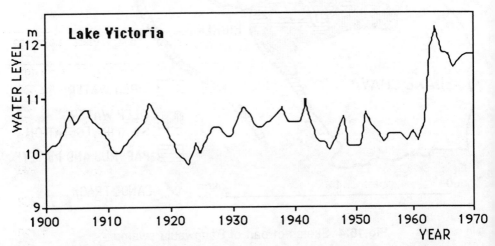

Fig.16.5 Fluctuation of the annual minimum level of Lake Victoria.

360

Three clearly marked zones can usually be found in lake areas:

- a lake zone, under water throughout the year;

- an intermediate zone, waterlogged for long periods;

- a marginal zone, flooded for only a brief part of the year.

Sometimes it is difficult to determine which part of an area can be considered as swamp and which part belongs to the lake. Equally difficult is to establish the area of influence of exogenous and endogenous processes.

Hydrology of swamps and lakes

Swamps.   With the variety of swamps and lakes found in the humid warm areas, it is difficult to classify their regimes. Three different types of African swamps have been described by Balek (1977): the Bangweulu type, Kafue Flats type and Lukanga type. Table 16.2 shows their water balance components.

Table 16.2

ANNUAL WATER BALANCE OF SOME AFRICAN SWAMPS

Parameter	Unit	Bangweulu	Kafue Flats	Lukanga
Drainage area	km^2	102 000	58 290	19 490
Area of swamp	km^2	15 875	2 600	2 600
Rainfall on drainage area	mm	1 190	1 090	1 250
Rainfall on swamp	mm	1 210	1 110	970
Evaporation from free surface	mm	2 340	2 070	1 710
Evapotranspiration outside swamps	mm	890	785	908
Additional evapotranspiration within swamps	mm	1 120-1 260	196	252
Total evapotranspiration in swamps	mm	2 000-2 180	1 000	1 120
Water loss as % of inflow	%	60	4	7.8

The Bangweulu type is formed by standing or slowly flowing water, deep swamp vegetation, papyrus and Matete reeds and numerous grassy islands. Bangweulu swamp itself transforms the regime of the Chambeshi river into an entirely different regime of the Luapula river in the headwaters of the Upper Congo. Evapotranspiration accounts for about 60% of the total inflow.

Another distinctive type is the swamps formed in the river floodplains, such as the area of Kafue Flats along the Kafue river, a tributary of the Zambezi. Before the construction of a dam, the swamps were saturated every year by the flooded rivers and dried out during the dry season. About 4% of the inflow becomes evapotranspiration from the swamps.

The Lukanga swamp in the Kafue basin has a complex hydrological regime (Macrae 1934; Vajner 1969). The swamp can be characterized as a sidestream reservoir. The Kafue river flows into it during the high flood season and passes by it in other months. Macrae showed that the channel connecting the river and the swamp can carry water in both directions, and there is a spill of water during the flood season of more than $500 \times 10^6 m^3 y^{-1}$. In the swamp 7.8% of the inflow is used for evapotranspiration.

As noted above, the hydrological regime in flatlands is very often subject to external infuences, so that flatlands should be analyzed together with their attached basins as a complex hydrological system. For instance, Okavango swamp in Botswana is located in a semi-arid region, but it is fed by the river Okavango, which has its headwaters in a wet/dry region. The river deposits all its sediment load in the swamp. As can be seen in Table 16.3, the main source of the inflow to the swamp is exogenous while the main water consumption due to evapotranspiration is endogenous.

As another example of the combined effect of areas with catchment response and flatlands, Fig.16.6 shows a rainfall-runoff relationship based on the percentage of flatlands in areas with catchment response.

Lakes. It is uncertain which lakes should be considered as belonging to the flatlands. Several deep lakes in humid warm regions, such as Tanganyika, are products of tectonic activity and they drain steep basins with a well developed river network. More typical of the flatlands are lakes such as Lake Victoria and Lake Chad, with a very high ratio of surface area to depth. Table 16.4 gives the basic characteristics of the largest lakes in the humid tropics, without morphological distinctions.

Table 16.3

ANNUAL WATER BALANCE OF OKAVANGO SWAMP
(from Dincer et al 1978)

Mean area of swamp	10 000 km^2
Max. area of swamp	13 000 km^2
Min. area of swamp	6 000 km^2
Mean active storage	$4 \times 10^9\ m^3$
Max. active storage	$7 \times 10^9\ m^3$
Min. active storage	$1 \times 10^9\ m^3$
Inflow	$10.5 \times 10^9\ m^3$
Precipitation	$5 \times 10^9\ m^3$
Evapotranspiration	$14.9 \times 10^9\ m^3$
Outflow - surface	$0.3 \times 10^9\ m^3$
Outflow - groundwater	$0.3 \times 10^9\ m^3$

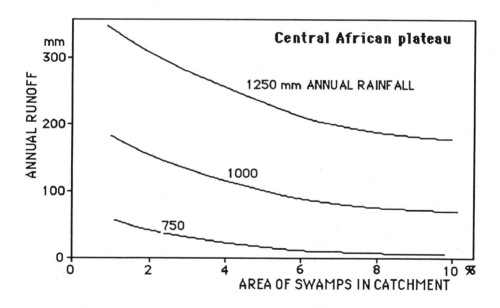

Fig.16.6 Annual runoff as a function of annual rainfall and proportion
of catchment covered by swamps (from Balek 1983).

The greatest African lake and the third largest in the world is Lake Victoria, also known as Ukereve. It is more than 400*km* long and 240*km* wide and has a shoreline 7000*km* long. Geologists believe that the lake once covered a much larger area and included Lake Kyoga. This would mean that the water level was at least 90*m* higher than today. The seasonal fluctuation of the lake level averages 650*mm*, but occasionally, for example in the 1960's, higher fluctuations have been observed.

Lake Tanganyika is the seventh largest in the world. Since the maximum depth of the lake is 1435*m*, the lake bed is far below sea level. In fact Tanganyika is the second deepest lake in the world, after Baikal. A subsurface ridge divides the lake into two parts. The only outflow from the lake, the River Lukuga, is intermittent, and lake evaporation often exceeds precipitation and inflow. An artificial outlet from the lake was made in 1878 by digging through the banks at Albertville and the lake level went down by 10*m*. The temperature in the first 480*m* varies between 25 and 27 ºC and from this depth it is fairly constant at 23.1ºC.

Lake Nyasa is a very long lake and like Lake Tanganyika it is still in the process of tectonic development. The contributing catchment has an area of about 100,000*km*^{2} and the lake is drained through the River Shire into the Zambezi. Seasonal fluctuation is about 1*m*.

Table 16.4

LARGEST LAKES IN THE TROPICS

Lake	Area (*km*^{2})	Max depth (*m*)	Volume (*km*^{3})	Elevation (*m*)
Victoria	66 400	92	2 656	1 135
Tanganyika	32 890	1 435	18 940	773
Nyasa	30 800	706	7 000	472
Chad	18 000	12	27	240
Maracaibo	13 000	250	-	0
Tonle Sap	10 000	12	40	15
Rudolph	8 660	72	-	427
Nicaragua	8 430	77	108	32
Titicaca	8 110	400	710	4 100
Mobutu Sese Seko	5 300	57	64	620

Lake Chad is located in a shallow depression, and owing to the flat relief the lake area varies greatly depending on the regime of the main inflows, the Logone and Chari. Its size can vary between 10 000 and 25 000 km^2, while the mean depth is only 1.5m. The highest level occurs in July, when the lake length increases to 250km and the width to 150km. The mean annual fluctuation of water level is below 0.8m, the maximum being 2.0m. Because both inflows carry a great amount of sediment, the water becomes brackish. In a northeasterly direction the lake is intermittently connected by surface runoff with the Bodel pan.

Maracaibo is a lagoon lake in Venezuela which fills a tectonic depression near Tablazo bay, connected with the ocean by a short shallow channel. In the southern part the lake water is fresh, while in the northern part it is brackish. The salinity varies from 400 to 2300$mg\ l^{-1}$, depending on the rate of exchange of water between the lake and the Caribbean Sea and on the hydrological regime of the lake inflows.

Tonle Sap is in the Mekong basin and stores excess flood water carried by the Mekong (see Chapter 19).

At the margins of the great Rift formations, south of the Ethiopian Highland, is Lake Rudolph, 300km long and 25 - 60km wide. The maximum depth is 70m. Most of the inflow is from the Omo River, and there is no outflow.

Nicaragua is a lake which fills another depression formed by volcanic activity in Central America. The San Juan river drains the lake into the Carribbean Sea.

Titicaca is the highest of the tropical lakes. It fills a tectonic depression and is a relict of an earlier, much larger lake. The lake drains through the River Desaguadero into Lake Poopo, which has no outflow. There are several inflows into the lake, the largest being the Ramis. The temperature at the surface fluctuates between 11 and 14 ^oC, and deeper there is a constant temperature of 11 ^oC. The water is moderately saline.

The size of many lakes, particularly in flatlands, varies seasonally and from year to year.

The salinity of the inland lakes is highly variable and in general, lakes and pans without surface drainage or a substantial contribution from base flow have a high saline content. For example, the salinity of the large African lakes varies between 65ppm for Lake Victora and 480ppm for Lake Mobutu Sese Seko.

In some lakes secondary effects are produced by the sediments. For instance in Lake Kivu large amounts of methane are generated by bacteria from the organic carbon of the sediments (Tietze and Geth 1980). In 1986 a lethal cloud of carbon dioxide from volcanic activity or decaying material in Lake Nyos, Cameroon, devastated three villages and killed almost 2000 people.

Table 16.5

ANNUAL WATER BALANCE OF AFRICAN LAKES, RELATED TO THE LAKE AREA

Lake	Area (km^2)	Inflow (*mm*)	Precipitation (*mm*)	Outflow (*mm*)	Evaporation (*mm*)
Victoria	66 400	241	1 476	316	1 401
Tanganyika	32 890	1609	950	141	2 418
Nyasa	30 800	472	2 272	666	2 078
Kariba	5 250	8 840	686	7 038	2 088

Impact of human activities

Lakes and swamps. Tropical lakes and swamps have always been one of the main sources of food, and the favorable conditions along the lake shores have been one of the positive factors in the development of tropical civilizations. With increasing population, eutrophication has been found to be vitally important in the tropical water economy. Eutrophication has been described as the process of aging of lakes, but such a description emphasizes the effects rather than the causes.

In the tropics, lakes in particular are only a temporary feature of the land surface. Most of the lakes and swamps have been formed by the joint action of rivers, wind, tectonic activity and earth movement, and they disappear from the earth through a process of natural eutrophication involving the filling of the depressions with nutrient-containing sediments. The process of man-made eutrophication is much faster than the slow rate of natural eutrophication, and the process is even more accelerated in the tropics. It has been established that phosphorus content is a determining factor in eutrophication, and the reduction of phosphate loadings and the removal of phosphates already present in lakes is one of the most effective means of eutrophication control. In addition, another effective means of control in warm humid areas is the removal of organic matter before decomposition, and the introduction of herbivorous fish or hippopotamus which graze on swamp vegetation.

Tropical lakes do not behave in the same way as the cool lakes of temperate regions. For instance, the great variability of rainfall in a single year has a negative effect on lake regimes. Tropical soils are highly erodible, and if the plant cover is damaged by over-grazing or drought then a large volume of sediments enriched with phosphates enters waterways and lakes.

Artificial storages.   The natural landscape of the tropics has been greatly changed during the past decades by the construction of reservoirs, which have influenced the ecology and economic and social life in extensive areas.   Table 16.6 gives a list of the largest artificial reservoirs in the tropics, but a great number of small reservoirs produce equally significant impacts.   The water loss from the reservoir surface can be very high in tropical regions, and the chemical and biological balance of the rivers flowing out of the reservoirs is changed. Transformation of riverine ecological conditions to lacustrine ones and the seasonal effect of the storage of floodwater result in thermal and chemical stratification, the effects of which are not always easily predictable. Fish distribution is almost always influenced by a changed river regime.

Perhaps the most serious problems result from resettlement of the population. They arise particularly when large schemes are designed, and the rural population has to leave extensive areas for towns, or is shifted to a strange or even hostile region where adaptation to existing conditions is difficult.

Table 16.6

LARGE ARTIFICIAL LAKES IN THE TROPICS

Lake and river	Country	Storage	Area	Dam height
Kariba, Zambezi	Zambia, Zimbabwe	160	5 250	100
High Aswan, Nile	Egypt, Sudan	157	5 120	95
Volta, Volta	Ghana	148	8 480	70
Itaipu, Parana	Brazil, Paraguay	129	1 400	190
El Mantecho, Caroni	Venezuela	111	-	136
Pa Mony, Mekong	Laos	107	-	115
Cabora Bassa, Zambezi	Mozambique	66	2 700	100

Agriculture.   Soils in humid warm flatlands are structurally unstable, and slake readily under the impact of raindrops (Lal 1983).   Alternate desiccation and flooding causes a surface crust to develop that drastically reduces the infiltration rate. This rapid deterioration of the soil structure and its decline in rainfall intake are due partly to low organic matter content.

Tropical rivers generally cannot create extensive rich alluvial plains because of their rather low content of dissolved solids and suspended matter. For example, even the Amazon flatlands are not suitable for crop production. Extensive areas of soils in the Amazon flatlands are under water five to seven months of the year, the depth of the water being over 4m ; attempts to build up a fertile soil profile by siltation have been ineffective.

Only those southeast Asian rivers such as the Red River, Mekong, Irrawaddy, Brahmaputra, Ganges and Indus carry a substantial amount of sediment and are capable of creating fertile plains.

Typical savanna soils are formed by a permeable layer superimposed on impermeable soil, so that a perched water-table is formed. Some savanna soils are derived from moist soils by degradation. Savanna land is therefore more suitable for transfer to cultivated grassland used for grazing, than for irrigation schemes.

Implications for hydrological data collection

Many water development schemes are planned in the flatlands. When dealing with the problems of water development in the tropics, the different conditions existing there have to be borne in mind, particularly the higher rate of energy flow through the tropical ecosystem, caused by higher temperature and greater potential productivity, and the higher consumption of nutritive salts and thus a greater need for their replenishment. Waste products and dead organisms rot more rapidly and require a higher consumption of oxygen. High evaporation and transpiration rates are another feature of the tropics.

Water resource planning should be part of an integrated approach to the planning of land use and management, but the program should be flexible with emphasis placed on priorities. Past experience strongly suggests giving priority to small and medium size development schemes and to projects in which benefits are most likely to follow immediately after their completion.

The main problems in planning water development schemes are the relation of water resources to the environment, finding the most rational way of using water resources, and finding the most rational way of developing them for that use.

An adequate solution of these problems requires the development, maintenance and operation of hydrological networks, including groundwater and water quality. Monitoring of lake and swamp levels, and their thermal and chemical stratification under various impacts need to be intensified. Any technological activities in marsh-ridden areas and their vicinity should be examined from the point of view of possible impacts on the aquatic ecosystem. Hydrological observations therefore need to be extended by monitoring of biological processes and changes in economic, human and social aspects.

Each water resource project must be based on adequate data, and if these are not available, a solid temporary network should be established as part of the preparatory work for the scheme.

Basic and applied research related to specific problems should be based on locally conducted experiments in intensively observed experimental and/or representative areas.

Problems of training, manpower and resettlement should also be clarified during the preparatory stage.

17. Dry warm flatlands

Occurrence and general characteristics

Definition. An idealized flatland could be represented by a horizontal area without any runoff. If it is impervious, the rainfall remains on the surface until it evaporates. If it is sufficiently pervious, all the water infiltrates, and in the arid and semi-arid zone is evaporated later, as the potential evaporation exceeds the annual rainfall. In favorable circumstances, part of the infiltrated water may recharge an aquifer.

The real world is not so simple, but the hydrological characteristics described above may occur at a macro scale from land areas which slope locally. More generally, flatlands are characterized by a macro-scale response which is far from the conventional one in a sloping catchment, and is evidenced by a degeneration of the surface drainage network (Rodier 1964, 1985).

In warm arid areas, stream flow is infrequent and sediment concentrations are very high, for endogenous streams influenced only by the local climate. In these conditions, small watercourses with a slope between 1 and $3m$ km^{-1} usually disappear after a few kilometers.

For exogenous rivers entering arid plains from humid areas, the degeneration of the drainage network begins at slopes of the order of magnitude of $0.1m$ km^{-1}. These two sets of limits, with intermediate situations, define an upper slope limit for flatlands in dry warm areas.

Occurrence. While the dry warm areas in general are shown in Fig.11.1, a separate presentation of dry warm flatlands on a world map is infeasible, because of the countless small plains more or less randomly distributed in the lower part of many watercourses in the hyper-arid, arid and semi-arid zones. To show these areas would require a scale of 1:200 000. The problem is similar to that of exclusion from the warm zone of mountainous areas with an annual potential evaporation below 1000mm.

The larger areas covered by this chapter are as follows:

Asia: The Dead sea (Israel, Jordan), the Tigris-Euphrates plain (Iraq), marshes and lakes of the Sistan area (Iran, Afghanistan), lower valley of Indus (Pakistan), great and little Rann of Kutch (India).

Africa: Lower valley of Nile (Egypt), Sudd marshes on upper Nile (Sudan), Lake Turkana (Rudolph) and its catchment (Kenya, Ethiopia), Lake Chad and its catchment (Chad, Cameroon, Nigeria, Niger), internal delta of Niger (Mali), lower valley of Senegal river (Senegal, Mauretania), Kalahari desert (Botswana).

North America: Central valley (California), Great Salt Lake (Utah).

South America: Marshes near Bermeja and Pilcomayo rivers, Gran Chao (Argentina, Paraguay).

Australia: Lake Eyre and its basin, the Simpson desert, the Great Sandy desert, the Nullarbor plain, the West Australian plateau, and several other large plains (see map

371

in Australian Water Resources Council 1972).

The above areas, though having a diversity of hydrological characteristics, do not include chotts, which are not so large but may be significant in relation to water resources. In Africa the most important chotts are chott Dierid (Tunisia), and chott Melrhir and chott ech Chergui (Algeria).

Morphological and hydrological characteristics. As noted above, the spatial scale of topography is important in the hydrological definition of flatlands. Most flatlands have micro depressions at areal scales of some cm^2 or dm^2. At the meso scale, very shallow depressions may cover $500 m^2$ up to $1 km^2$. At the macro scale, in playas, salt lakes and some plainlands, the depressions may cover many km^2. Several desert formations may be classified as flatlands at the macro or meso scale: the plains of regs with an impervious surface of mixed gravel and clay; the ergs; sand dunes, generally very pervious; and the hamada, a plain of blocks.

The general characteristics of plant distribution and growth in dry warm flatlands are similar to those described in Chapter 11 for areas with catchment response, but the dominant species differ, and perennial grasses in particular are more likely to occur on the flatlands.

For permeable soils, rainfall enters the plant root zone and is later transpired by the vegetation or evaporated from the soil surface. It remains in the plant root zone for a period related to the depth of the rainfall event, the aridity of the climate, and the season. A small part of a high rainfall may penetrate beyond the plant root zone and continue to move slowly downwards until it reaches the water-table. This is a feature of sand dunes, where the soil at the bottom of the dunes is observed to be damp for several weeks after a high rainfall. Groundwater recharge is typically of the order of $1 mm \, y^{-1}$, and the time taken for water to reach the water-table is very long, often of the order of hundreds or thousands of years (Allison et al 1985).

In the case of impermeable soils, most of the rainfall is stored in micro depressions. If the rainfall is sufficient, these small depressions overflow and water moves slowly towards the larger depressions. At the meso scale there is no overland flow to streams or lakes, but a small increase in the slope may induce overland flow and streamflow in the poorly organized channel network.

Impermeable soils are relatively frequent in tropical areas. The kinetic energy of the rainfall creates an impervious film (surface crust) at the soil surface, even in some sandy soils. This film inhibits further infiltration and allows water to pond on soils for which laboratory studies indicate a significant permeability. The surface of the soil dries very quickly after rainfall, and the small amount of water which has penetrated the surface film is also evaporated.

The hydrological characteristics of flatlands with exogenous inputs will be described later.

Water balance components

Precipitation.   With the exception of orographic effects, dry warm flatlands have the precipitation characteristics described in Chapter 11 for areas with catchment response.

Evaporation.   In arid areas, the long term actual evaporation E is far below the potential evaporation E_o, because of the lack of available water for most of the year. It is close to the potential evaporation for periods during which water is lying on the soil surface or the surface is close to saturation. These periods are generally short (hours to days), but may be much longer or even permanent when there are exogenous inputs, as in interior deltas and chotts.

In the arid and hyper-arid zones, the annual evaporation E is practically equal to the annual precipitation P_y.   In high rainfalls, small amounts of recharge can occur, as reported by Colombani (1978) and Vachaud et al (1981) for areas in Africa with no permanent vegetation, and by Chapman (1961), Allison et al (1985) and others for areas in Australia with various soils and types of natural vegetation.

Fig. 11.3 shows the variation with latitude of both potential and actual evaporation in central Africa. The interannual variability of the potential evaporation is relatively low.

In sub-tropical areas, the seasonal variation of potential evaporation is characterized by a maximum monthly value in summer which may exceed 200*mm* . For tropical areas, the maximum is just before the rainy season. At the boundary between the tropical and subtropical zones, often in desert, the rainy season is very short, and the maximum potential evaporation occurs in summer, often reaching 300*mm* per month.

Owing to the spatial variability of rainfall, water may pond on the surface in relatively small areas at a particular time. In these circumstances, the rate of evaporation may be considerably higher than the potential evaporation, due to advection (the "oasis effect").

Infiltration. Except in the sub-humid zone, infiltration into impermeable soils is 15-50% of a heavy rainfall (say 50-80 mm). Part of the water accumulated in micro depressions infiltrates, and part evaporates, at rates of several $mm\ d^{-1}$. The water consumption of plants increases with each input of infiltration, and the roots of trees and other deep-rooted plants may exploit the water to depths of 10m.

Impermeable soils cover large areas because of the formation of an impermeable film (Valentin 1981), which transforms soils, which in other climates would be relatively permeable, into soils which have low infiltration rates. This occurs mainly in areas with sparse vegetation and where there are some fine grain sizes in the composition of the upper soil layers. Even ploughed soils become impervious after several rains. The phenomenon does not occur in sub-humid areas where plant cover impedes the formation of a surface crust.

Except again in the sub-humid zone, there is more infiltration into pervious soils, but virtually all the water at the micro and meso scales is removed by evapotranspiration. The main opportunity for direct groundwater recharge is where there are fissured rocks (which occurs only in sloping areas) and in some subtropical soils where the rainfall occurs in the cold season. Opportunities for recharge are greatly enhanced where there is endogenous input from sloping land in the same climate, which concentrates the runoff before it reaches the flatlands, or where there is exogenous input from humid zones. In desert areas, this only occurs in exceptionally wet years.

A striking example of an endogenous input is the Korama basin in the semi-arid area of Niger. Here a perennial river is fed by an aquifer recharged by percolation from the pervious soils in the flatlands and the surrounding sloping areas. This is a transitional case of an endogenous interaction.

The situation is different in sub-humid areas, as the soil is often pervious and the higher rainfall increases the probability of groundwater recharge in favorable geological conditions, in spite of higher water consumption by the more dense plant cover.

In considering infiltration at the macro scale, it should be noted that, for most of the large flatland areas listed above, the morphology of the plains and surrounding areas corresponds to wetter conditions than have existed for the last 2000 - 3000 years (the Chad basin for instance was partly an internal sea). This accounts for the negligible slopes, large areas of clay soils, and the frequent presence of high concentrations of salt in the soil.

Surface storage.   For relatively pervious soils there can be significant storage of water on the surface during periods of high intensity rainfall, and for a short period after. For less pervious soils, the scale of the depressions is significant, as described earlier.

After rainfall, the micro depressions dry up very quickly. At the meso scale, depressions with water depths of 50-200mm dry less quickly but nevertheless rapidly. In tropical areas the water is warm, and as the proportion of the land surface covered by water is small, the advective effect is important. Evaporation rates are higher than from large deep reservoirs, and may reach 10$mm\,d^{-1}$.

In subtropical areas the water remains on the surface for a relatively long time in winter. This occurs frequently in semi-arid areas, and also in arid and even hyper-arid areas during exceptionally wet periods. As desert trails often follow the flat clayey land, travelling in these areas is very difficult at such times.

At the macro scale the depressions are larger and deeper, and the daily evaporation rate is lower. There may be some surface drainage during periods of exceptional rainfall, through depressions which are covered by vegetation.

However, as a general rule runoff in dry warm flatlands occurs only under endogenous or exogenous influences.

Groundwater.   Groundwater is a very important component of the water resources of dry warm areas (Jacobson and Lau 1983). There is no space here for a review of all the main aquifers of this very large region, but the characteristics of the main types will be described, with some examples.   Unesco/UNEP/FAO (1979) gives a short review which covers most of this area.

Some parts of what follows applies also to some areas with catchment response, as the same aquifer may underlie alternating sloping lands and flatlands. Except in the sub-humid zone, direct groundwater recharge from rainfall is infrequent, but important inputs to aquifers come from the intermittent rivers of areas with catchment response and exogenous rivers originating in the humid zone. Such inputs may also be significant to the equilibrium of fossil aquifers, which can be very extensive.

The five categories of aquifers described below follow the Unesco/UNEP/FAO (1979) classification, with slight modifications towards a classification developed by Margat (in press) for groundwater in Africa.

375

Shallow aquifers. Alluvial aquifers may be important when a watercourse overlies a recharge zone thick and permeable enough to constitute a groundwater reservoir. Particularly favorable conditions may occur in the distributary zones (deltas) frequent in the lower reaches of watercourses. Although these watercourses are dry most of the time, the aquifers may be recharged by one or two floods, particularly in exceptionally wet years. The smaller aquifers of this kind may be exhausted before the arrival of the next recharge event; in such cases, enhancement of the recharge by low surface dams and/or reduction of groundwater outflow by subsurface barriers may provide a perennial water resource.

Crystalline rock and Precambrian areas. Although these formations cover a considerable part of the arid zone, their water resources are poor because, under normal conditions, they are compact and devoid of porosity. Aquifers can be found only in the overlying weathered formations, in networks of fissures, and in faults and dikes. Sometimes, for example in India, there are horizontal zones of fractures which form significant aquifers. Generally however the yield is poor, as indicated in Table 17.1.

Primary areas. Considerable well production can be acheived from aquifers of interbedded limestone with a network of fissures, or dolomite or porous sandstone. In the Northern Territory of Australia, some wells supply up to $90\,m^3h^{-1}$. The water quality is often acceptable .

Volcanic and karstic areas and generally folded or dislocated chains. These very different formations may contain groundwater, principally under sloping land. Some volcanic rocks may contain considerable quantities of water in their vesicular cavities.

Karstic areas are relatively frequent in the subtropical zone. They sometimes feed springs, but there is typically a considerable diminution of flow in drought years. In most cases the aquifers are small and the recharge irregular in arid and hyper-arid areas. A particularly large system underlies much of the Nullarbor plain in Australia; the groundwater is discharged at the coast of the Great Australian Bight.

Large sedimentary basins. These underlie wide areas in western Africa and in Australia they contain continuous and often important aquifers.
The most extensive sedimentary basin in Africa is the interbedded continental aquifer of the Cretaceous period. It is continuous from Tasudeni to the Sudan, and extends across tropical and subtropical arid and hyper-arid zones. It is unconfined at some of the margins, but the central part is confined. Its clay sandstones have quite high permeability, and flow from wells is high. Although there is variable mineralization, the quality of the water is generally good. The main body of water is

376

mineralization, the quality of the water is generally good. The main body of water is fossil, but wadis on the margins occasionally recharge the aquifer to a small extent. While this resource is permanent and independent of droughts, overuse will of course reduce the head.

Another very substantial aquifer is the Maestrichtian sand, which extends over more than 100 000 km^2, and is several hundred meters thick in Senegal. Flows of 50-120m^3h^{-1} of good quality water can be obtained from wells. Similar aquifers are used in northern Nigeria (see Table 17.1) and in Chad.

Table 17.1

GROUNDWATER RESOURCES IN NORTHERN NIGERIA

Aquifer	Lithology	Thickness (m)	Area of outcrop (km^2)	Average SWL# (m)	Range in yield (ls^{-1})
Chad formation (Pliocene)	Clays, silts, fine sand with beds of sand and gravel.	100-200	9 200	36	1.7-5.0
Sokoto group (Paleocene)	Clays and shales with a fissured limestone formation.	90	n.d.[*]	n.d.	n.d.
Rima group (Maestrichtian)	Sandstones and siltstones with interbedded shales.	250	6 800	·.6	2.2-4.4
Gundumi formation (Maestrichtian)	Coarse sands and gravels. Some clays.	n.d.	8 200	14	1.6-3.3
Basement complex	Weathered basement mantle or jointed and fractured crystalline rocks.	100	12 200	13	0.0-0.3

#SWL - Static water level (below surface)   * not determined

In Australia the most important sedimentary basins are the Great Artesian basin, the Canning basin and the Carnarvon basin (Hahn and Fisher 1963).

The Great Artesian basin is a sandstone aquifer from the Triassic to the Cretaceous period, with a thickness up to $2100m$. It covers 1 700 000km^2, principally in the tropical and subtropical parts of Queensland. It is recharged from streams along the eastern and northern margins, where the annual rainfall is much higher than in the center of the basin. Its hydrology and use have been described by Habermehl (1987).

The Canning basin is less important; it consists of a thick sequence of formations in Western Australia, and is principally exploited near the coast. The Carnarvon basin, also in Western Australia, extends further south along the coast.

There are also large sedimentary basins in India, California and Peru, but not as extensive as the largest ones in Africa and Australia.

Endogenous interactions

This section will consider the influence of adjacent sloping areas on the hydrology of flatlands in the dry warm regions. On flatlands not subject to external influences, the accumulation of sufficient water will cause flow between surface depressions, first at the micro scale, then at the meso and macro scales if the water supply is sufficient. The flow paths between depressions may take the form of shallow unstable channels, or it may occur as sheet flow. This surface movement of water, which can be thought of as the beginning of runoff, is very sensitive to influences such as the spatial pattern of storm rainfall, wind direction, plant growth, the processes of erosion and sedimentation, and human influences.

When watercourses on sloping land, with a well defined channel network, flow on to flatlands with these characteristics, an intermediate situation develops between a fully organized drainage network and the flatland situation described above. This phenomenon has been termed "hydrographic degeneration" by Rodier (1964, 1985).

Hydrodraphic degeneration. The characteristic feature is that the continuity of runoff is no longer obvious, nor the continuity of the channel itself, and losses of water from the main channel are very important.

A simple case is that of a watercourse which, before it reaches the flatland, has distinct banks with often a narrow flood plain with thorn trees. On reaching a lower

378

slope the stream bed becomes unstable, producing a lowering of the banks and increasing the width of the flood plain, often with one or more depressions detached from the main stream bed. The runoff generally ends in a clay based depression which receives the water and sediments carried by the watercourse; sometimes a short delta is formed.

Before reaching this stage, the channel may join a major depression with very low slope, which collects the input of several such streams. With sufficient input the flows of these tributaries may travel in both directions along this main depression, filling a series of pools. After sufficient storm events to fill these pools, runoff from the whole area continues in a downstream direction.

If the slope of the main depression is low but significant, it may carry continuous runoff from the beginning of the rainy season and have a relatively large flood plain, several discontinuous channels and many small pools. The flow finally terminates at the point where it cannot be sustained by the total input from the tributaries. At the limit of the arid and semi-arid zones these conditions may permit the formation of a river which is fed successively from different parts of a large catchment with an area of the order of 50 000km^2. The Ba Tha river in the Chad basin is an example; for two to three months each year it flows into a small lake (Lake Fitri), and it dries up for the remainder of the year.

In the hyper-arid and arid zones, where there is a fossil channel network resulting from the earlier wetter period mentioned above, the situation is more complex; the river may use part of an old bed for one flood, and another one for the following flood.

Hydrographic degeneration is characteristic of low slopes in hyper-arid, arid and semi-arid areas, but is less common in sub-humid areas. It results from the following factors:

(i)   A long period without rainfall when the vegetation disappears and the soil on the slopes becomes bare, without any protection against erosion. The water reaching the flat lands has a high sediment concentration.

(ii)   Flows for a very short duration which may not be long enough for the maintenance of a continuous channel, particularly in flatlands where the potential energy is very low and the runoff has a large sediment load.

(iii)   Often a fossil drainage network in very flat areas where flow becomes more and more difficult.

379

Consequences of hydrographic degeneration. When a stream from sloping land enters a flatland, it develops a channel which results in some runoff from the flatland as well as the runoff from the sloping land. The resulting runoff however has little relation to catchment area, and values of specific discharge, measured in $m^3 s^{-1} km^{-2}$, are meaningless.

In addition, when the potential evaporation is high, the runoff becomes very low soon after leaving the sloping areas, even in favorable circumstances (for example in tropical areas where the rainfall is concentrated into a wet season). A general study of annual runoff in the tropical Sahel (Rodier 1975) gives an average annual runoff of 15 mm for an area of $2000km^2$ with an annual rainfall of $500mm$. In the very favorable conditions of the Ba Tha river the annual runoff is $10mm$ from a catchment area of $4500km^2$ and the same annual rainfall. As in sloping areas, the distribution of values of annual runoff is generally skewed.

Often the flow stops completely after a distance related to the magnitude of the discharge reaching the flatland area, and does not reach a drainage terminus in a lake or the sea. This is typical of the arid and particularly the hyper-arid zones.

The duration of surface storage on the flatlands remains short, though longer than it would be without the input from the sloping land. There is an increase in the depth of infiltration, which may result in groundwater recharge in arid and even some hyper-arid areas.

As a result of the input from the sloping land, the evaporation from the flatlands may be higher than the local precipitation.

Exogenous interactions

This section discusses the exernal inputs from rivers rising in wetter regions, such as the tropical humid zone, or from mountainous areas in the same general climate which have much higher rainfall due to the orographic effect. Most of the world's large flatlands listed earlier benefit from this external influence.

General hydrographic features. Because of their large mean discharges, most of the large rivers from humid areas are able to cross arid areas (Fig.17.1) and reach the sea. However in their passage through the drier areas they are affected by a different form of hydrographic degeneration, which results in very large flood plains and many channels which leave the main branch. These distributary channels often do not join the main branch again, and the water is evaporated from ponds, lakes and swamps. There are several different hydrographic patterns:

380

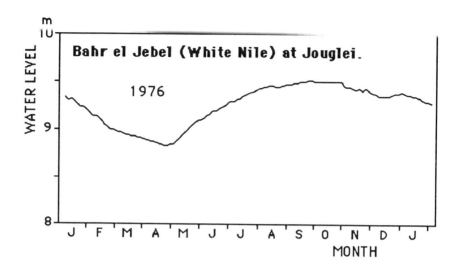

Fig.17.1 Water level of the Bahr el Jebel at Jouglei in a typical year,
showing the regulating effect of Lake Victoria on low flows.

- The rivers flow in a wide flood plain with secondary branches and lakes (e.g. Senegal river, Indus and lower Nile);

- The common flood plain of two or more rivers becomes very large, with widths reaching 200-400*km*. Many large channels are relatively stable, and there may be large lakes which fill or empty with long wet periods or droughts. Examples are the Tigris and Euphrates, the Chari and Logone in the Chad basin, and the internal delta of the Niger. The water of the Chari flows into an internal depression whose southern but not deepest part is Lake Chad. The lake area ranges from less than 2000 to 26000*km²*. In very wet years the water overflows eastwards towards a depression 400*km* away, but is stopped after 50*km* by evaporation and transmission (infiltration) losses.

- Chotts are flat depressions in the arid area of North Africa, which receive an important supply from groundwater. These chotts (e.g. Chott ech Chergui, Chott Tharsa, Chott el Hodna, Chott el Djerid) are very shallow and swampy, and contain high salt concentrations because of the geological conditions.

- The rivers flowing towards Lake Eyre in Australia exhibit a wide variety of drainage patterns (Kotwicki 1986), from single main channels to the complexity of the Channel Country, a large area of innumerable small watercourses which create a mosaic of small islands.

Influence on water balance components.   The water balance varies widely in space. There is a large difference between areas influenced by large permanent rivers with a flood for several months each year, and the margins of flatlands which receive only a small supplement of external water for two or three weeks once every five years or so. The annual evaporation from Lake Chad is about 2200*mm*, but it would be only 350*mm* for a flat area in the same climatological zone without external water input. For given geological conditions, the disparity for groundwater recharge is even greater.

The actual annual evaporation is generally higher in areas with exogenous inputs than for flatlands with only endogenous inputs. This is due to the longer duration of floods and generally shallow water-tables. The evaporation may approach the potential evaporation over large areas, and because of advective effects may exceed it over open water bodies.

The locally produced runoff may be increased in areas with exogenous inputs, as the rain falls on water or wet soil, particularly when the local rainfall occurs at about the same time as the flood in the large river. At the macro scale this effect is not significant compared with the external water supply. Losses by evaporation remain very high if we consider only the runoff measured upstream of the large basin. For instance, the Niger river loses about 48% of its runoff by evaporation while crossing its internal delta (Rodier 1985). While the volume of runoff recharging the aquifers is considerable, it is very small compared with the evaporation. The Bahr el Jebel (White Nile) loses 50% of its mean annual discharge of 840m^3s^{-1} while crossing the swamps of the Sudd (Sutcliffe 1974). The mean discharge of the Tigris in Iraq decreases from 1236m^3s^{-1} at Baghdad to 218m^3s^{-1} not far from the delta (Guilcher 1979).

These losses smooth the peaks of the annual hydrograph into a uniform curve (as for the Chari river in Chad), except in a very dry year. A flat maximum to the hydrograph is observed in some cases, e.g. the Logone, a tributary of the Niger (see Fig.17.2). Owing to the extreme width of the flood plain, an unusually high peak flow results in a very small change in water level in the main channel; the flood frequency curve for the Logone at Logone Birne has almost an asymptote at 950m^3s^{-1}. This explains the negative skew of many flood frequency curves in such areas. This is not as characteristic in distribution curves for annual runoff, because during wet years when the maximum discharge is not much different from the mean, the flat part of the hydrograph lasts for a longer time.

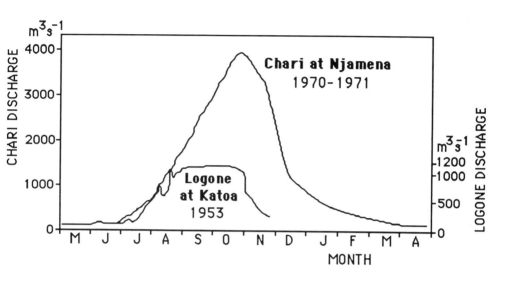

Fig.17.2 Discharges of the Chari in a typical year
and of the Logone in 1953.

In discussing frequency distribution curves it should be recognized that, while the channels of exogenous rivers are more stable than those of endogenous streams, there can be modifications of the pattern of channels or the longitudinal profile of the bed (Sutcliffe 1974), resulting in the distribution being shifted while retaining the same general character.

The runoff of exogenous rivers usually joins the sea by a delta, or in very arid zones a sebkha (a flat plain more or less connected with the sea, which is flooded only during the wet period).

Exogenous runoff brings with it much sediment, often very advantageous for agriculture. Under natural conditions, a large part of this sediment is trapped by the natural vegetation in swamps. Without any other external input, water often leaves flatland areas with little sediment but some organic matter; the resulting output is sometimes called the "black flood".

Exogenous runoff from humid tropical areas often originates in very old geological formations and therefore carries few solutes; this is the reason why the water in Lake Chad is relatively fresh. However, this is not true everywhere, particularly in subtropical areas, and the flatlands themselves may be a source of salt. Many flatland areas have salt problems. Because the external water supply increases interactions with groundwater, the phenomenon of salt rising towards the upper soil layers may be more frequent than in areas under endogenous influences only.

Other things being equal, the greater duration of depression storage and the large inundated area should provide excellent conditions for infiltration in areas with exogenous inputs. However, the deposition of fine-grained sediments may create impervious layers which impede downward water movement. On average, infiltration is higher than in areas only under endogenous influences, but for a given area it depends on the geological and morphological conditions.

All these aspects are very sensitive to human influence, as will be shown later.

Current pattern of land and water use

General considerations.   Before discussing particular forms of land and water use, it will be useful to consider the hydrological characteristics which determine the situation relating to the two main problems of water use:

-       lack of water (low flow or dry season period)

-       excess of water (disastrous floods, waterlogging)

These problems will be considered mainly in relation to areas with external inflows, which are frequent in flatland areas and allow much more economic development.

The relevant hydrological characteristics are:

-       the annual precipitation, its interannual variability (which determines the drought hazard), its seasonal distribution in relation to crop cultivation, and maximum daily values in relation to erosion flood hazards, particularly in areas subject to tropical hurricanes;

-       the annual volume of runoff produced locally in sloping areas or externally in humid areas. The first is less important and has a very irregular time distribution, with a series of floods on occasions and also the possibility of no runoff for some years. The second is much more important; all the flat areas may be flooded for a duration which has a statistical distribution with negative skew. The distribution of flood magnitudes also has a negative skew and almost an upper limit. The water-level regime is also ver important for irrigation, as are the characteristics of the minimum flow.

384

In order to manage the water resources in flatland areas it is essential to understand the operation of the complex hydrological system of rivers, channels with or without flow, lakes, reservoirs and swamps; the fluxes between components of this system under various conditions; and possible evolution of the system and its flows. In addition, consideration must be given to the possibilities of water supply from aquifers, taking account of aquifer depth, recharge rates and groundwater quality. Good knowledge of water quality, particularly salinity and chemical composition, is essential for water resource project management and operation. The nature and volume of sediments may also have a strong influence on the development of land and water use.

There is a wide range in the physical conditions in flatlands. In areas such as the lower Nile, the Tigris-Euphrates and the Indus, conditions were found to be extremely favorable for agriculture several thousand years ago, with flat fertile soils, ample water and the possibility of controlling it, easy water transport, and the possibility of establishing cities protected by water bodies and swamps. At the other end of the scale of land use are areas such as the lowest part of the Chad, a flat depression with brackish water near the surface, in which a few wisps of grass at intervals of $50m$ constitute a good pasture for camels.

Socio-economic conditions vary as widely as the physical conditions. The over-populated lower Nile valley with its agricultural traditions and transport facilities is very far from the Chari-Logonne system with a population density of $1-3p\ km^{-2}$, where the inhabitants generally do not practise agriculture and there are serious difficulties in the export of goods.

There are also great differences in the methods of water use, from maintaining natural hydrological conditions to flow modification by primitive structures and to the sophisticated methods of large projects. Each method may correspond to an optimal adjustment to the prevailing physical and socio-economic conditions. There are also large differences in the stage of development and operation of water systems; some have been used for so many years that most of the land is useless for cultivation, while in others the impact of man has been negligible up to this time.

In view of this diversity, only some examples of land and water use will be given. As the availability of water is more tied to the existence of external inputs than to geographical zones, water problems will be considered in relation to the different types of water use.

Stock grazing. Extensive stock grazing generally uses the driest areas of the dry warm flatlands, with the exception of the hyper-arid. It is also common in semi-arid areas, but is prevented by some diseases in parts of the sub-humid zone.

Stock need both food and water, and cannot graze beyond a certain range from watering points. If water is available at too few points, over-grazing and trampling in these areas makes them unproductive. Effective management therefore requires increasing the number of watering points, by diverting local runoff into surface ponds or cisterns and/or by constructing wells or bores. As evaporation losses from surface storages are considerable, increasing the water depth in at least part of the pond (as has been done in the Hafir area of Sudan) can be effective. Various ways are used to raise groundwater to the surface, from primitive methods using donkeys or oxen as a power source to windmills and electric pumps. Maintenance of mechanical and electrical equipment may be a problem in some areas.

The water requirement for stock grazing is less than that for agriculture, but during drought periods there is a tendency for stock to be moved towards more humid areas near permanent rivers or an internal delta.

Agriculture.   With the exception of dryland farming, which is not possible everywhere in the dry warm areas, agriculture in this region is determined by the need for water. In some depresssions in flood plains it may be possible to cultivate crops without modifying the water system, but irrigation of crops is the general rule. In some areas, water from mountainous areas may be used with simple techniques, as in the Mzab area in Algeria, but generally a more or less complex system of channels is necessary, and regulation of river flow upstream of the flatlands is also often required. In some areas, such as the Central Valley in California, water may be transported by artificial channels from more humid areas.

Water availability in irrigation areas may be a problem of water-level as well as one of volume of supply. For instance, in the recent drought in the Sahel the irrigation of rice failed because water levels were too low; although there was adequate river flow, no pumping plant was available.

Useful water supplies in irrigation areas can be improved by:

(i)   increasing the water supply by methods such as construction of storage reservoirs, diversion of water from other catchments, and using ground-water conjunctively with surface water, possibly with the provision of artificial recharge;

(ii)   reduction of wasteful use of water by good maintenance and the use of water- economizing irrigation techniques such as drip irrigation.

Navigation.   Where large rivers cross dry flatlands, there are often excellent opportunities for economic transport of goods, including agricultural products and stock.

Mining and tourist trade.   These have been briefly discussed in Chapter 11.

Urban water supplies.   Urbanization in flat areas presents two water-related problems: water supply and protection against floods.   During the drought in the Sahel, the city of Niamey in Niger used half of the flow of the Niger river for a few days in 1974, and the river flow stopped completely for a short period in 1985. Water supply systems have to be constructed that can cope with such situations, and this may involve diversion of water over long distances.   The system should also be designed to provide for the needs of industry in urban areas.

For flood protection, many villages in swamp areas have been built on hills, while others have constructed dikes.   Fortunately the difference between water levels in average and extreme floods is not very high in flatland areas, but protection may be required against floods which inundate plains that remain dry for several years at a time.

Hydrological impacts of human activities

Past effects.   Human activities may influence the natural hydrology of an area in three ways: construction of large structures (dams, large channels, roads etc), agricultural and grazing practices (including some small structures), and urbanization.   This influence may have a major impact on the natural system particularly when it continues for centuries on a large part of a flatland area.

Many flatland areas which some centuries ago were swamps of low economic value are now areas with a high agricultural yield, after development and proper management of a good network of drainage channels.   Some also make use of old irrigation schemes based on surface or groundwater resources.

Unfortunately, if the principles of soil and water conservation are not respected, the influence of these human activities may be disastrous.   Particularly in warm areas, many flatlands which were good pastures have been over-grazed, resulting in a change in the floristic composition towards less palatable and often less nutritious plants.   In more arid areas this has resulted in the onset of desertification, with changes in the hydrology at the micro scale, usually involving less infiltration, more surface runoff and, in sloping areas, more erosion.

Flatlands may suffer two serious consequences of irrigation, salinization of the soil and land subsidence. In the first case, evapotranspiration causes a capillary rise of groundwater through the unsaturated zone, leading to accumulation of salt at and near the surface and sevare constraints on plant growth and survival. The soil surface may become less pervious, reducing infiltration. The effect can be prevented by a well managed network of drainage channels, but these should be established a priori, as salinization effects are difficult to reverse. As a result of salinization, some parts of Mesopotania and the Indus valley can no longer be cultivated, and agricultural yields have decreased in many flatland areas that have been used for centuries.

In irrigation areas, extraction of groundwater at a higher rate than it can be recharged causes a lowering of the water-table. In fine-grained soils this leads to land subsidence, which may change the complex natural drainage system of flatland areas.

It is seldom that roads and railways, and particularly bridges over waterways, are constructed without having some effect on the complex hydrological system of flatlands, specially during floods. Although the effect may be beneficial, frequently there is an increase in the area affected by floods, and there may be a reduction in groundwater recharge.

The effect of urbanization is to increase the area of impermeable or less permeable land, and so reduce infiltration overall. This increases the volume of surface water which must be removed by drainage under conditions where there is little gravitational head to cause flow. These conditions also make it difficult to dispose of sewage, and frequently the groundwater in urban areas becomes polluted as a result.

Potential for future effects. In recent years the characteristic feature of flatlands, specially in warm areas, has been the increasing need for food resulting from an increasing population density. While this situation has been catastrophic for many areas in the subtropical zone, it has had less impact in tropical flatlands in Africa and Central and South America, with lower densities of population. Even in these areas, populations are increasing and it will be necessary to develop agriculture for more intensive production.

This section will cover five types of human activity which can have a serious impact on the hydrology of a flatland area: desertification, salinization, lowering of water-tables, pollution, and the effect of structures. Erosion has already been discussed in relation to sloping areas (see Chapter 11). Increased sediment, which is its consequence in flatlands, is not always a benefit, as in the case of a large flood from an eroding area leaving a meter or more of infertile sand and gravel on an irrigation area. Such problems are however more restricted in their areal impact than the five categories listed above.

Desertification is a consequence of over-grazing, often by goats, which is greatly accelerated by severe droughts. As the plant cover is reduced, the top soil is eroded and sorted by wind action, and the end result may be the development of sand dunes or clay plains which are often saline.

The process of salinization has been explained above and in Chapter 5. In warm dry flatlands the risk of salinization is high if one or more of the following factors is present: significant salinity in the irrigation water, a shallow saline aquifer, saline bedrock below the flatland sediments, or an inadequate drainage network. It is essential that there should be continuous monitoring of the salinity of both the irrigation water and the drainage water. In arid areas there is seldom the opportunity to remove soil salinity by application of large volumes of fresh water, as occurs in rice fields. The crops and cultural practices should suit the chemical composition of the soil water and the soil structure. The process of salt accumulation is a major risk for irrigation development in the future.

Increased use of groundwater for irrigation is developing rapidly, and there is an increasing danger of excessive lowering of water-tables. Some lowering of the water-table may be beneficial, as it may increase the natural recharge, but in general there will be a need for artificial recharge of groundwater which is being over-exploited.

In flatland urban areas, the problem of wastewater disposal, and the avoidance of pollution of groundwater and even surface water, is probably more critical than the problem of disposal of flood water. Increased individual water use and increases in industrial use of water can cause serious impacts even when the population density is not increasing. Treatment of wastewater to the point where it can be re-used, either locally or downstream after a period of natural purification, should be considered as an alternative to importing more fresh water into the urban area.

The possible effects of road and railway structures have already been mentioned. Hydraulic structures (dams, canals, irrigation systems etc) may have both immediate and subtle impacts on the hydrological system. Changes in the amount and nature of the sediment downstream from such structures can have long-term effects on the morphology of the channel system, and it may be many years before a new stable water and materials balance is achieved. Such long-term effects on the ecosystem are often difficult to predict, but an attempt should be made to identify possible consequences; if they cannot be ameliorated by changing the hydraulic design, monitoring systems should be established to give early warning of trends and their magnitude. Unfortunately, as a result of other conflicting interests, the best technical solution is not always adapted.

Implications for hydrological data collection

For rainfall and groundwater data, the main difference from sloping areas is the difficulty of access in flatland areas during wet periods. In arid areas, groundwater studies should place emphasis on water quality and the process of groundwater recharge.

Due to advective effects, the evaporation from pans will depend on the wetness of the surrounding area, so that information on local conditions is essential for analysis of pan data.

The main impact of flatlands on hydrological data collection however relates to measurements on surface water, which are completely different from those used in sloping areas with catchment response.

With no external inputs, the first surface water characteristic to be studied is the area which is flooded, the depth at selected points, and if possible the direction of flow under different conditions. More detailed flow measurements should be made at one or more characteristic depressions at the meso scale. Where flows occur at the macro scale in extreme events, measurements should attempt to identify the direction and magnitude of the flow.

With a significant external input, there will be a reasonably well defined channel crossing the flatland area, and due to its importance for water management the following observations should be made: downstream discharge in the main channel and floodplains, lateral flow (and its direction) between the channel and the floodplain, water-levels in depressions and lakes, slope of the water surface, changes in channel and floodplain morphology, and the influence of drought and vegetation on the hydrological regime. A long-term hydrometric station is required to monitor seasonal changes and long-term trends.

The surface water system should be sufficiently observed in low, normal and flood conditions, to enable the development and calibration of a hydraulic or mathematical model, possibly including sediment transport and water quality. The operation of such a network by traditional methods is difficult; there are problems of access, of having competent observers at the right location, of maintaining recorders, and of measuring channel losses. In an ORSTOM study on the Logone river, one set of loss measurements extended over a width of 30*km*. In spite of the low slope, the bed at even main hydrometric stations may be unstable, as in the upper Nile and the Logone. In these circumstances even slight movements due to tectonic activity can modify the natural water system.

Many of these problems are resolved by remote sensing and remote data transmission. The relative absence of clouds in dry warm areas provides favorable conditions for remote sensing, but the requirements for efficient maintenance may inhibit the use of remote data transmission.

Effective use of remote sensing for hydrological purposes in dry warm flatlands requires a spatial resolution not exceeding 50*m*, repetition of coverage at frequent intervals (preferably less than 15 days), and a suitable observation network for ground truth. The last is particularly important in flatland areas, because of the diversity of surface conditions. For example, on Lake Chad the following elements must be unambiguously identified: free water, fixed islands of papyrus, mobile islands of papyrus, water areas more or less covered by reeds, grass plains with saturated soil, etc.

Although the value of remote sensing and remote data transmission has been demonstrated in dry warm flatlands, few of the large areas listed earlier make operational use of remote sensing. There is some use of remote data transmission.

Management of land and water resources

In the drier areas or where the soil is too poor for agriculture, economic activity in dry warm flatlands is confined to stock grazing (and sometimes mining or tourism).

However these flatlands are often contiguous with sloping areas which pour their runoff on to the flatlands. In these circumstances, conditions are favorable for agriculture, with low soil erosion and a topography suitable for irrigation, provided there is sufficient water, good soil, and people who are prepared to undertake agriculture rather than be pastoralists.

A major difficulty is the variability from year to year of the runoff from sloping arid lands, and particularly the phenomenon of persistence of both wet and dry years. This can be ameliorated by the construction of dams where sites can be found in sloping areas, but this is not always possible or fully effective. To combat drought efficiently, it is necessary to have an integrated approach covering food supplies, land and agricultural management, and social and political issues (Tixeront 1979).

The situation is better in flatlands receiving runoff from more humid areas. The statistical distribution of the runoff is less skewed, and the consequences of droughts and extreme floods are not so serious. The difficulty is often to obtain sufficient understanding of the complex water system in flatland areas.

In managing land and water resources in dry warm flatlands, the hydrology is only one component, but it is an essential component. Disasters will occur in the future, as they have in the past, if this is ignored.

18. Small flat islands

Introduction

The origins of islands and broad global patterns of climate affecting maritime environments generally have been discussed in Chapter 12, which deals primarily with the hydrology of high islands. Low islands are predominantly of limestone and the main concentrations are in the tropical and subtropical regions due to the constraints on the growth of coral, the main reef-building organism, which requires warm sea temperatures. Individual islands are generally of small size, rarely exceeding 500 and frequently less than $100 km^2$. Limestone islands of more recent formation tend to have very low elevations, often no more than a few meters above sea level. Older islands may have been tectonically uplifted or affected by previous sea-level changes with elevations which may reach several tens of meters but rarely more.

Water balance components and characteristics

Rainfall, soils and vegetation. The absence of orographic control makes low islands dependent on the rainfall associated with the naturally occurring upward moving air systems such as occur at the ITCZ and with cyclonic depressions, both tropical and those derived from high latitudes. Periods of drought both seasonal and longer term are typical of low islands, increasing in degree and frequency with lower mean annual rainfall. The islands in the equatorial belt of the central Pacific region demonstrate the feature (Figs.12.4 and 12.5; Table 12.2). Convectional rainfall which develops in consequence of the heating of the air above land masses will tend in islands to fall either upon the leeward side of the prevailing wind direction or in the sea beyond. Convectional rainfall is usually very localized and variable in amount. The low elevations also make such islands very susceptible to the effects of tropical cyclones, both of the winds and the high seas. Apart from damage to vegetation and human habitations, extensive soil salination may occur which will require leaching to recover.

Low islands of the higher latitudes are mainly continental and formed of hard older rocks with soils, vegetation and climate corresponding to the adjacent mainland. In limestone islands, the rock is easily soluble in weak acid (rainfall wigh carbon dioxide in solution) and leaves little residual material. In consequence, other than where solution has continued for lengthy periods without removal of the residues, soil mantles over limestones are thin and stony or non-existent. Thicker residual soils are loamy and of a lateritic nature. In low latitudes, vegetation may range from tropical forest to desert scrub. For the Bahamas, Holdridge (1967) has developed an ecological-climatic classification which uses a combination of the temperature, rainfall and potential evaporation regimes. The vegetation of small low islands exhibits varying degrees of adaptation to high salinity environments, which may be associated with tidal flooding, intermittent sea spray or the presence of shallow brackish groundwater.

Surface runoff. The low land gradient constrains surface runoff and the effect is componded when the underlying formations are of high permeability such as recent limestones with thin soils. Sheet runoff is promoted during intense rainfall events and particularly when the water-table is close to the ground surface. In calculating the water balance of the island of Kiribati in the South Pacific, Lloyd et al (1980) found it necessary to attribute 90% of rainfall as sheet runoff during periods when the monthly rainfall exceeded 260 *mm*. Direct measurements of sheet flow cannot be made with any accuracy and can constitute a difficulty for water balance calculations.

Groundwater. The groundwater body below a low island is comparable in origin to the basal aquifer in a high island, but since it generally represents the only source of potable water supply, other than from rainwater catchments, knowledge of its occurrence and response characteristics is more critical. In a low island, the fresh groundwater occurs typically as a lens shaped body (Fig.18.1). The thickness of the lens, which is related to the freshwater head in accordance with the Ghyben-Herzberg relation, is constrained by the low land elevation. Groundwater flux is also restricted by the constraints on hydraulic head and there is some thickness of the transition zone (Peterson 1984 a).

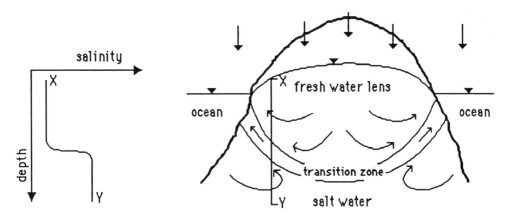

Fig.18.1 A fresh water lens in a homogeneous island
aquifer (from Barker 1984 b).

The importance of an understanding of the local controls to the movement of the
interface was referred to in discussing the basal aquifer in high islands. The
response is likely to be more critical still in the case of low islands where
groundwater is, in effect, the only naturally occurring freshwater source. In a
modelling study of the island of Anguilla in the West Indies (British Geological
Survey Internal Report Series) the lens system was simulated (by a mathematical
model) for a period of 280 days without recharge. The island has a maximum width
of some 5*km* and a maximum ground elevation of around 60*m*.   The lowering of
the water-table in the central lens area was 50-100*mm* which corresponds well with
observed data during similar periods.   The modelled rise of the interface during the
same period was very small, a few centimeters near the centre of the island and a
few meters near the coast. The lag in response depends on the permeability of the
aquifer within the lens and in the vicinity of the interface. In this case, the aquifer
was modelled as a two-layer system with a thin high permeability layer in the near
surface levels and low permeability below which extended to beyond the interface.
The presence of a layer of high permeability near the interface would substantially
increase the response time of the interface movement.   The importance of
monitoring the lens base, to assist prediction of changes in borehole water quality,
needs to be stressed.   An example of a progressive response to a longer term
trend is apparent in Fig.18.2 with data from an island in the Caribbean.   An
approximate position of the centre of the transition zone (taken as 50% sea water)
can be obtained by extrapolation of a probability plot of salinity versus depth for
which a dispersive distribution should result in a straight line.

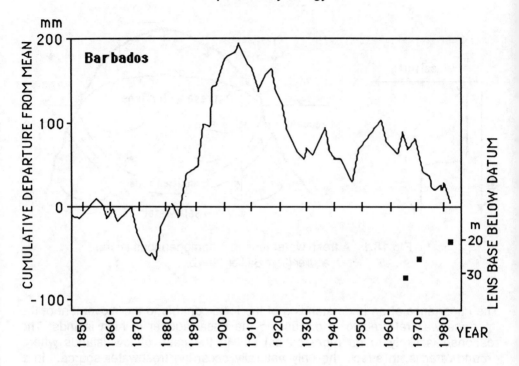

Fig.18.2 Cumulative departure from mean of annual island rainfall, and associated lens base movement (from Lloyd 1984).

The need for accurate information on recharge, storage and movements of the interface is particularly important in low limestone islands in view of the nature of the resource and limitations to its occurrence. The various methods of recharge calculation include a soil water budget, chloride balance of rainfall and recharge, baseflow analysis and flow net studies which include modelling. All these methods have constraints and limitations, some of which are emphasized in small island environments, and there is obvious value in attempting calculations by more than one method so that some indication of likely accuracy can be obtained by a comparison of results. Few very detailed studies on recharge have been made on small islands, mainly due to financial and institutional constraints, and even fewer have had the advantage of long term observed records. The need for good recharge data is particularly important for low limestone islands in view of the natural limitations on groundwater storage.

Estimation of groundwater recharge

Soil water balance. For any period Δt, the water balance for the unsaturated zone is

$$P - E_i - E_s - Q_s - R = \Delta V_u \qquad (18.1)$$

where
- P = rainfall in period Δt
- E_i = evaporation of water intercepted by vegetation
- E_s = evapotranspiration from the unsaturated zone
- Q_s = surface runoff
- R = groundwater recharge
- ΔV_u = change in soil water storage

The choice of balance period Δt is constrained by the availability of data for the various components, but if major seasonal variations are to be studied, monthly observations are minimal and daily would give improved reliability (Howard and Lloyd 1979).

Interception of rainfall by vegetation increases in significance in tropical regions (West and Arnold 1976) but it is difficult to quantify and it is logical to assume that some corresponding adjustment to the evapotranspiration term would also be needed. Values of 15 and 7%, which are of significant magnitude, have been used in studies in Male and Tarawa respectively (Lloyd 1984).

The standard methodology of a soil water balance is easily applied (Lloyd 1984) and an example is shown in Fig.18.3 . There are certain inherent weaknesses and difficulties in its application. The method assumes dispersed flow through the soil profile which commences and continues only when the soil water deficit is zero (at field capacity). Actual evaporation has to be adjusted at soil water levels below field capacity, and the various empirical methods used at present are not wholly compatible. The maximum soil water deficit will vary with soil type, thickness, and the type and root depth of the vegetation. Direct measurements to obtain these various parameters require careful observations at a sufficient number of sites to be representative of the catchment area. Small island soils, whether overlying volcanic rocks or limestone, tend to be of varying thickness and often discontinuous, features which reflect age and an irregular bedrock surface. Allowance may also

397

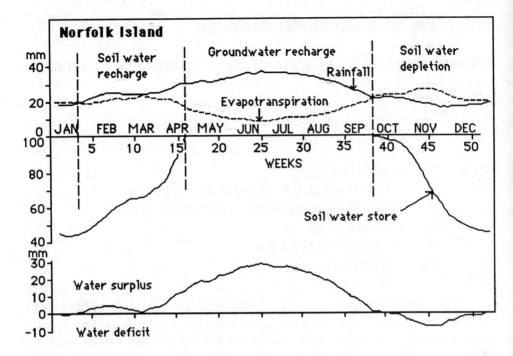

Fig.18.3 Annual water balance for grassland on Norfolk Island (after Abell 1976).

have to be made for an instantaneous recharge bypassing the main soil profile through exposed outcrops or wide channels in the soil such as desiccation cracks or the annuli of large root systems. The occurrence of this form of recharge is apparent in groundwater hydrographs but is normally difficult to estimate. Lloyd (1984) used a figure of 10% of effective rainfall $(P-E_i)$ as instantaneous recharge and similar inputs are quoted by Fox and Rushton (1976).

There is considerable degree of uncertainty in the soil water balance calculation under typical small island conditions. Where reasonable reliability can be asumed (such as when there are good comparisons with other recharge calculations), the method has the value of being sensitive to short-term vaiations in recharge and is capable of application over localized areas. In regions of low rainfall, it is not uncommon for negligible recharge to occur for several years. Table 18.1 lists a few results of studies made on small islands.

Table 18.1

RECHARGE RATIOS ON SELECTED LOW LIMESTONE ISLANDS IN PACIFIC OCEAN

Island	Rainfall (*mm*)	Water balance	Chloride balance	Flow net	Other
			Recharge ratio (R/P) as %		
Kwajalein [1]	2500	41-59			
Christmas [2] Island	867	(i) 29 (ii) 25 (iii) 18 (iv) 11			
Guam [3]	2100		38		
Male [4]	1897			41	
Peleliu [5]	3759				20

1. Peterson (1984 a). Range reflects assumed variations in soil thickness and moisture deficits. 2. Falkland (1984 a). (i)-(iv) 0,10, 20, 30 % coconut trees. 3. Ayers (1984 c). 4. Grimmelman (1984). 5. Grimmelman (1984). Well hydrograph analysis.

Chloride balance. The method assumes that the chloride in the unsaturated zone is constant, so that the input from rainfall equals the output in groundwater recharge, or

$$R\,C_r = P\,C_p \qquad (18.2)$$

where C_p and C_r are the chloride content in the rainfall and groundwater recharge respectively. A major source of error in the application to small islands is the variability of the chloride content in maritime rainfall, which changes seasonally and more irregularly depending on the amount of aerosols entrained. The chloride content of shallow phreatic groundwater may not be representative of the composition of recharge due to a variety of factors, which include enhancement by recycling irrigation or waste water, evaporation losses from the aquifer by phreatophytic vegetation and dispersive mixing with sea water at shallow levels in a thin lens. In high islands there is a difficulty due to the time lag of movement through a thick unsaturated zone and the consequent uncertainty of correlation with a specific seasonal recharge event. Recharge calculations by the chloride balance are more readily applicable on low islands or where the annual rainfall variations are small. The method provides broad estimates of recharge which are likely to be on the low side. A fuller discussion of its application on small islands has been given by Vacher and Ayers(1980).

Groundwater flow models. The general water balance equation for the groundwater in an island aquifer is

$$R - E_g - Q_o - Q_t - Q_p + Q_r = \Delta V_g \qquad (18.3)$$

where

R	=	recharge
E_g	=	evapotranspiration from the groundwater
Q_o	=	outflow of fresh groundwater as base flow or spring discharge
Q_t	=	outflow by dispersion into the transition zone
Q_p	=	abstraction by pumping, infiltration galleries etc
Q_r	=	return flows, e.g. by recycling
ΔV_g	=	change in groundwater storage

Flow modelling is used mainly in low limestone islands where more precise knowledge of a small resource in a thin lens is essential to development planning. The terms Q_o, Q_t and ΔV_g require information on the hydraulic conductivity (permeability) and storage coefficient of the aquifer. These are mainly determined by pump testing, which presents particular problems in application in small island environments (Barker 1984 a). These include the complications resulting from the tidal and other well water-level fluctuations, the difficulties in testing a heterogeneous layered aquifer system (Fig.18.4) with partially penetrating wells, and constraints on durations of pumping to avoid excessive drawdowns and upconing. The diffusivity (ratio of transmissivity to storage coefficient) can be obtained by correlation of water-level changes with tidal cycles (Carr and Van der Kamp 1969) but these are again subject to difficulties of interpretation in an island situation and with variable ground layering. The storage coefficient is difficult to measure in thin transmissive aquifers and information on the diffusivity can prove helpful in this respect.

The strict modelling of the transition zone requires the use of equations expressed in pressure rather than observed head in order to take account of the density

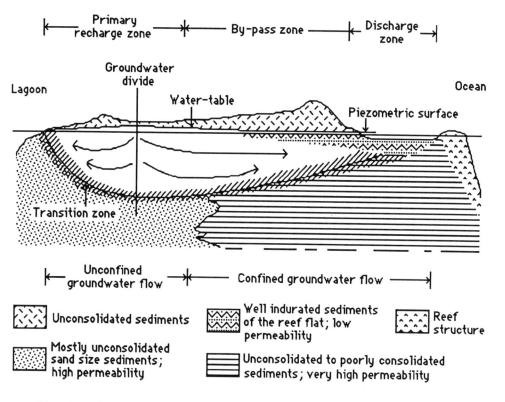

Fig.18.4 Conceptual model of the occurrence and movement of fresh
groundwater within the hydrogeologic framework of an atoll
island (from Ayers 1984 b).

variations which occur. Data on the density variations are not easy to obtain. More
commonly, a sharp interface approximation is used and the equations are more
simply expressed in terms of fresh water and salt water heads from which the
interface can be inferred.

Groundwater models, whether partial or comprehensive, can be utilized in a variety
of ways in addition to their more generally accepted role as a predictive
management tool. A model can provide information on recharge by calibrating the
assumed inputs to the aquifer response in accordance with the observed physical
parameters. In the Anguilla model referred to earlier, recharge estimated by
modelling suggested a value several times larger than that obtained by the chloride
balance.

Hydrochemistry and water quality

The basic controls to the hydrochemistry and water quality of groundwater in small islands have been discussed in relation to small high islands (Chapter 12), where a greater range of rock types including limestones occur. The chemical compositions of fresh carbonate groundwaters are mainly distinguished by contents of calcium and bicarbonate (Table 18.2). The other main changes in chemical composition which occur are the result of mixing with sea water.

The freshwater lens systems in low limestone islands are particularly vulnerable to surface pollution, both biological and chemical, and to increasing salinity as a result of over-development. Water as a domestic supply source can be affected (Table 18.3), and also groundwater used by phreatophytic plants. Protective measures to avoid biological pollution of shallow aquifers are difficult and in densely populated islands almost impracticable. Legislation is particularly critical to ensure water protection and conservation and restrictions in use of dangerous chemicals. Some low limestone islands are highly populated. Male in the Maldive group, for example, has 37 000 people in a land area of $3.8 km^2$. As a result of over-development, the thickness of the fresh water lens has decreased in recent years from 20 to $8m$.

Table 18.2

COMPOSITION OF SOME CARBONATE ROCK GROUNDWATERS[*]
(concentrations in $mg\ l^{-1}$)

Source	pH	Na^+	Ca^{2+}	Mg^{2+}	K^+	HCO_3^-	SO_4^-	SiO_2	Cl^-
Florida	8.0	3.2	34	5.6	0.5	124	2.4	12	4.5
Nikunau (Kiribati)	7.3	7.2	105	16.8	0.2	132	n.d.	n.d.	17
Niue	7.7	7.0	49	8	0.2	175	13	< 1	13
Tarawa (Kiribati)	7.7	9.2	74	17	<0.4	293	19.2	n.d.	14

[*] Reproduced from Morrison et al (1984)

Table 18.3

COLIFORM LEVELS IN ISLAND DRINKING WATER[*]

Island	Number of bores/wells tested	Total coliforms per 100ml	Faecal coliforms per 100ml
Vaitupu (Tuvalu)	15	200 -> 5000	0 - 300
Niue	18	0 - 500	<10
Tongatapu (Tonga)	12		Mostly 0
Savo (Solomons)	55	10 -> 6000	
Nikunau (Kiribati)	6	< 100	
Chrismas Island (Kiribati)	3	3800 - 4400	50 - 1000

[*] Reproduced from Morrison et al (1984).

Water availability and development

Development of the lenses below low limestone islands has problems comparable to those described in relation to the development of the basal aquifer in high islands. The lenses in low islands are generally much thinner, which compounds the likelihood of lateral encroachment of the sea-water wedge and upconing. Critical information to assist development programs should include the ground layering (Fig.18.4) and the position of the interface and as far as possible details of the varying hydraulic parameters including anisotropy. Techniques which are utilized to obtain such information include surface geophysics (Ayers 1984 b), drilling and aquifer testing. Partial penetration of test boreholes in a multi-layered aquifer system with a transitional zone to sea water presents particular problems to hydraulic analysis.

Skimming wells and infiltration galleries have an obvious applicability. The wells should be dispersed as widely as possible and pumping rates should be low. It will rarely be feasible to abstract more than 30% of total recharge from thin lens systems, and this amount may also need to be reduced in accordance with any other withdrawals, notably by phreatophytic vegetation.

Conclusion

Some general comparisons with the hydrology of low coastal areas of larger land masses can be made, but the more pervasive influence of the oceanic environment which relates to geology in addition to current hydrology results in significant differentiation of small low islands.

19. Deltas and coastal areas

Introduction

Deltas are formed where a river debouches into a standing body of water (ocean, sea or lake). The sediments carried by the river are deposited on the sea (or lake) floor and in the estuaries, raising their bed levels. As a result, the delta expands into the sea.

Most rivers nearing the coast split into a number of branches, and in the context of this paper the deltaic area is defined as that part of the river basin which is situated downstream from this point (the "apex"). This definition implies that only "modern" deltas will be considered, as distinct from "fossil" deltas. Inland deltas will not be covered.

The hydrological features and technical problems of land and water development in deltas are similar to those in other low-lying coastal or lacustrine areas, such as coastal marshes, lagoons and other embayments, and can be considered together.

Deltas have a high potential for intensive economic development, and in a large number of deltas this development has taken place. Four ancient civilizations originated in the deltas of the Nile, the Euphrates-Tigris, the Indus and the Yellow river.

This high potential is due to a number of natural advantages, such as the natural fertility and high water retention capacity of the soil, the shallow water-table, the flatness of the land, the network of watercourses and the relative abundance of water, and the location as the outlet of the river basin. These natural advantages are accompanied by inherent drawbacks: waterlogging of the land, flooding by the river and the sea, salt water intrusion, silting of harbors, bad foundation soils etc. Special measures requiring high investments are necessary to obtain the full benefits of the high potential and a certain threshold has to be overcome. Such measures include artificial drainage, flood protection by embankments, river diversion and channel improvement, and closure of estuaries. Measures of this scope create a man-made environment and profoundly affect the hydrological conditions as perhaps nowhere else. Therefore the hydrology of deltas and other low-lying coastal areas cannot be discussed without considering the effect of human intervention and the degree of hydraulic development.

Both nature and man affect the hydrological conditions of deltas, creating new environments. Deltas belong to the most rapidly changing parts of the crust of the earth: the river course is lengthening, there is a rhythm in delta building because the river is shifting its activities from one part of a delta to another, and all this under the natural conditions prevailing in "virgin" deltas.

General characteristics and diversity

Specific hydrological features of deltas. In the upstream parts of a river basin the river flows and river stages are governed by the rainfall on the catchments. In the watercourses of a delta the flows and water levels depend on the flows and river stages at the apex and the water levels at sea. These factors together form the boundary conditions of the flow system. For this reason consideration must be given to sea levels and their variations.

The variations due to tides are predictable, and the morphology of the deltas and their utilization by man have been influenced by these regular variations (Unesco 1969).

The tides propagate into open estuaries. The mechanics of this propagation belongs to the field of tidal hydraulics (shallow water tides), but some knowledge of the processes is essential for the understanding of delta hydrology. The tidal reach is the reach where the flows and water levels are affected by the tides and where

406

alternating currents occur. The river reach is beyond the tidal effect; the direction of flow is always the same and the flow depends only on the upland river discharge (Fig.19.1). In the upstream part of the tidal reach the flow is also unidirectional but varies in magnitude with the tidal phase.

The length of the tidal reach in a given estuary depends on the upland discharge. With increasing upland discharge the effect of the tidal variation is pushed back (Fig.19.2).

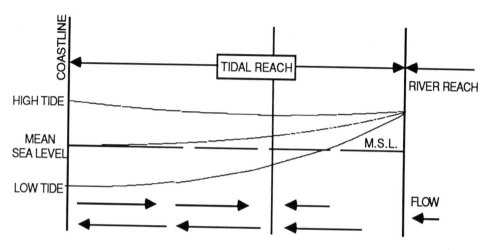

Fig.19.1 Propagation of tides in an estuary.

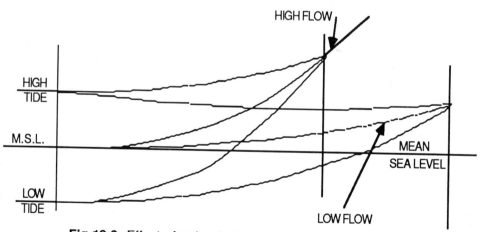

Fig.19.2 Effect of upland discharge on tidal propagation.

Sea levels along coasts are also affected by winds. Landward winds blowing over coastal water produce a rise (set-up) of the sea level. This rise is superimposed on the tidal variation (Fig.19.3). When the winds are strong and the coastal waters are shallow this set-up may of the magnitude of several meters.

According to the place of occurrence, sea level variations as affected by winds are described as storm surges, cyclones, typhoons and hurricanes (Table 19.1).

Like the tides, the storm surges penetrate into the estuaries (Fig.19.4). When the land areas are not embanked extensive flooding occurs. The propagation of the high sea levels depends among other things on the duration of the surge; for storms of short duration (up to 12 hours) the time available is not sufficient to fill the more inland areas to the maximum level at sea.

In 1953 in the Netherlands, in 1959 in Japan and in 1970 in Bangladesh the existing sea dikes were breached and thousands of people lost their lives.

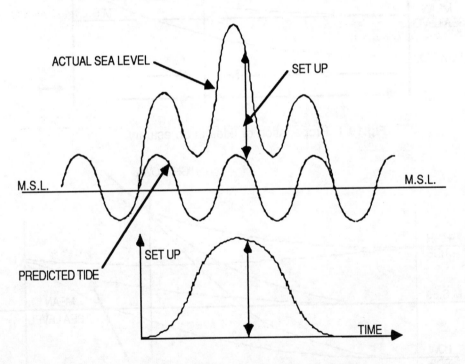

Fig.19.3 Effect of storm surges on tides.

Table 19.1

EFFECT OF WIND ON SEA LEVEL

Description	Location	Set-up (*m*)
Storm surges	North Sea (UK, FRG, the Netherlands)	4 - 5
Cyclones	Bay of Bengal (Bangladesh, West Bengal)	6 - 9(?)
Typhoons	Pacific Ocean (Japan, China)	3 - 4
Hurricanes	Gulf of Mexico	7 - 8

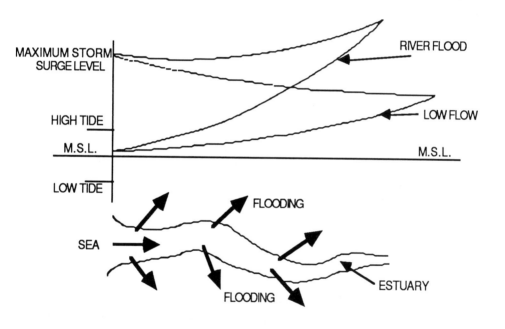

Fig.19.4 Coincidence of river and sea floods.

Build-up of deltas. The build-up of the land areas in deltas and the growth of the deltas into the sea depend on many factors which are related to the river basin and the recipient basin.

The flow and its variability, the sediment transport and the ice conditions of the river are important genetic or delta formation factors. The relevant factors of the recipient basin are salinity, tides and tidal currents, storm surges, foreshore topography and aeolian effects. Variations in the relative elevation of land and sea (transgressions and regressions in the geological history) also govern delta growth. Finally lateral confining boundaries in the recipient basin (high areas and barriers) determine the shape of the delta.

The build-up of the land areas in a delta is related to the extension of the delta into the sea (Fig.19.5). The deposition of the forebed lengthens the river course, decreasing the hydraulic gradient and consequently the sediment carrying capacity. The river is no longer capable of carrying all the sediment to the sea and the river bed silts up. This causes more frequent flooding of the land areas.

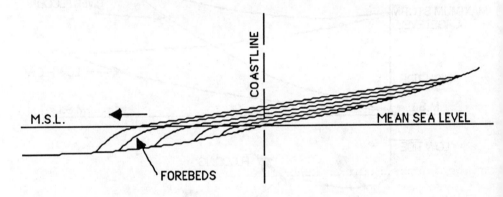

Fig.19.5 Delta extension and rise of the river bed.

In deltas with no flood protection, this flooding leads to typical geomorphological features like natural levees and backswamps (Fig.19.6). There is a longitudinal and lateral sorting of sediments; the coarser sediments are deposited first and the finer sediments are carried further. This explains why coarse sediments prevail in the natural levees while finer particles are found in the backswamps. The natural levee may be as much as 3 to 4*m* higher than the backswamp, and the width of the system ranges from several hundreds to a few thousand meters.

The longitudinal sorting of the sediments explains why near the coast the natural levees are lower and consist of finer material than nearer the apex (Fig.19.6).

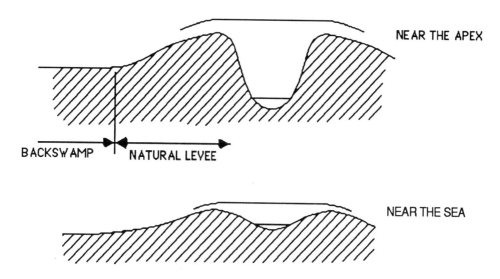

Fig.19.6 Natural levees and backswamps.

Because most of the land area in a delta consists of backswamps, these areas are typically level. There are no permanent water divides; the flow direction of excess water from rainfall depends on variable factors like the vegetation, the areal distribution of rainfall and the wind direction. Artificial drainage basins are created by impoldering, as will be discussed later.

In the subsoil of almost all the world's deltas, unconsolidated sediments are found often to a depth of several hundreds of meters. These sediments consist of pervious material (gravel and sand) forming aquifers, and semi-pervious material (clay and loam) forming aquicludes. This structure is due to the occurrence of geosynclinal subsidence which keeps pace with the deposition of river sediments. During periods of marine transgressions saline water intruded into the aquifers with the result that saline and brackish water is found in them.

Delta types. Deltas exhibit a great variety with respect to their external shape, the number of river branches and ramifications, the development of natural ridges and the soil conditions. Attempts have been made to classify deltas according to these features. Most meaningful are the genetic classifications which relate delta features to delta formation factors (Unesco 1964). The intricacy of the processes in deltas forms an obstacle in establishing classifications which are more than descriptive.

A well-known classification according to the shape of the sea front is that of Gulliver (1899) who distinguished bay deltas (infilled drowned valleys like the Irrawaddy and Chao Phraya deltas), unilobate deltas (Ebro), multi lobate (bird-foot) deltas (Mississippi, Volga, Po, Rhone), cuspate-lobate deltas (Danube, Nile), cuspate deltas (Tiber), rounded deltas (Niger, Mekong, Ganges) and blocked deltas (Senegal, Vistula). This sequence can be related to the ratio of the constructive forces of the river and the destructive forces of the sea.

A more comprehensive classification has been presented by Samojlov (1956). He considers various delta features of the sea front and sea bed and the land areas, and attempts to relate these features with the stage of morphologic genesis.

Bates (1953) classified the channels and gullies in the submarine delta according to the relative densities of river and sea water. The general case is that of a river with fresh water debouching into sea water (hypopycnal inflow). When the density of the river water is higher than that of the recipient basin, hyperpycnal inflow occurs (deltas in fresh water lakes and the delta of the Yellow river). There may also be cases of equal densities (homopycnal inflow).

Volker (1966) classified deltas according to their gross or overall slope, which is defined as the ratio between the land elevation of the apex above mean sea level and the shortest distance to the sea. This parameter shows a surprisingly wide range, from 1×10^{-5} to 5×10^{-4} (Unesco 1964). Together with the channel characteristics this slope governs the length of the tidal reach and the length over which the sea water intrudes into the open estuaries.

Silvester and De La Cruz (1970) applied regression analysis to quantitatively relate features like distance of the apex from the sea, delta area, maximum delta width and ratio to number of mouths with river discharge, tidal range, river slope, slope of the continental shelf etc.

Drowning deltas are deltas where the process of further delta build-up has been reversed because of decrease of sediment supply, rise of the sea level and land subsidence. A typical example is provided by the deltaic area of the Rhine and Meuse rivers in the Netherlands, which at present would no longer exist if man had not stopped the process of destruction (Volker 1979).

Hydrology and water resources

Surface water. As described earlier, the network of interconnected watercourses in a delta exhibits an intricate flow pattern of alternating flows varying with the tidal phase. Important independent variables are the upland discharge and the distribution of this discharge over the various river branches of distributaries. Knowledge of these flows is necessary to analyze the sea water intrusion in the open estuaries and to predict the effects of embanking, closure of estuaries, diversion of water and changes in the discharge from the upstream catchment.

Where the river flow is not substantially affected by the tidal variation, the upland discharge can be directly measured. Where an alternating current exists, the upland discharge has to be derived from the difference between the volume of the ebb and the volume of the flood. Since both determinations are subject to error, the derivation of the upland discharge may be inaccurate if the tidal volumes are large compared with the volume of upland discharge.

Hence the determination of the distribution of this discharge is a delicate matter, the more so as it depends not only on the hydraulic characteristics of the river branches but also on the tidal variations in the mouths of the estuaries, which may not be synchronous and in phase with the ocean tides.

Models which simulate the actual conditions are a powerful tool in analyzing the flow processes in deltaic waters and in making the predictions mentioned above. These models may be hydraulic scale models or mathematical models. An essential point is that it is not sufficient to have data on the hydraulic characteristics of the channel and the boundary conditions (the upland discharge and the tidal variation at sea). For the calibration of the model and in particular the determination of the hydraulic roughness it is necessary to have the results of synoptic measurements of flows and levels at a number of stations and under different conditions of tidal range and upland discharge, including floods. These data collection periods involving many people, vessels and instruments may be quite expensive.

Because of the low velocities and the possible reversal of the current, the acoustic method of discharge measurement may be quite suitable in watercourses which are not too wide. The moving boat method has been found suitable in wide estuaries.

Groundwater. Groundwater plays an important role in the hydrological conditions of deltaic areas, not so much as a contribution to river flow in the

413

upstream parts of the catchment but as seepage water in low-lying areas and, locally, as a source of useful water (ESCAP 1963).

From the viewpoint of groundwater hydraulics, the structure of the subsoil is quite simple since it consists of a alternation of horizontal aquifers and semi-pervious layers. If the transmissivities and the resistances are known, the groundwater flows can easily be computed for given boundary conditions. These geotechnical parameters can be determined from pumping tests or, preferably, from the analysis of existing large-scale groundwater flows.

Groundwater under large parts of a delta is usually saline or brackish, specially near the coast. Further inland the groundwater may be fresh, due to the effect of seepage from higher adjacent recharge areas where part of the rainfall infiltrates.

The general picture of the salinity distribution is of a gradual increase with depth and a gradual decrease from the coastal areas to the adjacent higher areas. Differences over small distances may occur below and above semi-pervious layers. In the deltas of the Mekong and the Ganges fresh groundwater is found near the coastline at depths of 200-300m below semi-pervious layers, whereas above the layers the gorundwater is very saline. The fresh groundwater is connected with remote recharge areas.

The determination of the salinity destribution of the water in the aquifers, both horizontal and vertical, can best be done by a combination of test borings with water sampling and the electrical resistivity method of geophysical prospecting. Geophysical well-logging provides useful additional data. The resistivity method has been applied for the determination of groundwater salinity down to a depth of 250 -300m.

Salt water intrusion in watercourses. In many deltas surface water is abundant in quantity, but as a resource it is often useless because of its quality, in particular the high salt content. This is due to several sources of salt: sea water intrusion into open estuaries, seepage of groundwater with a high salinity, sea water admitted during the locking of ships, and a possible salt discharge of the river itself (from saline soils and waste salts from industries, as in the Rhine).

Sea water instrusion into open estuaries occurs wherever a river with fresh water debouches into the sea. By virtue of its higher density, the sea water penetrates into the river mouth and moves in an upstream direction in spite of the river flow in the opposite direction. The intrusion may assume the shape of a wedge as shown in Fig.19.7 if there are no tides or other conditions furthering a mixing of the saline and fresh water.

The process of sea water intrusion is still imperfectly understood and there is a proliferation of formulae and models all containing coefficients which have no general validity, but require field data for their determination in a particular situation.

For the case of the simple stratified estuary illustrated in Fig.19.7, Thijsse (1954) gave the semi-empirical formula

$$L = \frac{c^2 B^2 h^4}{Q^2} \frac{\Delta \rho}{\rho} \qquad (19.1)$$

where

L	= length of salt water wedge (*m*)
B	= width of a rectangular cross section (*m*)
h	= depth (*m*)
Q	= upland discharge (*m^3 s^{-1}*)
$\Delta \rho / \rho$	= relative density difference
c	= modified Chezy coefficient.

From measurements in various drainage canals exposed to sea water intrusion Thijsse found $c \approx 15 m^{0.5} s^{-1}$.

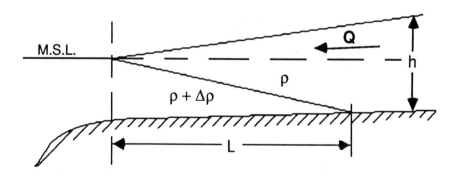

Fig.19.7 Definition diagram for a sea water wedge in a stratified estuary.

The formula shows the dominating effect of the channel depth h. This is in agreement with the experience in many estuaries where dredging for harbors produced an invasion of sea water.

Mixed or partly mixed estuaries are generally analyzed by considering a concentration c which is defined as the average concentration during a complete tidal cycle and over the cross-sectional area. Steady conditions for a given upland discharge Q require that the convective flow of salt in the downstream direction is balanced by a flow in the upstream direction generated by the concentration gradient (Fig.19.8):

$$Q \cdot c = - DA\frac{dc}{dx} \qquad (19.2)$$

where A = cross-sectional area
       D = transport coefficient.

The transport coefficient D varies not only from delta to delta but also with the upland discharge, the tidal flood volume, and the location in the estuary. For this reason it can only be estimated if extensive field data are available.

There is an IHP project to make an inventory and evaluation on a global basis of the available approaches, formulae and models in this field.

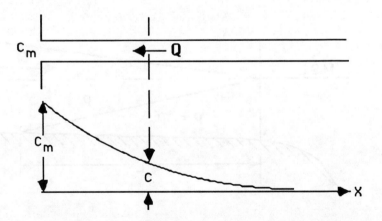

Fig.19.8 Definition diagram for salinity distribution in a mixed estuary.

The sea water intrusion into open estuaries forms a major obstacle in meeting the needs for fresh water in deltas. In the Gambia river in West Africa, the saline effect from the sea extends during low flow to at least $235km$ from the mouth.

Measures to prevent or reduce the intrusion include augmentation of low flows (by releases from upstream reservoirs), filling deep channels to reduce the depth (e.g. Rotterdam waterway in the Netherlands), and construction of submerged low sills. The most drastic way of arresting sea water intrusion into estuaries consists in closing them at the mouth by a dam equipped with drainage sluices.

Salt water intrusion into aquifers. In those deltaic aquifers where the groundwater is fresh there may be intrusion of salt water from the sea or from interconnected aquifers where the ground water is saline or brackish.

When land below mean sea level is reclaimed, a seepage flow will carry sea water into the aquifers and finally bring seepage water with a high salinity to the surface channels. This may create problems for the water supply to crops and even endanger the technical feasibility of the reclamation project. The sea water may also come from low river reaches where sea water has intruded.

Abstraction of fresh groundwater by wells may draw in neighboring groundwater with a high salinity, so that the pumped water becomes unfit for use. This may happen in fresh water lenses floating on the underlying saline groundwater, as occurs under coastal dunes and inland sandy ridges.

There has recently been much research on the hydraulics of groundwater with different densities and distinct interfaces (Unesco 1987 b). When the geotechnical factors are known, a prediction of changes in the salinity distribution under the effect of hydraulic gradients can be made with greater accuracy than in the case of sea water intrusion in estuaries.

Seepage flows of saline water are difficult to prevent because of the generally considerable thickness of the aquifers through which the seepage flow is taking place.

Land use and water resources

Factors determining land use. Early settlements of fishermen and farmers were located on the natural levees and coastal ridges where the depth of flooding is small, the soils are well-drained and fresh water is found at shallow depths. On these levees houses were built and various crops and fruit trees were grown. This

417

settlement pattern can still be recognized in all deltas, even in those where later a high level of flood protection and drainage was reached and all the land was occupied.

With a growing population, the backswamps were also gradually cultivated to supply the necessary food.

In the temperate zone the high marshes along the coasts were used as grazing lands. In deltas exposed to storm surges artificial hillocks were erected to serve as emergency mounds. Later on these mounds were connected by dikes and a more active method of defence was introduced. Along the rivers, crops were grown outside the river flood season and in certain spots low dikes were erected which would constitute the beginning of large scale protection against floods.

In the deltas in the humid tropical zone where wetland rice is the staple crop, flood protection was less essential than in the temperate zone. Rice can grow in water and local varieties have been developed that can grow in water with an increasing depth. Rice can therefore be cultivated in the absence of flood protection (Volker 1983) provided that the rate of rise does not exceed 150$mm\ d^{-1}$. An extreme case of adaptation of the crop variety to prevailing conditions is that of the so-called floating rice which can grow in water depths up to 3-4m. This requires the river floods to be of the mild type, which is generally the case in large river basins with pervious soils and low slopes. When the floods are flashy, nothing can be grown during the flood season. This explains why dikes have been built in the delta of the Red river in northern Vietnam, which has flash floods, while there are no river dikes in the delta of the Mekong in the southern part of the country, which has gentle floods. However, flood protection during the rainy season is an absolute necessity to cultivate the high-yield varieties of rice which require perfect water control, and also to grow dryland crops.

Embankments, simple as they are as hydraulic structures, produce a number of negative side effects and may profoundly influence the hydrological conditions. The isue of embanking is often a controversial matter.

There are two other reasons why wetland rice is the most widely spread agricultural crop in deltas in the humid tropical zone. The first lies in the soils which in large areas are acid or potentially acid. The best treatment of such soils is to keep them wet and to grow a crop like rice with a shallow root system. The second reason is that controlled drainage in these zones involves the removal of considerable volumes of water and requires the lay-out of costly artificial drainage systems.

The hydrology of polders.   Polders are man-made hydrologic units in coastal and upland wetlands. A polder can be defined as a well circumscribed level area which has been isolated from the surrounding hydrological system. In the polder the levels of the surface water and groundwater are artificially controlled according to the requirements of the agricultural or urban areas. Thus a polder (Fig.19.9) is a catchment area with an artificial drainage system and a controlled outflow through a structure. Depending on the water levels inside and outside, the outflow is by gravity or pumping. Isolation from the surrounding hydrological system and protection against flooding are mostly achieved by the embankments.

Embanking level areas which are exposed to extensive flooding entails a number of environmental side effects, many of which may be negative, in the fields of hydrology, morphology, water management and health.

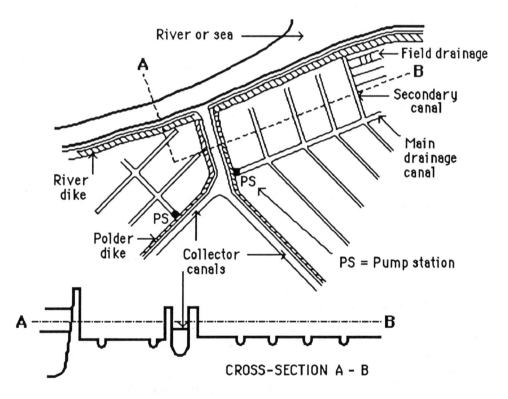

Fig.19.9 Polder drainage system.

Because of the elimination of longitudinal overland flow and overbank storage, embanking produces a rise in flood levels. This effect is very pronounced in the case of flash floods and may then amount to a few meters in major floods. Accurate predictions can be made by applying the physical or mathematical models mentioned previously. The effect has been observed in many flood plains and has led to controversies on the desirability of embankments. The strip on the natural levee between the channel and the embankment is exposed to deeper flooding. This is precisely the place where people live, and there are cases where dikes were deliberately breached to lower the flood level.

Under certain conditions embanking may cause a rise of the river bed, increasing flood levels still further. This is followed by increasing the height of the river dikes (e.g. the Rhine between 1700 and1850, the Yellow river and the Red river) so that a vulnerable situation is created.

Embanking increases the tendency of the rivers to meander, and bank erosion may occur, threatening the dikes. Finally river training may become necessary, which is a very expensive operation compared with diking.

By preventing flooding, embankments also eliminate the beneficial effects of the floods. These consist in the first place of a flushing of the land areas after a prolonged dry period, removing accumulated human and animal wastes, bacterial growth sites, and brackish water. In the second place, in areas where rice is grown, the floods often form a supplemental irrigation when the local rainfall falls short of meeting the demand. Here, too, there are cases of farmers cutting the dikes which were designed to protect them.

When areas which were exposed to flooding are embanked, it becomes necessary to make provisions for the evacuation of excess water produced by local rainfall. It is not sufficient to make outlets consisting of sluices with gates in the encircling embankment. A system of artificial main and subsidiary drainage canals is required to convey excess water to the outlet. This implies a loss of water conservation. As a result the water management has to be completely modified and, in addition to the drainage system, provision may have to be made for irrigation.

Thus embanking and impoldering entail a completely different environment, which offers new possibilities for intensive exploitation of the natural resources.

Land reclamation by impoldering has been carried out in many deltas and low-lying areas all over the world. In some cases (Denmark, Egypt, India, Japan and the Netherlands) water areas have been impoldered which were previously submerged by several meters of water. The minimum elevation of the new land in these cases is around 7*m* below mean sea level.

Water management. Water management in deltaic areas has some distinctive features which are not found in other parts of a river basin.

The hydrological conditions in a deltaic area are greatly affected by natural and man-made changes to the conditions in the upstream parts of the catchment. Obvious examples are the effect of climatic variations and of erosion in the upper reaches of a basin.

Of particular significance is the effect of upstream storage reservoirs on water management in the delta. Major multiple storage schemes on the river can produce a reduction of flood flows and an augmentation of dry season flows, provided the reservoirs are operated appropriately. Classical examples are the High Aswan dam on the Nile, and the Bhumiphol and Sirikit dams on the Chao Phraya river in Thailand. The former dam made it possible to apply perennial irrigation in the Nile delta and in upper Egypt, and also reduced the floods. In Thailand the upstream storage development enabled a gradual transformation of the system of supplemental irrigation by flooding during the flood season, into a system with controlled supplemental irrigation and irrigation during the dry season. In this way the disadvantages of flood protection by embanking could be avoided. Upstream reservoirs may also have negative effects on the conditions in the delta in that they reduce the supply of suspended sediment to the delta, and erosion of the delta lands may set in. This occurred in the delta of the Mississippi, where in the period 1965-1978 more than 50% of the marshes and forested wetlands in the lower delta disappeared. Erosion along the delta front of the Nile increased considerably after the construction of the High Aswan dam.

A well-balanced development of the deltaic and upstream parts of a river basin is seldom possible. There may be conflicting interests with respect to the use of the river water and its quality. The two parts of the river basin may be situated in different provinces, states or countries. In many cases the use of the deltaic areas preceded the development of the other parts of the delta, and the inhabitants of the deltas did not wait until upstream development had taken place.

A second particular feature of water management in deltaic areas is the fact that an intervention in the hydrological regime in one part of the area affects the conditions in other parts of the area. This is due to the interconnected system of watercourses and the delicate dynamic balance between the distribution of the upland discharge over the river branches, the siltation and building up of the land area, the sea water intrusion and the maximum and minimum water levels. For this reason stepwise development projects for land and water use in a deltaic area

421

should fit into a master plan for the ultimate utilization of the total potential of the area.

A third special aspect of water management of deltaic areas results from the problem of sea water intrusion into open estuaries. The intrusion is most pronounced during dry periods, when the upland discharge is small and the need for fresh water is a maximum. To repulse the saline intrusion, high volumes of fresh water are required, which cannot be used for other purposes. In the Netherlands more than half of the total water demand consists of water that is used to keep the sea water intrusion within certain limits and to rinse or flush the canals.

As mentioned earlier, a drastic way of combating saline intrusion is the enclosure of estuaries and other tidal inlets (Fig.19.10). A dam is built in the mouth together with sluices to evacuate flood waters from the river. Because of the inflow of fresh water and drainage of excess water to the sea, the enclosed area is gradually transformed into a fresh water reservoir. This form of coastal storage is the alternative to the upstream storage mentioned earlier. Examples of coastal storages can be found in the Netherlands, France, Japan, India (Kerala state) and on a smaller scale in Bangladesh and in the Mekong delta.

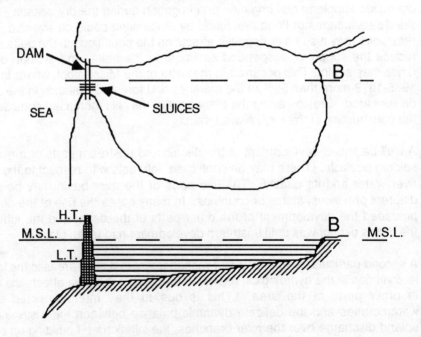

Fig.19.10 A coastal reservoir.

422

The hydrological design of such reservoirs is based on two balances, the water balance, which is similar to the balance of upstream reservoirs, and the salt balance which indicates the salinity of the water in the reservoir.

The water balance for monthly or annual periods contains the following items:

Inflow	Outflow
Inflow from the river	Drainage to the sea
Rainfall on the reservoir	Abstraction of water from the reservoir
Drainage from land adjacent to the reservoir	Evaporation

Change in storage

In the case of gravity drainage of excess water to the sea, the volumes that can be discharged depend on the tidal levels. When the sea levels are abnormally high during storm surges the discharge to the sea may be hampered for one or more days.

The salt balance reads:

Inflow	Outflow
Salt water load in river	Drainage to the sea
Drainage of saline water from adjacent land	Abstraction of water from the reservoir
Diffusion of salt from the polder bed	
Salt admitted by locking of ships	
Leakage of sluices and gates	

Change in storage

Hydrological characteristics of particular deltas

Delta of the Rhine and Meuse rivers. As mentioned earlier, the delta of the Rhine and Meuse is a "drowning" delta whose survival depends on the system of primary and secondary dikes and the system of artificial drainage. A large part of the Netherlands (41 160km^2) is situated below mean sea level (Fig.19.11) and more than 60% of the country would be flooded by storm surges and river floods if the dikes did not exist. Polders with a land elevation 4-9m below sea level consist of reclaimed lakes. Polders 0-3m below mean sea level are due to natural and man-induced land subsidence; about a thousand years ago, these areas were at the level of the normal high tide. Natural land subsidence or relative rise of the sea level amounts to 150-200 *mm* per century; man-induced subsidence is the

423

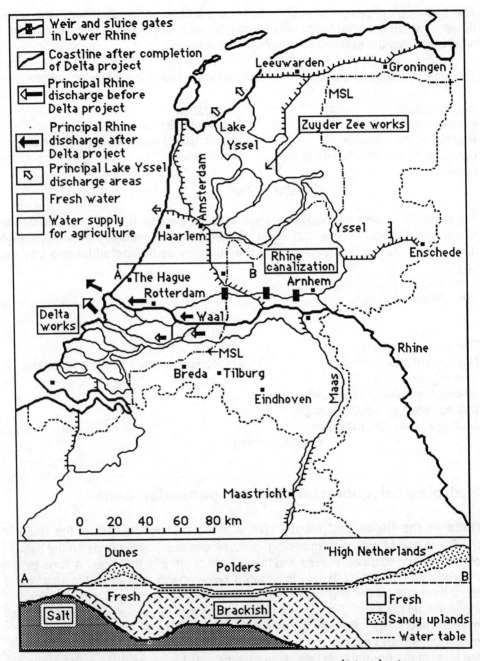

Fig.19.11 The delta of the Rhine and Meuse (Maas) rivers.

consequence of agricultural land drainage and abstraction of groundwater and may be of the order of 1-2*m*.

The map shows the two major water management projects of this century: the Zuiderzee works and the Delta works. The Zuiderzee works were the response to the storm surge disaster of 1916. Essential elements are the enclosing dam 32*km* long (1927-1932), the Ijssel lake, a fresh water reservoir of 1200*km²* and five deep polders with a total area of 2200*km²*. The Delta works were implemented after the disaster of 1953 which ravaged the southwestern part of the country. These works comprise the closure of the estuaries of the Rhine and the Meuse leaving open the entrance to the port of Rotterdam.

Delta of the Mekong river.   The delta of the Mekong river is partly situated in the southern part of Vietnam and partly in Kampuchea. The total area is some 55000*km²* (Fig.19.12). The delta as a whole is densely populated and quite productive. The staple crop is rice grown during the wet monsoon, which is also the flood season.

The apex of the delta is situated at Phnom Penh where the Mekong splits into a number of branches ("Quatre Bras"). Upstream of Phnom Penh the Great Lake is connected with the Mekong by the Tonle Sap (Fig.19.12). This lake acts as a natural regulation reservoir: when the Mekong is in spate, the lake is filled by the Tonle Sap and extends in area; when the flood recedes the lake empties and water flows back to the Mekong. Thus the floods are attenuated and the dry weather flow is sustained by water from the Great Lake.

Except in some coastal areas there are no embankments and most of the area is flooded to different depths depending on the distance from the coast. Thus human intervention in the hydrology of the delta has been limited. There is however a dense network of canals. These canals are interconnected and the tides and the river floods can freely penetrate into these watercourses.

Delta of the Irrawaddy river.   The delta of the Irrawaddy river in Burma (35 000*km²*) provides an interesting example of the controversy in the southeast Asian region on the issue of embanking as a means of flood protection. As in the other large deltas of this region, lowland rice is the staple crop and is grown during the rainy season.   Except in very dry years rainfall during this season is sufficient to meet the demands of the crop. Hence there is no need for supplemental irrigation by the flood waters as is the case in the Central Plain in Thailand where the rainfall is lower. The floods in the Irrawaddy delta caused more damage than benefits, and as early as 1861 the first dike along the west bank of the main stem was erected. Embanking was extended specially in the period 1880-1920. This met a growing

425

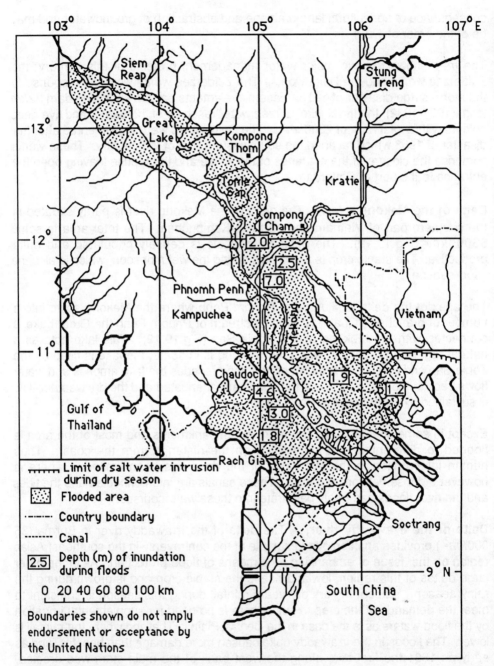

Fig.19.12 Delta of the Mekong river.

426

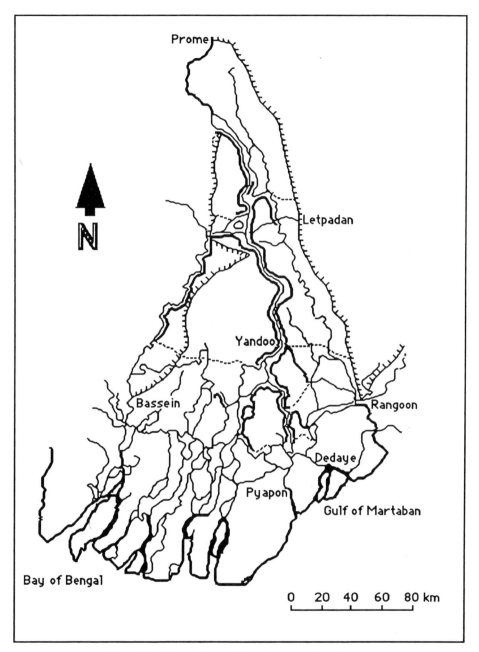

Fig.19.13 Delta of the Irrawaddy river.

opposition as it was feared that the flood levels would rise considerably. With the elimination of the floods the lands were also deprived of the supply of fertilizing silt. A compromise was adopted consisting of horse-shoe shaped embankments which are open at the downstream ends (Fig.19.13). With high floods a partial flooding occurs so that part of the storage effect is preserved.

In 1926 a high flood occurred which caused much damage specially to the railways. The next year in the Mississippi the highest flood levels were measured, which were attributed to the effect of double embanking of the river. Following these events it was proposed in 1929 to gradually demolish the dikes on the east bank of the Irrawaddy at a rate of 16*km* per year. In view of the costs of compensation for acquisition this suggestion could not be put into practice. The matter came again to the fore after the big flood of 1939. Again it was not put into effect because the farmers did not want to be drowned out for the sake of future generations.

References

Abell R.S. (1976): A groundwater investigation of Norfolk Island, Bureau Mineral Resour. Record No. 1976/62 (Aust. Gov't. Publ. Serv., Canberra).

ACAH (1980): *Water for Agriculture: Future Needs* (Advisory Council for Agriculture and Horticulture in England and Wales, London).

Aellen M. (1985): Die Gletscher der Schweizer Alpen im Jahr 1983/84, *Die Alpen (J. of Swiss Alpine Club)* **61**: 188-213.

Albergel J. (1987): Genèse et prédétermination des crues du Burkina Faso, These, Université Pierre et Marie Curie, Paris.

Alley W.M.(1977): *Guide for collection, analysis, and use of urban stormwater data* (American Society of Civil Engineers, New York).

Allison G.B. (1987): Estimation of groundwater discharge and recharge with special reference to arid areas, in "International Conference on Groundwater Systems Under Streaa", Aust. Water Resour. Council Conf. Series No.13, pp.231-238, Aust. Govt. Publishing Serv., Canberra.

Allison G.B., Stone W.J. and Hughes M.W. (1985): Recharge in karst and dune elements of a semi-arid landscape, as indicated by natural isotopes and chloride, *J.Hydrol.* **76**:1-25.

Ameghino F. (1884): *Las Inundaciones y Sequias en la Provincia de Buenos Aires* (Buenos Aires, Argentina).

Amorocho J. and Orlob G.T. (1961): *Non-linear analysis of hydrologic systems,* Univ. California, Sanitary Eng.Res.Lab., Berkeley.

Anderson M.G. and Burt T.P. (1982): The contribution of throughflow to storm runoff: an evaluation of a chemical mixing model, *Earth Surf. Proc. & Landforms* **7**:565-574.

Anderson M.T. and Hawkes C.L. (1985): Water chemistry of northern Great Plains strip mine and livestock water impoundments, *Water Resour. Bull.* **21** (3): 499-505.

Ando Y., Takahashi Y. and Kuan M.F. (1984): Relationship between land use and final infiltration rate in urban areas, Proc.3rd Internat. Conf. on Urban Storm

Drainage (ed. P.Balmer, P.A.Malmqvist and A.Sjoberg), pp.1029-1036 (Chalmers Univ. of Tech., Goteborg, Sweden).

Andrews F.M. (1962): Some aspects of the hydrology of the Thames basin, *Proc. Inst. Civil Eng.* **21**: 55-90.

Archibald G.G. (1983): Forecasting water demand, *J. Forecasting* **2**: 181-192.

Arenas A.O. (1983): Tropical storms in Central America and the Caribbean: characteristic rainfall and forecasting of flash floods, in *Hydrology of Humid Tropical Regions* (ed. R. Keller), IAHS Publ. No. 140: 39-51.

Aschwanden H. and Weingartner R. (1985): *Die Abflussregimes der Schweiz,* Geographisches Institut der Universitaet Bern, Publikation Gewaesserkunde 65.

Askew A.J. (1987): Climate change and water resources, in *The Influence of Climate Change and Climatic Variability on the Hydrologic Regime and Water Resources,* IAHS Publ.No. 168: 421-430.

Australian Water Resources Council (1969): The representative basins concept in Australia, *AWRC Hydrol. Ser. No. 2* (Dept. National Development, Canberra).

Australian Water Resources Council (1972): Hydrology of smooth plainlands of arid Australia, *AWRC Hydrol. Ser. No. 6* (Aust. Gov't Publishing Service, Canberra).

Ayebotele (1979): Water resources, Report to the UN Conference on Science and Technology, IHP document SC/214/IHP/ICIII/Inf.ACAST, Unesco, Paris.

Ayers J.F. (1984 a): Geological construction of atolls, Proc. Regional Workshop on Water Resour. of Small Islands, Suva, Fiji (Commonwealth Science Council Tech. Publ. No. 154), pp. 69-78.

Ayers J.F. (1984 b): Groundwater occurrence beneath atoll islands, Proc. Regional Workshop on Water Resour. of Small Islands, Suva, Fiji (Commonwealth Science Council Tech. Publ. No. 154), pp. 157-179.

Ayers J.F. (1984 c): Estimate on groundwater recharge - the chloride balance approach, Proc. Regional Workshop on Water Resour. of Small Islands, Suva, Fiji (Commonwealth Science Council Tech. Publ. No. 154), pp. 344-352.

Bagchi A.K. (1982): Orographic variation of precipitation in a high-rise Himalayan basin, IAHS Publ. No. 138: 3-9.

Baghirathan V.R. and Shaw E.M. (1978): Rainfall depth-duration-frequency studies for Sri Lanka, *J.Hydrol.* **37** (3-4): 223-239.

Balek J. (1977): *Hydrology and Water Resources in Tropical Africa* (Elsevier).

Balek J. (1983): *Hydrology and water resources in tropical regions,* Developments in Water Science No.18 (Elsevier).

Barabas S. (1981): Eutrophication can be controlled, *WHO Water Quality Bull.* **6**:94 and 155.

Barbagello J. (1984): Las areas anegables de la "Pampa Deprimida", un plan teo hidrologico para su solucion, in *Hydrology on large flatlands* (ed. M.C. Fuschini Mejia), Proc. Olavarria Symp. **2**: 787-864, Unesco, Republica Argentina.

Barker J.A. (1984 a): Aquifer properties and pumping tests: an introduction, Proc. Regional Workshop on Water Resour. of Small Islands, Suva, Fiji (Commonwealth Science Council Tech. Publ. No. 154), pp. 115-123.

Barker J.A. (1984 b): Freshwater - saltwater relations, Proc. Regional Workshop on Water Resour. of Small Islands, Suva, Fiji (Commonwealth Science Council Tech. Publ. No. 154), pp. 124-130.

Barker J.A. (1984 c): Coastal aquifer modelling, Proc. Regional Workshop on Water Resour. of Small Islands, Suva, Fiji (Commonwealth Science Council Tech. Publ. No. 154), pp. 243-254.

Barry R.G. and Chorley R.J. (1976): *Atmosphere, weather and climate,* 3rd edn. (Methuen).

Bastug A. (1984): Mockus triangle hydrograph method for design flood recurrences, Papers of Unesco-VITUKI International Postgraduate Course on Hydrological Methods, Budapest, Hungary.

Bates C.C. (1953): Rational theory of delta formation, *Bull. Amer. Assoc. Petr. Geol.* **37**: 2119-2162.

Bates B.C. & Pilgrim D.H. (1983): Investigation of storage-discharge relations for river reaches and runoff routing models, *Trans. Inst. Eng. Aust.* **CE25**(3):153-161.

Baumgartner A. and Reichel E. (1975): *The world water balance* (Elsevier, Amsterdam).

Baumgartner A., Reichel E. and Weber G. (1983): *Der Wasserhaushalt der Alpen* (R.Oldenbourg Verlag, Munich & Vienna).

Bay R.R. (1969): Runoff from small peatland watersheds, *J. Hydrol.* **9**: 90-102.

Beable M.E. and McKerchar A.I. (1982): Regional flood estimation in New Zealand, National Water and Soil Conservation Organisation, Water & Soil Tech. Publ. 20.

Beard J.S. (1964): Savanna, in *Ecology of Man in a Tropical Environment* (Mogens, Switzerland).

Beran M.A. (1986): The water resources impact of future climate change and variability, in *Effects of Changes in Stratospheric Ozone and Global Climate* (ed. J.G. Titus), vol.1 (UNEP/EPA, Washington).

Beran M.A. and Rodier J. (1985): Hydrological Aspects of Drought, *Studies and Reports in Hydrology 39* (Unesco/WMO, Paris).

Berdanier C.R., Asmussen L.E. and Ferriera V.A. (1984): Application of computerized map analysis system tp hydrologic study of flatland watersheds, in *Hydrology on large flatlands* (ed. M.C. Fuschini Mejia), Proc. Olavarria Symp. **1**: 495-526, Unesco, Republica Argentina.

Beskow G. (1947): Soil freezing and frost heaving with special application to roads and railroads, Swedish Geol. Soc., Ser. C. No 375, 26th yearbook No.3.

Beven K. (1984): Infiltration into a class of vertically non-uniform soils, *Hydrol. Sci. J.* **29** (4): 425-434.

Beven K. and Clarke R.T. (1986): On the variations of infiltration into a homogeneous soil matrix containing a population of macropores. *Water Resour. Res.* **22** (3): 383-388.

Beven K.J. and Germann P.F. (1982): Macropores and water flow in soils, *Water Resour. Res.* **18** (5): 1311-1325.

Biswas A.K. (1984): Climate and development, in *Climate and development* (ed. A.K.Biswas),Vol.13 (Tycooli Internat. Publ.).

Bjornsson H. (1975): Subglacial water reservoirs, jokelhlaups and volcanic eruptions, *Jokull* **25**: 1-14.

Bjornsson H. (1976): Marginal and supraglacial lakes in Iceland, *Jokull* **26**: 40-51.

Blackie J.R. and Newson M.D. (1986): The effects of forestry on the quantity and quality of runoff in upland Britain, in *Effects of Land Use on Fresh Waters* (ed. J.F.G. Solbe), pp. 398-412 (Ellis Horwood, Chichester).

Blake G.J. (1975): The interception process, in *Prediction in catchment hydrology* (eds. T.G. Chapman & F.X. Dunin), pp. 59-81 (Aust. Acad. Sci., Canberra).

Bonell M., Gilmour D.A. and Cassells D.S. (1983): Runoff generation in tropical rainforests of northeast Queensland, Australia, and the implications for land use management, in *Hydrology of Humid Tropical Regions* (ed. R. Keller), IAHS Publ. No.140: 287-297.

Bonell M., Hendriks M.R., Imeson A.C. and Hazelhoff L. (1984): The generation of storm runoff in a forested clayey drainage basin in Luxembourg, *J. Hydrol.* **71**: 53-77.

Boorman D.B. (1985): A review of the Flood Studies Report rainfall-runoff model parameter estimation equations, Institute of Hydrology Report 94, Wallingford, Oxon.

Bordas M.P. and Canali G.E. (1980): The influence of land use and topography on the hydrological and sedimentological behaviour of basins in the basalt region of South Brazil, Symposium on the Influence of Man on the Hydrological Regime with Special Reference to Representative and Experimental Basins, Helsinki, IAHS Publ. No. 130: 55-60.

Bouchet R.J. (1963): Evapotranspiration réelle et potentielle, signification climatique, IAHS General Assembly (Berkeley), Publ.No. 62: 134-142.

Boulad A.O., Miller J.P. and Bocquire G. (1977): Determination of the age and rate of alteration of a ferralitic soil from Cameroon, *Sci. Geol. Bull.* **30**: 175-178.

Bouma J. (1986): Using soil survey information to characterise the soil water state, *J. Soil Sci.* **37**: 1-7.

Bowles D.J., Grant J.L., Humphries W.E. and O'Hayre A.P. (1985): Design and impact analysis for diversion at Coal Creek mine, *Water Resour. Bull.* **21**(6): 995-1003.

Bree T. and Cunnane C. (1980): The effect of arterial drainage on flood magnitude, in *Casebook of Methods of Computation of Quantitative Changes in the*

Hydrological Regime of River Basins due to Human Activities (Unesco, Paris).

Brodie J.E., Prasad R.A. and Morrison J.R. (1984): Pollution of small island water resources, Proc. Regional Workshop on Water Resour. of Small Islands, Suva, Fiji (Commonwealth Science Council Tech. Publ. No. 154), pp. 378-386.

Brown J. (1978): Influence of climate and terrain on ground temperature in the continuous permafrost zone of Northern Manitoba and Keewatin District, Canada, Research Report 809, Divn. Building Res., Nat. Res. Council Canada, Ottawa, Ontario.

Brown J., Dingman S.L. and Lewellen R.I. (1968): Hydrology of a drainage basin on the Alaskan coastal plain, CRREL Research Report 240, U.S. Army Corps of Engineers.

Brown L., Chandler W.U., Flavin C. et al (1986): State of the world 1986, a Worldwatch Institute report on progress towards a sustainable society, Norton, New York.

Bruce R.B., Myrhe D.L. and Sanford J.O. (1968): Water capture in soil surface microdepressions for crop use, 9th Internat. Congr. Soil. Sci. : 325-330.

Bruijnzeel L.A. (1983): Evaluation of runoff sources in a forested basin in a wet monsoonal environment: a combined hydrological and hydrochemical approach, in *Hydrology of Humid Tropical Regions* (ed. R. Keller), IAHS Publ. No. 140: 165-174.

Brunig E.F. (1975): Tropical ecosystems: state and targets of research into the ecology of humid tropical systems, *Plant Res. and Development.* 1: 22-38.

Brutsaert W. (1982): *Evaporation into the atmosphere* (Reidel, Boston).

Bucher B. and Demuth S. (1985): Vergleichende Wasserbilanz eines flurbereinigten und eines nicht flurbereinigten Einzugsgebietes in Ostkaiserstuhl fur den Zeitraum 1977-1980, *Deutsche Gewasserkundliche Mitteilungen* **29** (1): 1-4.

Budyko M.I. (1956): *The heat balance of the earth's surface* (Leningrad). (English translation by N.I. Stepanova, Washington, US Weather Bureau).

Budyko M.I. (1971): *Climate and life* (Hydrometeoizdat, Leningrad).

Budyko M.I. (1977): *Global ecology* (Globahaya ekologia, in Russian) (Mysl Publishers, Moscow).

Budyko M.I. (1986): *The evolution of the biosphere* (Reidel).

Buishand T.A. and Velds C.A. (1980): Neerslag en Verdamping, Koninklijk Nederlands Metereorologisch Institut.

Bundesanstalt fur Gewasserkunde (1986): Jahrbuch, Bundesrepublik Deutschland und Berlin (West), 1986 (Koblenz).

Buning R.A. (1985): Flood control and drainage project, city core of Bangkok, Thailand, *Land and Water Internat.* **55**: 3-16.

Burden R.J. (1982): Hydrochemical variation in a water-table aquifer beneath grazed pastureland, *J. Hydrol. (N.Z.)* **21**: 61-75.

Burt T.P. and Butcher D.P. (1985): On the generation of delayed peaks in stream discharge, *J. Hydrol.* **78**: 361-378.

Cailleux A. and Tricart J. (1959): Le probleme de la classification des faits géomorphologiques, *Ann. Geog.* **65**: 162-186.

Camus H. and Bourges J. (1983): Recherches en milieu mediterranéen aride. Analyse du ruissellement sur un bassin versant du sud tunisien, ORSTOM-DRE, Tunis.

Canterford R.P., Pierrehumbert C.L. and Hall A.J. (1981): Frequency distribution for heavy rainfalls in north-west Australia, WMO Seminar on Hydrology of Tropical Regions, Miami.

Card R.J. (1979): Synthesis of streamflow in a prairie environment, in *The hydrology of areas of low precipitation*, IAHS Publ. No.128: 11-21.

Carr P.A. and Van der Kamp, G.S. (1969): Determining aquifer characteristics by the tidal method, *Water Resour. Res.* **5** (5): 1023-1031.

Cahuepe M., Leon R.J.C., Sala O. and Soriano A. (1982): Pastizales naturales y pasturas cultivadas, dos systemas complementarios y no opuestos, *Revista Faculdad de Agronomia* **3** (1): 1-11 (Buenos Aires, Argentina).

Chambers R. (1978): Water management and paddy production in the dry zone of Sri Lanka, Occasional Publ. No.8, ARTI, Colombo.

Champa S. and Boonpirugsa S. (1979): Water resources development of the Mae

Klong River basin, IWRA 3rd World Congr. on Water Resour., Mexico City 8: 3975-4010.

Chandler R.L. and McWhorter D.B. (1975): Upconing of the saltwater-freshwater interface beneath a pumping well, *Ground Water* 13: 354-359.

Chapman B.M. (1982): Numerical simulation of the transport and speciation of nonconservative chemical reactants in rivers, *Water Resour. Res.* 18: 155-167.

Chapman T.G. (1961): Hydrology survey at Lorna Glen and Wiluna, W.A., CSIRO Aust. Divn. Land Res. & Reg. Surv., Tech. Paper No.18 (CSIRO, Canberra).

Chapman T.G. (1968): Catchment parameters for a deterministic rainfall-runoff model, in *Land Evaluation* (ed. G.A. Stewart), pp. 312-323 (Macmillan, Australia).

Chapman T.G. (1984): Australian approach to hydrology of smooth plainlands, in *Hydrology on large flatlands* (ed. M.C. Fuschini Mejia), Proc. Olavarria Symp. 1:545-551, Unesco, Republica Argentina.

Chapman T.G. (1985): The use of water balances for water resource estimation, with special reference to small islands, *Bull. No .4*, Pacific Regional Team, Aust. Development Assistance Bureau (Sydney, Australia).

Chapman T.G., Bliss P.J. and Smalls I.C. (1982): Water quality considerations in the hydrological cycle, in *Prediction in water quality* (eds. E.M. O'Loughlin and P. Cullen), pp. 27-68 (Aust. Acad. Sci., Canberra).

Chernogaeva G.M. (1974): Analytical model of the water balance structure, *Water Resources* (5): 31-37.

Chorley R.J. (1978): The hillslope hydrological cycle, in *Hillslope hydrology* (ed. M.J. Kirkby), pp. 1-42 (Wiley-Interscience).

Chow V.T. (1964): Runoff, in *Handbook of Applied Hydrology* (ed. V.T. Chow), pp.14.6-14.8 (McGraw-Hill, New York).

Christian C.S. (1958): The concept of land units and land systems, Unesco Symposium on Climate, 9th Pacific Sci. Congr., 20: 74-80.

Clarke R.T. (1977): A review of research on methods for the extrapolation of data and scientific findings from representative and experimental basins, Tech.Doc. in Hydrol.(Unesco, Paris).

Coleman F. (1972): Frequencies, tracks and intensities of tropical cyclones in the Australian region, 1909-1969, Met.Summary, Bureau of Meteorology, Melbourne, Australia.

Colenbrander H.J. (ed.) (1986): *Water in The Netherlands* (TNO Comm. on Hydrological Research, The Hague).

Colombani J. (1978): Utilisation de la sonde à neutrons pour la caracterisation d'un profil de sol du Sud Tunisien, simultanement à l'emploi de mésures tensiométriques et de tracages isotopiques et salins, Journées Sci. du Groupe Français d'Humidité Neutronique CEA CADARACHE.

Colombani J. (1979): Pluviométrie, in *Etudes des Potentialités du Bassin Conventionnel du Lac Tchad* (ORSTOM, Paris).

Colombani J., Kallel R. and Eoche Duval J.M. (1972): *Les précipitations et les crues exceptionelles de l'automne 1969 en Tunisie* (Tunis)

Colombani J., Olivry J.C. and Kallel R. (1984): Phénomènes exceptionelles d'érosion et de transport solide en Afrique aride et semi aride, in *Challenges in African Hydrology and Water Resources,* Proc. Harare Symposium, IAHS Publ. No. 144: 295-300.

Cominetti M.E. (1986): Problems of water quality associated with the mining and extraction industries of south west England - some case histories, in *Effects of Land Use on Fresh Waters* (ed. J.F.G. Solbe), pp. 147-161 (Ellis Horwood, Chichester).

Commission of the European Communities (1982): *Groundwater Resources of the European Communities* (Schafer, Hanover).

Cook H.L. (1946): The infiltration approach to the calculation of surface runoff. *Trans. Amer. Geophys. Union* 27: 726-747.

Cornejo J., Salinas H.Y., Ocena C.L., Brack A.E. and Iglesias W. (1979): Modification of a desert environment. The San Lorenzo irrigation project, Peru, in *Water Management and Environment in Latin America*, pp. 139-159 (Pergamon Press, Oxford).

Cosby B.J., Whitehead P.G. and Neale R. (1986): A preliminary model of long term changes in stream acidity in south western Scotland, *J. Hydrol.* 84: 381-401.

Croll B.T. (1986): The effect of the agricultural use of herbicides on fresh waters, in

Effects of Land Use on Fresh Waters (ed. J.F.G. Solbe), pp. 201-209 (Ellis Horwood, Chichester).

Czelnay R. (1969): On the accuracy of estimated areal averages, *Jdojaras* 73:340-350.

Czelnay R. (1971): Minimum error interpolation by structure functions, *Jdojaras* 75:318-325.

Dabin B. (1957): Note sur le fonctionnement des parcelles expérimentales de l'érosion à la station d'Adiopodioume (Cote d'Ivoire), mimeographed report, Decret Permanent Bureau Sols AOF, Dakar.

Dacharry M. (1974): *Hydrologie de la Loire en amont de Gien*, Publications du Department de Géographie de l'Université de Paris-Sorbonne (Nouvelles Editions Latines, Paris: 2 vols.).

Dagg M. and Pratt M.A.C. (1962): Relation of stormflow to incident rainfall, *E. Afric. Agric. For. J.* **27**: 31-35.

Dahlstrom B. (1978): Methods of computation, space-estimation and graphical display of precipitation, Unesco Workshop on Water Balance of Europe, Varna, Bulgaria, pp. 45-65.

David L. (1981): Theory of processes in catchment-development and the application in regional water management (in Hungarian), *Vizugyi Kozlemenyek* **62** (3):383-410.

David L. (1985): Environmentally sound management of freshwater resources, IWRA 5th World Congr. on Water Resour., Brussels **2**: 841- 850.

Davidson M.R. (1985): Numerical calculation of saturated-unsaturated infiltration in a cracked soil, *Water Resour. Res.* **21** (5): 709-714.

Dale W.R. (1984): Water utilisation in the Pacific Islands, Proc. Regional Workshop on Water Resour. of Small Islands, Suva, Fiji (Commonwealth Science Council Tech. Publ. No. 154), pp. 45-54.

Debenham F. (1952): *Study of an African swamp* (Colonial Office, London).

De Bruin H.A.R. and Kohsiek W. (1977): Evaporation from grass on clay soil in the summer of 1976 compared with the Penman formula, Scientific Report WR 77-10 KNMI, De Bill.

De Martonne E. (1927): Regions of interior-basin drainage, *Geog. Rev.* **17**: 397-414.

Demmark A. (1980): *Contribution à l'étude des écoulements solides des cours d'eau d'Algérie* (DMRH, Algiers).

Denmead O.T., Nulsen R. and Thurnell G.W. (1978): Ammonia exchange over a corn crop, *Soil Sci. Soc. Amer. J.* **42**: 840-842.

D'Hoore J.D. (1961): Influence de la mise en culture sur l'évolution des sols dans la forêt dense de basse et moyenne altitude, Unesco Symp. on Humid Tropical Zone, Abidjan, pp. 49-58.

Dillon P.J. and Liggett J.A. (1983): An ephemeral stream-aquifer interaction model, *Water Resour. Res.* **19** (3): 621-626.

Dincer T., Hutton L.G. and Kupee B.B.J. (1978): Study, using stable isotopes of flow distribution, surface groundwater relations and evapotranspiration in the Okavango swamp, Botswana, in *Isotope Hydrology*, pp. 3-25 (IAEA, Vienna).

DOCTER (1987): *European Environmental Yearbook 1987*, DOCTER - Instituto di Studi e Documentazione per il Territorio: Milano (DOCTER-UK, London).

Dominguez O. and Carballo E. (1984): Uso de la imagen satelitaria en el estudio de los procesos de anegamiento y/o inundacion en grandes llanuras, in *Hydrology on large flatlands* (ed. M.C. Fuschini Mejia), Proc. Olavarria Symp. **2**: 1089-1136, Unesco, Republica Argentina.

Dooge J.C.I. (1983): On the study of water, *Hydrol. Sci. J.* **28**(1): 23-48.

Doorenbos J. and Pruitt W.O. (1977): Crop water requirements, *FAO Irrig. Drainage Pap. 24* (FAO, Rome).

Doornkamp J.C., Gregory K.J. and Burn A.S. (1980): *Atlas of Drought in Britain, 1975-1976* (Institute of British Geographers, London).

Downes C.J. (1981): Water quality in the Pacific Islands, in *Pacific Island Water Resources* (ed. D. Dale), South Pacific Tech. Inventory 2:93-110 (DSIR, New Zealand).

Dreyer N.N. (1978): *Water balance in North America* (Nauka, Moscow).

Dreyer N.N., Nikoleyeva G.M. and Tsigelnaya I.D. (1982): Maps of streamflow resources of some high-mountain areas in Asia and North America, IAHS Publ. No. 138: 11-20.

Dubreuil P. and Vuillaume G. (1970): Etude analytique du ruissellement et de l'érosion en region tropicale sur bassins versants de quelques hectatres a Koutkouzout (Niger), Proc. Internat. Water Erosion Symposium, Prague.

Dubreuil P. and Vuillaume G. (1975): Influence du milieu physico-climatique sur l'écoulement de petits bassins intertropicaux, in *The Hydrological Characteristics of River Basins*, IAHS Publ. No. 117: 205-215.

Duckstein L., Fogel M. and Bogardi I. (1979): Event-based models of precipitation for semiarid lands, in *The hydrology of areas of low precipitation*, IAHS Publ. No.128: 51-64.

Dunne T. (1978): Field studies of hillslope flow processes, in *Hillslope Hydrology* (ed. M.J. Kirkby), pp. 277-293 (Wiley, Chichester).

Dunne T. (1979): Sediment yield and land use in tropical catchments, *J. Hydrol.* 42: 281-300.

Dunne T. and Black R.D. (1970): Partial area contributions to storm runoff in a small New England watershed, *Water Resour. Res.* 6: 1296-1311.

Dunne T., Moore T.R. and Taylor C.H. (1975): Recognition and prediction of runoff producing zones in humid regions, *Hydrol. Sci. J.* 10: 305-327.

Duran D. (1982): *La alternancia de sequias e inundaciones (un problem clave de la Pampa Deprimida)* (OIKOS, Buenos Aires).

Duran D. (1986): Simultaneidad de sequias e inundaciones, unpublished, Buenos Aires.

Dutt G.R., Shaffer M.V. and Moore W.J. (1972): Computer simulation model of dynamic biophysicochemical processes in soils, Tech. Bull. 196, Dept. Soil, Water & Eng. (Univ. Arizona, Tucson).

Eagleson P.S. (1970): *Dynamic hydrology* (McGraw-Hill, New York).

Eagleson P.S. (1982): Ecological optimality in water-limited natural soil-vegetation systems 1. Theory and hypothesis, *Water Resour. Res.* 18 (2): 325-340.

Eagleson P.S. (1986): The emergence of global-scale hydrology, *Water Resour. Res.* **22** (9): 6S-14S.

Elias V. and Cavalcante A.J.S. (1983): Recent hydrological and climatological activities in the Amazon basin, Brazil, in *Hydrology of Humid Tropical Regions* (ed. R. Keller), IAHS Publ. No. 140: 365-373.

Ellis J.B. (1985): Urban runoff quality and control, in *Advances in Water Engineering* (ed. T.H.Y. Tebbutt)pp. 234-240 (Elsevier, London).

Emmenegger C. and Spreafico M. (1979): La station hydrométrique fédérale de la Massa-Blatten au front du glacier d'Aletsch,Mitteilung Nr. 41, VAW ETH Zurich, pp. 23-38.

Engel H. (1977): *Anthropogene Einflusse auf die Abflussverhaltnisse des Oberrheins zwischen Basel und Worms* , Kurzdokumentation der Arbeiten der Hochwasser- studienkommission fur den Rhein, Koblenz.

Erhard-Cassegrain A. and Margat J. (1979): Introduction à l'économie générale de l'eau, Bureau de Recherches Géologiques et Minières, Orléans Cedex.

Ermakov Y.G. and Ignatiev E.M. (1971): Physical and geographical analogies and their causes, in *Natural Resources and Cultural Landscapes of Continents* (Moscow University).

ESCAP (1963): The development of ground-water resources with special reference to deltaic areas, Water Resour. Ser. No. 24 (ESCAP, Bangkok).

Evans D. (1983): Agricultural demand for water in lowland areas *J. Inst. Wat. Eng. and Sci.* **37**: 513-521.

Falkenmark M. (1981): Transfer of water know-how from high to low latitudes. Some problems and biases, *Geophysica* **17** (1-2): 5-20.

Falkenmark M. (1983): Urgent message from hydrologists to planners: water a silent messenger turning land use into river response, Symposium on Scientific Procedures Applied to the Planning, Design and Management of Water Resources Systems, Hamburg, IAHS Publ.No.147: 61-75.

Falkenmark M. (1984): New ecological approach to the water cycle: ticket to the future, *Ambio* **13** (3): 152-160.

Falkenmark M. (1986 a): Macro-scale water supply/demand comparisons on the global scene - a hydrological approach, *Beitrage zur Hydrologie* **6**:15-40 (Hydrology and Earht Sciences, Festschrift Prof. Dr. R. Keller).

Falkenmark M. (1986 b): Fresh water - time for a modified approach, *Ambio* **15** (4): 192-200.

Falkland A.C. (1984 a): Assessment of groundwater resources on coral atolls: case studies of Tarawa and Christmas Islands, Republic of Kiribati, Proc. Regional Workshop on Water Resour. of Small Islands, Suva, Fiji (Commonwealth Science Council Tech. Publ. No. 154), pp. 261-276.

Falkland A.C. (1984 b): Assessment of surface water runoff and determination of groundwater recharge on small high islands: a case study of Norfolk Island, Proc. Regional Workshop on Water Resour. of Small Islands, Suva, Fiji (Commonwealth Science Council Tech. Publ. No. 154), pp. 277-289.

Falkland A.C. (1984 c): Development of groundwater resources on coral atolls: experiences from Tarawa and Christmas Island, Republic of Kiribati, Proc. Regional Workshop on Water Resour. of Small Islands, Suva, Fiji (Commonwealth Science Council Tech. Publ. No. 154), pp. 436-452.

FAO (1974): *World Soil Map* (FAO/Unesco, Paris and Rome).

FAO/UNEP/IIASA (1983): Potential population supporting capacities of lands in the developing world, Land resources for population of the future, FAO Tech. Report FPA/INT/513.

Fatt C.S. (1985): Sediment problems and their management in peninsular Malaysia, *Water International* **10**: 3-6.

Ferguson J. (1952): The rate of natural evaporation from shallow ponds, *Aust. J. Sci. Res.* **5**: 315-330.

Finlayson B.L., McMahon T.A., Srikanthan R. and Haines A. (1986): World hydrology: a new data base for comparative analyses, Inst. Eng. Aust., Hydrol. and Water Resour. Symp., Brisbane, pp. 288-296.

Fioriti M.J. and Fuschini Mejia M.C. (1985): Los efectos de la ocupacion humana en el funcionamiento fisico del ambiente llanura. Casa de estudio, Pampa Bonaerense, GAEA, Contribuciones Cientificas, Buenos Aires.

Fleming P.M. (1964): A water budgeting method to predict plant response and irrigation requirements for widely varying evaporation conditions. *Proc. 6th Int. Cong. Agric. Engg.*, **2**: 66-77.

Fleming P.M. and Smiles D.E. (1975): Infiltration of water into soil, in *Prediction in Catchment Hydrology* (eds. T.G. Chapman & F.X. Dunin), pp. 83-110 (Aust. Acad. Sci., Canberra).

Forkasiewicz J. and Margat J. (1980): *Tableau mondial de données nationales d'économie de l'eau*, Bureau de Recherches Géologiques et Minières, Département Hydrogéologie, 79 SGN 784 HYD, Orleans.

Fournier F. (1969): Transports solides effectués par les cours d'eau, *Bull. Int. Assoc. Sci. Hydrol.* **14** (3): 7-47.

Fox F.A. and Rushton K.R. (1976): Rapid recharge in a limestone aquifer, *Ground Water* **14**: 21-27.

Framji K.K., Garg B.C. and Luthra S.D.L. (1982): *Irrigation and drainage in the world: a global review*, 3rd edn. (ICID, New Delhi).

Fuschini Mejia M.C. (ed.) (1984): *Hydrology on large flatlands* , Proc. Olavarria Symp., Conclusions **1**:71-95 (Unesco, Republica Argentina: 3 vols.).

Fuschini Mejia M.C. (1985): Hidrologia de las llanuras, 5th IWRA World Congr. on Water Resour., Brussels **3**: 1603-1614.

Gallo V. (1985): The role of thermal expansion factors in the variation of water level in evaporation pans (in Hungarian), *Jdojaras* **89**:167-172.

Gandin L.S. (1960): Objective analysis of meteorological fields, *Gidrometeorologisheskoe Izdatelstoo* (Leningrad).

Garcia M.L. (1983): Hydrology of Central America, *Beitrage zur Hydrologie* **9** (1): 23-40.

Gardiner V. (1980): Water supply, in *Atlas of Drought in Britain 1975-1976* (eds.J.C. Doornkamp, K.J. Gregory, and A.S. Burn), pp. 69-70 (Institute of British Geographers, London).

Gardner C.M.K. (1981): The soil moisture data bank: moisture content data from some British Soils, Institute of Hydrology Report 76, Wallingford, Oxon.

Gardner C.M.K. (1986): The present soil moisture regime on Yarnton Mead, Institute of Hydrology, Informal Report, Wallingford, Oxon.

Gash J.H.C. and Stewart J.B. (1977): The evaporation from Thetford Forest during 1975, *J. Hydrol.* **35**: 385-396.

Germann P.F. and Beven K.J. (1985): Kinematic wave approximation to infiltration into soils with sorbing macropores, *Water Resour. Res.* **21** (1): 33-44.

Gilmour D.A., Cassells D.S. and Bonell M. (1982): Hydrological research in the tropical rainforests of North Queensland: some implications for land use management, Proc. 1st Nat. Symp. Forest Hydrol., Melbourne, pp.145-151.

Giraldez J.V. and Sposito G. (1985): Infiltration in swelling soils, *Water Resour. Res.* **21**(1): 33-44.

Giusti E.V. (1978): Hydrogeology of the karst of Puerto Rico, *USGS Prof.Paper No.1012.*

Godz P., Costa J.L., Gonzalo Belo R., Vidal N. and Lazovich M. (1984): La Pampa Deprimida de la Provincia de Buenos Aires, Argentina, in *Hydrology on large flatlands* (ed. M.C. Fuschini Mejia), Proc. Olavarria Symp. **2**: 939-972, Unesco, Republica Argentina.

Gong Shiyang and Mou Jinze (1985): Methods of land and water conservation in the Wuding river basin, in *Strategies for river basin management* (eds. J. Lundqvist, U. Lohm and M. Falkenmark), pp.81-89 (Reidel).

Gras R.A. (1985): Metropolitan areas and water quality, Proc. Internat. Symp. on Water Management in Metropolitan Areas, Sao Paolo: 61-78.

Gray D.M., Granger R.J. and Landine P.G. (1986): Modelling snowmelt infiltration and runoff in a prairie environment, AWRA Cold Regions Hydrol. Symp.: 427-438.

Green F.H.W. (1980): Current field drainage in northern and western Europe, *J. Env. Managem.* **10**: 149-153.

Green J., King A. and Bowden K. (1975): Economics of sewerage design, R. Inst. Public Admin. Report No. C218.

Green W.H. and Ampt G.A. (1911): Studies on soil physics. 1. The flow of air and water through soils, *J. Agric. Sci.* **4**: 1-24.

444

Gregory K.J. and Walling D.E. (1973): *Drainage Basin Form and Process* (Edward Arnold, London).

Grigoriev A.A. (1970): Types of geographical environment, Selected Theoretical Papers, Mysl, Moscow.

Grimm F. (1968): Das Abflussverhalten in Europa - Typen und regionale Gliederung, Wissenschaftliche Veroffentlichungen des Deutschen Instituts fur Landerkunde, NF 25/26 Leipzig, pp. 18-180.

Grimmelman W.F. (1984): Initial assessment of groundwater resources on small islands, Proc. Regional Workshop on Water Resour. of Small Islands, Suva, Fiji (Commonwealth Science Council Tech. Publ. No. 154), pp. 290-304.

Grove A.T. (1971): The dissolved and solid load carried by some West African rivers, *J. Hydrol.* **16**: 287-300.

Guilcher A. (1979): *Précis d'hydrologie marine et continentale*, 2nd edn. (Masson, Paris).

Guillen J.A. (1984): Hydrogeological facts about dyke aquifers and underground water circulation in Tahiti, Proc. Regional Workshop on Water Resour. of Small Islands, Suva, Fiji (Commonwealth Science Council Tech. Publ. No. 154), pp. 455-492.

Guiscrafet J., Klein J.C. and Moniod F. (1976): Les ressources en eau de surface de la Martinique, Monogr. hydrol. No. 3, ORSTOM, Paris.

Gulliver (1899): Shoreline topography, *Proc. Amer. Acad. Arts & Sci.* **34**: 151-258.

Gustafsson J.-E. (1984): Water resources development in the People's Republic of China, Meddelande Trita-Kut 1035, Royal Institute of Technology, Stockholm.

Gustafsson J.-E. (1985): Soil and water conservation on the Chinese loess plateau - the example of the Xio Shi Guo brigade, in *Strategies for river basin management* (eds. J. Lundqvist, U. Lohm and M. Falkenmark), pp. 91-97 (Reidel).

Gustard A. (1983): Regional variability of soil characteristics for flood and low flow estimation, *Agric. Water Managem.* **6**: 255-268.

Habermehl M.A. (1987): The Great Artesian Basin - its groundwater development, and natural and artificial discharge, in *Groundwater Systems under Stress*, Aust. Water Resour. Council Conf. Ser. No. 13 (Aust. Gov't. Publishing Serv., Canberra).

445

Hadley R. (1975): Classification of representative and experimental basins, Tech. Doc. SC/75/WS/66, Unesco, Paris.

Haeberli W. (1983): Frequency and characteristics of glacier floods in the Swiss Alps, Ann. Glaciol. 4: 85-90.

Hahn G.W. and Fisher N.H. (1963): Review of available ground water data of the Great Artesian Basin, in Water resources, use and management, pp.167-187 (Macmillan, Melbourne).

Hall A.J. (1984): Hydrological networks on small islands, Proc. Regional Workshop on Water Resour. of Small Islands, Suva, Fiji (Commonwealth Science Council Tech. Publ.No. 154), pp. 305-321.

Hall D.K. and Martinec J. (1985): Remote sensing of ice and snow (Chapman and Hall).

Hatfield J.L., Reginato R.J. and Idso S.B. (1984): Evaluation of canopy temperature -evapotranpiration models over various crops, Agric. & Forest Meteorol. 32: 41-53.

Havlik D. (1969): Die Hohenstufe maximaler Niederschlagssummen in den Westalpen, Freiburger Geographische 7.

Henderson-Sellars A. (1981): Climate sensitivity variations in vegetated land surface albedos, Proc. 6th Annual Climate Diagnostics Workshop, Columbia Univ., pp.135-144.

Hengeveld H. and De Vocht C. (1982): Role of water in urban ecology, Urban Ecol. 6 (1-4): 1-362.

Henin S. (1986): Water quality - the French problem, in Effect of Land Use on Fresh Waters (ed. J.F.G. Solbe), pp. 210-220 (Ellis Horwood, Chichester).

Hessell J.W.D. (1981): Climatology of the south-west Pacific islands. 2. Climatological summaries, South Pacific Tech. Inventory 2: 35-46.

Heusch B. (1985): Cinquantes ans de banquetes de DRS-CES en Afrique du Nord, Bulletin Erosion No.5, Grenoble.

Hewitt K. (1982): Natural dams and outburst floods in the Karakoram Himalaya, in Hydrological aspects of alpine and high mountainous areas, IAHS Publ. No. 138: 259-269.

Hewlett J.D. and Hibbert A.R. (1967): Factors affecting the response of small watersheds to precipitation in humid areas, in *Forest Hydrology* (eds. W.E. Sopper and H.W. Lull), pp. 275-290 (Pergamon, New York).

Hibbert A.R. (1967): Forest treatment effects on water yield, in *Forest Hydrology* (eds. W.E. Sopper and H.W. Lull), pp. 537-543 (Pergamon).

Higgins G.M., Dieleman P.J. and Abernethy C.L. (1987): Trends in irrigation development and their implications for hydrologists and water resources engineers, *J. Hydraul. Res.* **25**: 393-406.

Hindson J.R.E. (1955): Protection of dambos by means of contour seepage furrows, Ministry Agric. Internal Report, Lusaka.

Hodge C.A.H., Burton R.G.O., Corbett W.M., Evans,R. and Seale R.S. (1984): Soils and their use in Eastern England, Soil Survey of England and Wales, Bull.13, Harpenden.

Hoinkes H. (1969): Surges of the Vernagtferner in the Oetztal Alps since 1959, *Canadian J. Earth Sci.* **6**:853-861.

Holdridge L.R. (1967): *Life zone ecology*, Revised edition (San Jose, Costa Rica: Tropical Science Center).

Hornbeck J.W., Pierce R.S., Likens G.E. and Martin C.W. (1975): Moderating the impact of contemporary forest cutting on hydrologic and nutrient cycles, in *The Hydrological Characteristics of River Basins*, IAHS Publ. No. 117: 423-433.

Hornung M., Reynold B., Stevens P.A. and Neal C. (1987): Increased acidity and aluminium concentrations in streams following afforestation: causative mechanisms and processes, in *Acidification and Water Pathways*, Norwegian Nat. Comm. for Hydrology/Unesco/WMO **1**: 259-268.

Horton R.E. (1919): Rainfall interception, *Mon. Weather Rev.* **47**: 603-623.

Horton R.E. (1933): The role of infiltration in the hydrologic cycle, *Trans. Amer. Geophys. Union* **14**: 446-460.

Hotes F.L. and Pearson E.A. (1977): Effects of irrigation on water quality, in *Arid Land Irrigation in Developing Countries* (ed. E.B.Worthington), pp.127-158 (Pergamon Press, Oxford).

Howard K.W.F. and Lloyd J.W. (1979): The sensitivity of parameters in the Penman evaporation equation and direct recharge balance, *J. Hydrol.* **39**: 355-364.

Hsueh L.T. (1979): Groundwater over pumpage and land subsistence in Taipei basin, Taiwan, Proc. 3rd IWRA World Congr. on Water Resour., Mexico, pp. 3809-3818.

Humbert J. (1982): Cinq années de bilans hydrologiques mensuels sur un petit basin versant des Hautes Vosges, in *Structure et Fonctionnement du Milieu Naturel en Moyenne Montagne* (Recherche Géographiques à Strasbourg 19-21), pp.105-122 (Assoc. Geograph. d'Alsace).

Hindson J.R.E. (1955): Protection of dambos by means of contour seepage furrows, Ministry of Agric. Interim Report, Lusaka.

Hurst H.E. (1954): *Le Nil* (Payot, Paris).

Hvitved-Jacobsen T. (1986): Conventional pollutant impacts on receiving waters, in *Urban Runoff Pollution* (eds. H.Torno, J.Marsalek and M.Desbordes), pp. 345-378 (Springer-Verlag, Heidelberg).

Hwang D. (1981): Beach changes on Oahu as revealed by aerial photographs, Hawaii Inst. of Geophysics, H1G - 81/3.

Hydrologisches Jahrbuch der Schweiz (1965-1982)(Landeshydrologie, Bern).

Ichim I., Surdeanu V. and Radoane N. (1984): Landsliding, slope development and sediment yield in a temperate environment: north east Romania, in *Catchment Experiments in Fluvial Geomorphology.* (eds. T.P. Burt and D.E. Walling), pp.289-298 (Geobooks, Norwich).

Ikuse T., Mimura A., Takeuchi S. and Matsishita J. (1975): Effect of urbanisation on runoff characteristics, in *The Hydrological Characteristics of River Basins*, IAHS Publ. No. 117: 377-385.

Imeson A.C., Vis M.A. and Duysings J.J.H.M. (1984): Surface and subsurface sources of suspended solids in a forested drainage basin in the Keuper region of Luxembourg, in *Catchment Experiments in Fluvial Geomorphology* (eds. T.P. Burt and D.E. Walling), pp. 219-234 (Geobooks, Norwich).

Institute of Hydrology (1980): *Low Flow Studies* (IH Wallingford, Oxon.).

Institute of Hydrology / British Geological Survey (1985): Hydrological Data UK, 1982 Yearbook, IH/BGS Wallingford, Oxon.

Institute of Hydrology (1985): *Research Report 1981-84* (Natural Environment Research Council, Wallingford).

INTA (1973): Inventario del Recurso Suelo de la Provincia de Buenos Aires, (escala 1:500 000), unpublished, Instituto Nacional de Tecnologia Agropecuaria, Buenos Aires.

INTA (1977): *La Pampa Deprimida, condiciones de denaje de sus suelos. Mapa geomorfologico CGA (Tricart) y de la Red Hidrografica del rio Salado,* INTA No.154, Instituto Nacional de Tecnologia Agropecuaria, Buenos Aires.

Jackson I.J. (1971): Problems of throughfall and interception assessment under tropical forest, *J. Hydrol.* **12**: 235-253.

Jacobson G. and Lau J.E. (1983): The importance of groundwater in Australia - a national perspective, in *Groundwater and man,* Aust. Water Resour. Council Conf. Ser. No. 8, **3**: 109-121.

Jacobson G. (1984 a): Niue Island: an example of a raised atoll, Proc. Regional Workshop on Water Resour. of Small Islands, Suva, Fiji (Commonwealth Science Council Tech. Publ. No. 154), pp. 79-85.

Jacobson G. (1984 b): The assessment of groundwater resources on small oceanic islands, Proc. Regional Workshop on Water Resour. of Small Islands, Suva, Fiji (Commonwealth Science Council Tech. Publ. No. 154), pp. 322-343.

Jamieson D.J. and Nicolson N.J. (1984): Water resources of the Thames Basin: quantitative and qualitative aspects, *J. Inst. Wat. Eng. and Sci.* **38**: 379-391.

Jaworska M. (1968): Erozja chemiczna: denudacja zlewni rzek Wieprza i Pilicy, *Prace Panstwoweqo Instytuto Hydrologiczno-Meteorologicznego* **95**: 29-47.

Johnson A.I., Prill R.C. and Morris D.A. (1963): Specific yield - column drainage and centrifuge moisture content, Water-supply Paper 1662-A, U.S. Geol.Surv., Washington D.C..

Jones J.A.A. (1987): The effects of soil piping on contributing areas and erosion patterns, *Earth Surf. Proc. & Landforms* **12**: 229-248.

Junge C.E. (1963): *Air chemistry and radioactivity* (Academic Press, New York).

Kamarck A.M. (1976): *The tropics and economic development* (Johns Hopkins Univ. Press).

449

Kang Ersi (1985): A preliminary glacio-hydrological comparison between some glaciers in the Swiss Alps and the Chinese Tianshan, Arbeitschaft Nr. 7, Laboratory of Hydraulics, Hydrology and Glaciology, ETH Zurich, pp. 1-26.

Kasser P. (1973): Influence of changes in the glacierized area on summer runoff in the Porte du Scex drainage basin of the Rhone, IAHS Publ. No. 95: 221-225.

Kasser P. (1981): Rezente Gletscherveranderungen in den Schweizer Alpen. Gletscher und Klima, Jahrbuch der Schweizerischen Naturforschenden Gesellschaft, wissenschaftlicher Teil 1978, pp. 106-138, Birkhauser Verlag, Basei.

Kasser P., Aellen M. and Siegenthaler H. (1983): Die Gletscher derSchweizer Alpen 1975/76 und 1976/77, in *Glaziologisches Jahrbuch der Gletscherkommission der SNG*, Chs.97 and 98.

Kats N.Y. (1971): *Marshes of the Globe* (Nauka, Moscow).

Keller R. (1979): *Hydrologische Atlas der Bundesrepublik Deutschland* (Deutsche Forschungsgemeinschaft, Bonn).

King F.H. (1898): Principles and conditions of the movements of ground water, 19th Annual Report, U.S. Geol. Surv., Washington D.C..

Kinoshita T., Takeuchi K., Musiake K. & Ikebuchi S. (1986): Hydrology of warm humid islands, Paper presented at Workshop on Comparative Hydrology, Budapest, 11-12 July, 1986.

Kirkby M.J. and Chorley R.J. (1967): Throughflow, overland flow, and erosion, *Bull. IAHS* **12** (3): 5-21.

Kirkby M.J. and Morgan R.P.C. (1980): *Soil Erosion* (Wiley, New York).

Klige R.K. (1981): Man's impact on water resources, *Water International* **6** (3): 117-121.

Kling H., Furch K., Irmler U. and Junk W.S. (1981): Fundamental ecological parameters in Amazonia in relation to the potential development of the region, in *Tropical Agricultural Hydrology* (ed. R. Lal and E.W. Russell), pp. 19-36 (Wiley).

Koeppen W. (1931): *Die Klimate der Erde* (Berlin, Walter de Gruyter).

Koeppen W. (1936): Das geographische System der Klimate, in *Handbuch der Klimatologie*, by W. Koeppen & G. Geiger, v.1, part C (Gebr. Borntraeger, Berlin).

Kotwicki V. (1986): *Floods of Lake Eyre* (Engineering and Water Supply Department, Adelaide).

Kovacs G. (1977): Human interaction with groundwater, *Ambio* **6** (1): 22-26.

Kovacs G. (1978 a): Hydrology and water control on large plains, *Hydrol. Sci. Bull.* **23** (3): 305-332.

Kovacs G. (1978 b): Modernization of the principles of water control and storage in large plains, *Vizugyi Kozlemenyek* **60** (2): 208-226.

Kovacs G. (1980): The interpolation of hydrological data, Symposium on the Influence of Man on the Hydrological Regime with Special Reference to Representative and Experimental Basins, Helsinki, IAHS Publ.No.130: 389-397.

Kovacs G. (1981): *Seepage Hydraulics* (Akademiai Kiado, Budapest).

Kovacs G. (1984 a): General principles of flat-lands hydrology, in *Hydrology on large flatlands* (ed. M.C. Fuschini Mejia), Proc. Olavarria Symposium **1**: 297-355 (Unesco, Republica Argentina).

Kovacs G. (1984 b): Proposal to construct a coordinating matrix for comparative hydrology, *Hydrol. Sci. J.* **29** (4): 435-443.

Kovacs G. (1984 c): Determination of average runoff by applying continuum approach (in Hungarian), *Vizugyi Kozlemenyek* **64** (3):

Kovacs G. (1986 a): Time and space scales in the design of hydrological networks, Symposium on Integrated Design of Hydrological Networks, Budapest, IAHS Publ.No.158: 283-294.

Kovacs G. (1986 b): Decision support system for managing large international rivers, Manuscript, International Institute for Applied Systems Analysis, Laxenburg, Austria.

Kovacs G. (1987): Estimation of areal evapotranspiration (in preparation).

Kovacs G. and Associates (1981): *Subterranean hydrology* (Water Resour.Publ., Littleton, Colorado)

Kovacs G. and Molnar G. (1973): Determination of the snow water equivalent thickness of snow cover data, Symposium on Design of Water Resources Projects with Inadequate Data, Madrid **2**: 205-216, *Studies and Reports in Hydrology 16* (Unesco, Paris).

Kulikov Y.N. (1976): Special features of the water and heat balance, and ways to transform the nature of the Vasjugen area, *Water Resources* :95-110.

Kunkle S.H. (1972): Effects of road salt on a Vermont stream, *J. Amer. Waterworks Assoc.* **64**: 290-294.

Kushid Alam C.F. (1973): Distribution of precipitation in mountainous areas of West Pakistan, WMO No. 326, **2**: 290-306 (WMO, Geneva).

Kuusisto E. (1986): The energy balance of a melting snow cover in different environments, in *Modelling Snowmelt-Induced Processes* (ed. E.M. Morris), IAHS Publ. No. 155: 37-45.

Kuprianov V.V. (1974): Hydrological effects of urbanization in the Union of Soviet Socialist Republics, in *Hydrological Effects of Urbanization,* Studies and Reports in Hydrology 18, Unesco, Paris.

Lai Zuming (1982): A study on the variation coefficient of annual runoff of the rivers in Northwest China, IAHS Publ. No. 138: 285-294.

Lal R. (1981): Deforestation of tropical rainforest and hydrological problems, in *Tropical Agricultural Hydrology* (ed. R.Lal and E.W. Russell), pp. 131-140 (Wiley).

Lal R. (1983): Soil erosion in the humid tropics with particular reference to agricultural land development and soil management, in *Hydrology of Humid Tropical Regions* (ed. R. Keller), IAHS Publ. No. 140: 221-239.

Landsberg H.E. (1975): Weather, climate and settlements, Background paper for UN Conf. on Human Settlements, Vancouver, Report A/CONF. 70/8/1, United Nations, New York.

Landsberg H.E. (1981): *The Urban Climate* (Academic Press, New York).

Lang H. (1981): Is evaporation an important component in high alpine hydrology? *Nordic Hydrol.* **12**: 217-224.

Lang H. (1984): A view on our knowledge of water resources in relation to climate in the Alps, *Zurcher Geogr. Schriften* **14**: 53-58.

Lang H. (1986): Hydrological aspects of high mountain areas above the forest line,

Paper presented at Workshop on Comparative Hydrology, Budapest, 11-12 July 1986.

Langbein W.B. and Schumm S.A. (1958): Yields of sediments in relation to mean annual precipitation, *Trans. Amer. Geophys. Union* **39**: 1076-1084.

Laski A. (1977): Long term national water resources development planning in Poland, UN Water Conf., Mar del Plata, Argentina, 14-25 March 1977.

Lauer W. (1976): Klimatische Grundzuge der Hohenstufung tropischer Gebirge, Tagungsbericht 40, Deutscher Geographentag, Wiesbaden, pp.76-90.

Laurenson E.M. (1964): A catchment storage model for runoff routing, *J. Hydrol.* **2**: 141-163.

Lauscher F. (1976): Weltweite Typen der Hohenabhangigkeit des Niederschlags, *Wetter u. Leben* **2**: 80-90.

Laut P., Austin M.P., Goodspeed M.J., Body D.N. and Faith D.P. (1984): Hydrologic classification of sub-basins in the Macleay valley, New South Wales, *Trans. Inst. Eng. Aust.* **CE26** (3): 218-135.

Law F. (1956): The effect of afforestation upon the yield of water catchment areas, *J.Br.Waterworks Assoc.* **38**: 484-494.

Lawson T.L., Lal R. and Oduro-Afiriye K. (1981): Rainfall redistribution and microclimatic changes over a cleared watershed, in *Tropical Agricultural Hydrology* (ed. R.Lal and E.W. Russell), pp.141-151 (Wiley).

Lazaro T.R. (1979): *Urban Hydrology* (Ann Arbor Science Publishers, Michigan).

Lennett D.J. (1980): Handling of hazardous wastes, *Environment* **22** (8): 6-15.

Leon R.J.C. (1980): *Las comunidades herbaceas de la region Castelli-Pila* (Buenos Aires, Argentina).

Lettau H., Lettan K. and Molion L.C.B. (1979): Amazonia's hydrological cycle and the role of atmospheric recycling in assessing deforestation effects, *Mon. Weather Rev.* **107** (3): 227-238.

Limbrey S. (1983): Archaeology and palaeohydrology, in *Background to Palaeohydrology* (ed. K.J.Gregory), pp. 189-212 (Wiley, Chichester).

Lindh G. (1979): Socio-economic aspects of urban hydrology, *Studies and Reports in Hydrol.* 27, Unesco, Paris.

Lindh G. (1983): *Water and the City* (Unesco, Paris).

Lindh G. (1985): The planning and management of water resources in metropolitan regions, Report 3105, Dept. Water Resour. Eng., Lund Univ..

Linsley R.K. (1957): The hydrologic cycle in relation to meteorology, in *Compendium of Meteorology* (ed. T.F. Malone), pp.1048-1054 (Amer. Meteorol. Soc.).

Liu Changming, Zuo Dakang and Xu Yuexian (!985): Water transfer on China: the East Route Project, in *Large scale water transfers* (eds. G.N. Golubev and A.K. Biswas), pp. 103-118 (Tycooly, Oxford).

Lloyd J.W. (1984): A review of some of the more important difficulties encountered in small island hydrogeological investigations, Proc. Regional Workshop on Water Resour. of Small Islands, Suva, Fiji (Commonwealth Science Council Tech. Publ. No. 154), pp. 180-210.

Lloyd J.W., Miles J.C., Chessman G.R. and Bugg S.F. (1980): A groundwater resources study of a Pacific island atoll - Tarawa, Gilbert Islands, *Water Resour. Bull.* **16** (4): 646-653.

Lockwood J.G. (1976): *The physical geography of the tropics: an introduction* (Oxford Univ. Press, London).

Lomme O. (1961): Effets de caractères du sol sur localisation de la vegetation en zones equatoriales humide, Unesco Symp. on Humid Tropical Zone, Abidjan, pp.25-39.

L'vovich M.I. (1971): *Rivers of the USSR* (Mysl, Moscow).

L'vovich M.I.(1979): *World water resources and their future* (English translation ed. R.L.Nace)(American Geophysical Union).

L'vovich M.I. and Chernogaeva G.M. (1974): Man and environment, in *Studies of Geography in Hungary II*, pp.131-149 (Akademiai Kiado, Budapest).

Macdonald G.A., Abbott A.I. and Peterson F.L. (1983): *Volcanoes in the sea*, 2nd ed. (University of Hawaii Press, Honolulu).

Macrae F.G. (1934): The Lukanga swamp, *Geogr. J.* **83**: 213-218.

Major P. (1975): Etude des processus de l'évaporation de la nappe d'eau phréatic et de l'infiltration efficace au terrains d'étude. Possibilités de généralisation des résultats, in *The Hydrological Characteristics of River Basins*, IAHS Publ. No. 117: 35-46.

Major P. (1984): Generalisation des elements de bilan d'eaux sur les regions plates a l'aide d'un reseau de stations hydrologiques representatives, in *Hydrology on large flatlands* (ed. M.C. Fuschini Mejia), Proc. Olavarria Symp. **1**: 527-544, Unesco, Republica Argentina.

Maller R.A. and Sharma M.L. (1984): Aspects of rainfall excess from spatially varying hydrological parameters, *J. Hydrol.* **67**: 115-127.

Malmqvist P.A. and Hard S. (1981): Groundwater quality changes caused by stormwater infiltration, Proc. 2nd Internat. Conf. on Urban Storm Drainage, pp. 89-97, Univ. Illinois, Urbana-Champaign.

Margat J. (in press): *Les eaux souterraines de l'Afrique*, 2nd.edn. (UN, New York).

Margat J. and Saad K.F. (1984): Deep-lying aquifers: water mines under the desert, *Nature & Resour.* **20** (2): 7-13.

Marlenko N., Piatti L.M. and Redondo F. (1984): Problemas de drenaje e inundaciones en los bajos submeridionales santafesinos, in *Hydrology on large flatlands* (ed. M.C. Fuschini Mejia), Proc. Olavarria Symp. **2**: 611-640, Unesco, Republica Argentina.

Marsalek J. (1984): Caracterisation du ruissellement de surface issu d'une zone urbaine commerciale, *Sci. et Tech. de l'Eau* **17** (2): 163-167.

Marsh P. and Woo M-K. (1985): Meltwater movement in natural heterogeneous snow covers, *Water Resour. Res.* **21** (11): 1710-1716.

Marsh T.J. and Davies P.A. (1983): The decline and partial recovery of groundwater levels below London, *Proc. Instn. Civil Eng. Part 1* **74**: 263-276.

Marsh T.J. and Lees M.L. (1985): The 1984 drought. Hydrological data U.K., Institute of Hydrology / British Geological Survey, Wallingford.

Matalas N. C. (1987): *Hydrology in an Island-Continent Context*, Proc. Dahlem

Workshop on Resources and World Development, Part B: Water and Land, Berlin 1986 (Wiley).

Matthai H.F. (1979): Hydrologic and human aspects of the 1976-77 drought, USGS Prof. Paper 1130 (US Printing Office, Washington D.C.).

McIntyre A.K. (ed.)(1977): *Water: planets, plants and people* (Aust. Acad. Sci., Canberra).

McMahon T.A. (1979): Hydrological characteristics of arid zones, in *The hydrology of areas of low precipitation*, IAHS Publ. No. 128: 105-123.

McMahon T.A. (1982): World hydrology: does Australia fit?, Inst. Eng. Aust., Nat. Conf. Publ. 82/3: 1-7.

McNutt M. and Menard H.W. (1978): Lithospheric flexure and uplifted atolls, *J. Geophys. Res* . **83** (B23): 1206-1212.

McPherson M.B. (1973): Need for metropolitan water balance inventories, *J. Amer. Soc. Civil Eng.* **99** (HY10): 1837-1848.

McPherson M.B. and Zuidema F.C. (1977): Urban hydrological modelling and catchment research: International summary, Report IHP 77-4, Amer. Soc. Civil Eng., New York.

McPherson M.B. and Schneider W.J. (1974): Problems in modelling urban watersheds, *Water Resour. Res.* **10** (3): 434-440.

Meier M.F. (1983): Snow and ice in a changing hydrological world, *Hydrol. Sci. J.* **28** (1-3): 3-22.

Mein R.G. and Larson C.L. (1973): Modeling infiltration during a steady rain, *Water Resour. Res.* **9**: 384-394.

Miller D.H. (1977): *Water at the surface of the earth* (Academic Press, New York).

Ministry of Construction (1985-1986): IHP Representative Basins in Korea, Progress 1985-1986, Ministry of Construction, Republic of Korea.

Mink J.F. (1983): Groundwater hydrology in agriculture of the humid tropics, in *Hydrology of Humid Tropical Regions* (ed. R. Keller), IAHS Publ. No. 140: 241-247.

Monteith J.L. (1985): Evaporation from land surfaces: progress in analysis and prediction since 1948, *Proc. National Conf. on Advances in Evapotranspiration,* Chicago, Dec. 1985, pp. 4-12 (Amer. Soc. Agric. Eng., St. Joseph, Michigan).

Mooney E. (1976): Inferno verde, *Defenders* **51** (1): 24-27.

Mohr E.C.J. and Van Baren F.A. (1958): *Tropical Soils* (Interscience, Amsterdam).

Morel-Seytoux H.J. and Billica J.A. (1985): A two-phase numerical model for prediction of infiltration: applications to a semi-infinite soil column, *Water Resour. Res.* **21**(4): 607-615.

Morgenschweis G. (1980): Erfassung und Simulation des Boden Wasserhaushaltes am Biespiel eines Loss-Einzugsgebietes, unpublished PhD thesis, Univ. Freiburg, Fed. Rep. Germany.

Morrison R.J., Prasad R.A and Brodie J.E. (1984): Chemical hydrology on small tropical islands, Proc. Regional Workshop on Water Resour. of Small Islands, Suva, Fiji (Commonwealth Science Council Tech. Publ. No. 154), pp. 211-233.

Morton F.I. (1965): Potential evaporation and river basin evaporation, *Proc. Amer. Soc. Civil Eng.* **91** (HY6): 67-97.

Morton F.I. (1983): Operational estimates of areal evapotranpiration and their significance to the science and practice of hydrology, *J. Hydrol.* **66**: 1-76.

Morton F.I., Ricard F. and Fogarasi S. (1985): Operational estimates of areal evapotranspiration and lake evaporation - program WREVAP, NHRI Paper No. 24, Inland Waters Directorate, Ottawa.

Mosely P. (1985): Upstream - downstream interactions as natural constraints to basin-wide planning for China's river Huang, in *Strategies for river basin management* (eds. J. Lundqvist, U. Lohm and M. Falkenmark), pp. 131-140 (Reidel).

Motovilov Y.G. (1986): A model of snow cover formation and snowmelt processes, in *Modelling Snowmelt-Induced Processes* (ed. E.M. Morris), IAHS Publ. No 177: 47-57.

Mudiare O.J., Gray D.M. and McKay G.A. (undated): Influence of cloud cover on evapotranpiration demand.

Mueller-Dombois D. (1979): Classification of plant communities and methods of mapping primary production: tropical ecosystems, SCOPE *Publications on the role of terrestrial vegetation in the global carbon cycle* .

Mukammel E.I. and Neumann H.H. (1977): Advective effects influencing the evaporation of a Class A pan, WMO Seminar on Areal Evapotranspiration, Budapest, Hungary.

Musiake K., Inokuti S. and Takahasi Y. (1975): Dependence of low flow characteristics on basin geology in mountainous areas of Japan, in *The hydrological characteristics of river basins*, IAHS Publ. No. 117: 147-156.

Nakagawa S. (1984): Study on evaporation from pasture, Environmental Research Center Paper No. 4 , Univ. of Tsukuba, Japan.

Napierkowski J.J. and O'Kane P. (1984): A new non-linear conceptual model of flood waves, *J. Hydrol.* **69**: 43-58.

National Water and Soil Conservation Authority (1984): An index for base flows, Streamland 24, Water Directorate, Wellington, New Zealand.

Nemec J. and Rodier J.A.(1979): Streamflow characteristics in areas of low precipitation (with special reference to low and high flows), in *The hydrology of areas of low precipitation,* IAHS Publ. No. 128: 125-140.

NERC (1975): *Flood Studies Report* (HMSO, London: 5 vol.).

Newson M.D. (1980): The erosion of drainage ditches and its effect on bedload yields in mid-Wales: reconnaissance case studies, *Earth Surf. Proc.* **5**: 275-290.

Nielsen D.R., Van Genuchten M.T. and Biggar J.W. (1986): Water flow and solute transport processes in the unsaturated zone, *Water Resour. Res.* **22** (9): 89S-108S.

Nikolaeva G.M. and Chernogaeva G.M. (1977): *Water balance of Asia*, pp.26-44 (Moscow Radio).

Northcliff S. and Thornes J.B. (1981): Seasonal variations in the hydrology of a small forested catchment near Manaus, Amazonas, and the implications for its management, in *Tropical Agricultural Hydrology* (ed. R.Lal and E.W. Russell), pp.37-57 (Wiley).

Notodihardjo M. and Zuidema F.C. (1982): Integral quantitative and qualitative planning and management of water resources in the river basins in the

Gerban-Kerto-Susila region, Indonesia, in *Optimal allocation of water resources,* IAHS Publ. No. 135: 203-209.

Novikov S.M. (1981): The impact of draining amelioration on the runoff and elements of water balance of the ameliorated territories, in *Some Problems of Modern and Practical Hydrology,* Part 1, pp. 72-78 (Moscow University).

Nyquist H. (1924): Certain factors affecting telegraph speed, *Bell Systems Tech. J.* **3** (2): 324-346.

Odum E.P. (1971): *Fundamentals of ecology,* 3rd edn. (W.B. Saunders, Philadelphia).

OECD (1986): *Water pollution by fertilisers and pesticides* (OECD, Paris).

Oguntoyinbo J.S. and Akintola F.O. (1983): Rainstorm characteristics affecting water availability for agriculture, in *Hydrology of Humid Tropical Regions* (ed.R. Keller), IAHS Publ. No. 140: 63-72.

O'Loughlin E.M. (1981): Saturation regions in catchments and their relations to soil and topographic properties, *J. Hydrol.* **53**: 229-246.

O'Loughlin E.M. (1986): Prediction of surface saturation zones in natural catchments by topographic analysis, *Water Resour. Res.* **22** (5): 798-804.

Orimoyegun S.O. (1986): Influence of forest litter and crop residue on soil erosion, Nat. Workshop on Soil Erosion, Federal Univ. of Tech., Owerri, Nigeria, 8-12 Sept..

Ovington J.D. (1968): Some factors afecting nutrient distribution within ecosystems, in *Functioning of terrestrial ecosystems at the primary productivity level* (ed. F.E. Eckardt), pp. 95-105 (Unesco, Paris).

Owens L.B. and Watson J.P. (1979): Rates of weathering and soil formation on granite in Rhodesia, *Soil Sci. Soc. Amer. Proc.* **43**: 160-166.

Oyebande L. (1975): Water resource problems in Africa, *African Environment* (Special Issue), pp. 40-53 (Internat. African Inst., London).

Oyebande L. (1978): Urban water supply planning and management in Nigeria, *GeoJournal* **2** (5): 403-412.

Oyebande L. (1981): Sediment transport and river basin management in Nigeria, in *Tropical Agricultural Hydrology* (ed. R. Lal and E.W. Russell), pp.201-225 (Wiley).

Oyebande L. (1986): Socio-economic factors and hydrological applications in Nigeria, Unesco Workshop on Comparative Hydrology, 11-12 July, 1986.

Oyebande L. (1987): Effect of tropical forest on water yield, in *Forest, Climate and Hydrology - Regional Impacts*, Proc. UN Univ. Workshop, Oxford Univ., 26-30 March 1984.

Packman J.C. (1980): The effects of urbanisation on flood magnitude and frequency, Institute of Hydrology Report 63, Wallingford, Oxon.

Pankow J.F. and Morgan F.M.M. (1981): Kinetics for the aquatic environment, *Environ. Sci. Technol.* **15**: 1155-1164.

Parde M. (1947): *Fleuves et Rivières*, 3rd Edition (Paris).

Parfait J-A. and Lallmahomed H. (1980): The effects of change in land use on the hydrological regimes of three small basins in Mauritius, Symposium on the Influence of Man on the Hydrological Regime with Special Reference to Representative and Experimental Basins, Helsinki, IAHS Publ. No. 130: 351-358.

Parodi L. (1947): La Estepa Pampeana, la vegetacion de la Argentina, in *Geografia de la Republica Argentina,* vol.8 (GAEA, Buenos Aires).

Paterson J. (1984): Exploitation of natural systems and the existence of a sustainable yield solution, *Water* **11** (4): 5.

Patzelt G. (1973): Die neuzeitlichen Gletscherschwankungen in der Venedigergruppe (Hohe Tauern, Ostalpen), *Z. Gletscherkunde u. Glazialgeologie* **9** (1-2): 5-57.

Pearl R.T. (ed.) (1954): The calculation of irrigation need, Min. of Agric. and Fisheries Tech. Bull. 4.

Peck A.J. (1983): Response of groundwaters to clearing in Western Australia, International Conference on Groundwater and Man **2**: 327-335, Aust. Water Resour. Council Conf. Series No. 8, Aust. Govt. Publishing Serv., Canberra.

Peczely T. (1977): Some observations concerning the changes in the hydrological cycle caused by irrigation, in *Arid Land Irrigation in Developing Countries* (ed. E.B.Worthington), pp.159-170 (Pergamon Press, Oxford).

Pelleray (1957): *Etude des bassins versants experimentaux du Mayo-Kereng* (ORSTOM, Yaounde).

Pels S. and Stannard M.E. (1977): Some environmental changes due to irrigation development in semi-arid parts of New South Wales, Australia, in *Arid Land Irrigation in Developing Countries* (ed.E.B.Worthington), pp.171-183 (Pergamon Press, Oxford).

Penman H.L. (1956): Evaporation: an introductory survey, *Neth. J. Agric. Sci.* **4**: 9-29.

Pereira H.C. (1974): *Land Use and Water Resources* (Cambridge Univ. Press, London).

Peterson F.L. (1984 a): Groundwater recharge storage and development on small atoll islands, Proc. Regional Workshop on Water Resour. of Small Islands, Suva, Fiji (Commonwealth Science Council Tech. Publ. No. 154), pp. 422-430.

Peterson F.L. (1984 b): Hydrogeology of high oceanic islands, Proc. Regional Workshop on Water Resour. of Small Islands, Suva, Fiji (Commonwealth Science Council Tech. Publ. No. 154), pp. 431-435.

Petts G.E. (1984): *Impounded Rivers* (Wiley, Chichester).

PFRA (1983): The determination of gross and effective drainage areas in the prairie provinces, Hydrol. Report 104, Prairie Farm Rehabilitation Administration, Regina, Saskatchewan.

Philip J.R. (1957 a): The theory of infiltration: 4. Sorptivity and algebraic infiltration equations, *Soil Sci.* **84**: 257-264.

Philip J.R. (1957 b): Evaporation, and moisture and heat fields in the soil, *J. Meteorol.* **14**: 354-366.

Philip J.R. (1967): Sorption and infiltration in heterogeneous media, *Aust. J. Soil Res.* **5**: 1-10.

Philip J.R. (1969 a): Theory of infiltration, *Adv. in Hydroscience* **5**: 215-296.

Philip J.R. (1969 b): Moisture equilibrium in the vertical in swelling soils, *Aust. J. Soil Res.* **7**: 99-141.

Pierce R.S., Hornbeck J.W., Likens C.C. and Bormann F.H. (1970): Effects of vegetation elimination on stream water quantity and quality, in *Results of Research in Representative and Experimental Basins* , IAHS Publ.No. 96: 311-328.

461

Pilgrim D.H. (1983): Some problems in transferring hydrological relationships between small and large drainage basins and between regions, *J. Hydrol.* **65**: 49-72.

Pokshishevsky V. (1974): *Geography of the Soviet Union* (Progress Publishers, Moscow).

Poland J.F. (1984): Guidebook to studies of land subsidence due to groundwater withdrawal, *Studies and Reports in Hydrol.* 40, Unesco, Paris.

Pompe C.L.P.M. (1982): Design and calculation of rainwater collection systems, in *Rain Water Cistern Systems* (ed. F.N. Fujimura), pp.151-157 (Water Resour. Res. Center, Univ. Hawaii).

Post A. and Mayo L.R. (1971): Glacier dammed lakes and outburst floods in Alaska, USGS Hydrological Investigations Atlas HA-455, Washington D.C..

Postel S. (1984): Water: rethinking management in an age of scarcity, Worldwatch Paper 62, Worldwatch Inst., Washington D.C..

Postel S. (1985): Conserving water: the untapped alternative, Worldwatch Paper 67, Worldwatch Inst., Washington D.C..

Prasad R. and Coulter J.D. (1984): Hydrogeology of high oceanic islands, Proc. Regional Workshop on Water Resour. of Small Islands, Suva, Fiji (Commonwealth Science Council Tech. Publ. No. 154), pp. 431-435.

Preuss E.F. (1975): Groundwater recharge under humid climatic conditions in north Germany, in *The Hydrological Characteristics of River Basins*, IAHS Publ. No. 117: 305-312.

Prevot L., Bernard R., Taconet O. and Vidal-Madjar D. (1984): Evaporation from a bare soil evaluated using a soil water transfer model and remotely sensed soil moisture data, *Water Resour. Res.* **20** (2): 311-316.

Priestley C.H.B. and Taylor R.J. (1972): On the assessment of surface heat flux and evaporation using large-scale parameters, *Mon. Weather Rev.* **100**: 81-92.

Probst J. (1986): Dissolved and suspended sediment matter transported by the Girou River (France): mechanical and chemical erosion rates in a calcareous molasse basin, *Hydrol. Sci. J.* **31**: 61-79.

Randle P.H. (1981): *Atlas del Desarrollo Territorial de la Argentina* (OIKOS, Madrid).

Rango A., Salomonson V.V. and Foster J.L. (1977): Seasonal streamflow estimation in the Himalayan region employing meteorological satellite snowcover observations, *Water Resour. Res.* **13** (2): 109-112.

Ranjitsinh M.K. (1979): Forest destruction in Asia and the South Pacific, *Ambio* **8** (5): 192-201.

Raudkivi A.J. (1979): *Hydrology* (Pergamon, Oxford).

Richards L.A. (1931): Capillary conduction of liquids through porous mediums, *Physics* **1**: 318-333.

Richards P.W. (1964): *The Tropical Rain Forest: an Ecological Study* (Cambridge Univ. Press).

Richardson M. (1968): *Translocation in plants* (Arnold, London).

Riordan E.J., Grigg N.S. and Hiller R.L. (1978): Measuring the effects of urbanization on the hydrological regimen, Proc. 1st Internat. Conf. on Urban Storm Drainage (ed. P.R.Helliwell), pp.496-511 (Pentech Press, London).

Riou C. (1975): La détermination pratique de l'évaporation. Application à l'Afrique Centrale, Mémoirs ORSTOM 80, Bondy.

Roberts G. and Marsh T.J. (1987): The effects of agricultural practices on the nitrate concentrations in the surface water domestic supply sources of western Europe, in *Water for the Future: Hydrology in Perspective*, IAHS Publ. No. 164: 365-380.

Robinson M. (1986): Changes in catchment runoff following drainage and afforestation, *J. Hydrol.* **86**: 71-84.

Robinson M. and Beven K. (1983): The effect of mole drainage on the hydrological response of a swelling clay soil, *J. Hydrol.* **64**: 205-223.

Robinson M. and Blyth K. (1982): The effect of forest drainage operations on upland sediment yields: a case study, *Earth Surf. Proc. and Landforms* **7**: 85-90.

Robinson M., Ryder E.L. and Ward R.C. (1985): Influence on streamflow of field drainage in a small agricultural catchment, *Agric. Water Managem.* **10**: 145-158.

Roche M.A. (1981): Watershed investigations for development of forest resources of the Amazon region in French Guyana, in *Tropical Agricultural Hydrology* (ed. R.Lal and E.W. Russell), pp.75-82 (Wiley).

Rodda J.C., Downing R.A. and Law F.M. (1976): *Systematic Hydrology* (Newnes-Butterworths, London).

Rodier J.A. (1964): Regimes hydrologiques de l'Afrique Noire à l'Ouest du Congo, Memoirs ORSTOM 6, pp. 50-68, ORSTOM Bondy.

Rodier J.A. (1975): Evaluation de l'écoulement annuel dans le Sahel tropical africain, Travaux et Documents de l'ORSTOM No.46, ORSTOM Paris.

Rodier J.A. (1985): Aspects of arid zone hydrology, in *Facets of Hydrology II* (ed.J.C.Rodda), pp.205-247 (Wiley, Chichester).

Rodier J.A. (1986): Caracteristiques des crues des petits bassins versants réprésentatifs au Sahel, Cahiers ORSTOM, Série Hydrologie, Vol.21, No.2, 1984-85.

Rodier J.A., Colombani J., Claude J. and Kallel R. (1981): Le bassin de la Mejerdah, Monographies Hydrologiques No.6, ORSTOM, Paris.

Rodriguez-Iturbe I. (1986): Scale of fluctuation of rainfall models, *Water Resour. Res.* **22** (9): 15s-37s.

Rodriguez-Iturbe I. and Valdes J.B. (1979): The geomorphological structure of hydrologic responses, *Water Resour. Res.* **15** (6): 1409-1420.

Rothlisberger H. (1981): Eislawinen und Ausbruche von Gletscherseen, in *Gletscher und Klima* (ed. P. Kasser), Jahrbuch der Schweizerischen Naturforschenden Gesellschaft, wissenschaftlicher Teil, pp.170-212 (Birkhauser Verlag, Basel).

Rothlisberger H. and Lang H. (1986): Glacial hydrology, in *Glacio-Fluvial Sediment Transfer* (eds. Gurnell and Clark) (Wiley).

Rougerie G. (1960): Le façonnement actuel des modèles en Côte d'Ivoire forêtiaire, Mem. Inst. Francais d'Afrique Noire No. 58.

Roulet N.T. and Woo M.K. (1986): Hydrology of a wetland in the continuous permafrost region, *J. Hydrol.* **89** (1-2): 73-91.

Rudas J. (1973): Autocorrelation functions of discharges (in Hungarian), *Hidrologiai Kozlony* **53** (6): 282-288.

Sala O, Soriano A. and Perelman S. (1981): Relaciones hidricas de algunos componentes de un pastizal de la depresion del Salado, *Revista Faculdad de Agronomia* **2** (1): 1-10, Buenos Aires.

Salati E., Olio A.D. and Matsui E. (1979): Recycling of water in the Amazon basin: an untopic study, *Water Resour. Res.* **15** (5): 1250-1258.

Salati E. and Vose P.B. (1983): Analysis of Amazon hydrology in relation to geoclimatic factors and increased deforestation, *Beitrage zur Hydrologie* **9** (1): 11-22.

Samojlov I. (1956): *Die Flussmundungen* , transl. from Russian (Gotha).

Sartor J.D., Boyd G.B. and Agardy F.J. (1974): Water pollution aspects of street surface contaminants, *J. Water Pollution Control Fed.* **46** : 458-467.

Scarf F. (1970): Hydrologic effect of cultural changes at Montere experimental basin, *J. Hydrol. (N.Z.)* **9**: 142-162.

Schickedanz K. and Ackermann W.C. (1977): Influences of irrigation on precipitation in semi-arid climates, in *Arid Land Irrigation in Developing Countries* (ed.E.B.Worthington), pp.185-196 (Pergamon Press, Oxford).

Schiller E.J. and Latham L. (1982): Computerised methods in optimising rainwater collection systems, in *Rain Water Cistern Systems* (ed. F.N. Fujimura), pp. 92-101 (Water Resour. Res. Center, Univ. Hawaii).

Schmorak S. and Mercado A. (1969): Upconing of freshwater-seawater interface below pumping well, field study, *Water Resour. Res.* **5**: 1290-1311.

Sengele N. (1981): Estimating potential evapotranspiration from a watershed in the Loweo region of Zaire, in *Tropical Agricultural Hydrology* (ed. R. Lal and E.W. Russell), pp.83-95 (Wiley).

Showers V. (1979): *World facts and figures* (Wiley , New York).

Silvester R. and de la Cruz C.R. (1970): Pattern forming forces in deltas, *Proc. Amer. Soc. Civil Eng., J. Waterways, Harbors & Coastal Eng. Divn.* **96**: 201-217.

Silvestro J. and Guilen J.A. (1984): Optimisation of rainfall storage facilities, Proc. Regional Workshop on Water Resour. of Small Islands, Suva, Fiji (Commonwealth Science Council Tech. Publ. No. 154), pp. 473-492.

Simmons D.L. and Reynolds R.J. (1982): Effects of urbanization on base flow of selected south-shore streams, Long Island, New York, *Water Resour. Bull.* **18** (5): 797-806.

Simonffy K. (1979): Loss of information considering the peak values of hydrological events due to discrete fixed time-point sampling, Presentation at Unesco-VITUKI Postgraduate Course on Hydrological Methods, Budapest, Hungary.

Singh V.P. and McCann R.C. (1980): Some notes on Muskingum method of flood routing, *J. Hydrol.* **48**: 343-361.

Sklash M.G. and Farvolden R.N. (1982): The use of isotopes in the study of high-runoff episodes in streams, in *Isotope Studies of Hydrologic Processes* (eds. E.C. Perry and C.W. Montgomery), pp. 65-73 (Illinois Univ. Press, Dekalb, Ill.).

Slaymaker O., Hungr O., Desloges J., Lister D., Miles M. and Van Dine D. (1987): Debris torrents and debris flood hazards, Lower Frazer, Nicolum and Coquihallon Valleys, British Columbia. Excursion guide, 19th IUGG General Assembly, Vancouver.

Smart J.S. (1978): Analysis of drainage network composition, *Earth Surf. Proc.* **3**: 129-170.

Smiles D.E., Knight J.H. & Perroux K.M. (1982): Absorption of water into soil: the effect of a surface crust, *Soil Sci. Soc. Amer. J.* **46**: 76-81.

Smith D.D. and Wischmeier W.H. (1972): Rainfall erosion, *Adv. Agronomy* **14**: 109-148.

Sokolov A.A. and Chapman T.G. (eds.) (1974): Methods for water balance calculations, *Studies & Reports in Hydrol.* 17 (Unesco, Paris).

Solbe J.F.G. (ed.) (1986): *Effects of Land Use on Fresh Waters* (Ellis Horwood, Chichester).

Solomon S. (1967): Relationship between precipitation, evaporation and runoff in tropical-equatorial regions, *Water Resour. Res.* **3** (1): 163-172.

Solomon S. (1979): Feasibility study on the application of the grid-square technique to the Amazon basin, WMO, Geneva.

Soriano A. et al (1977): Ecologia de los pastizales de la Depresion del Salado, *Academia Nacional de Agronomia y Veterinaria* **31** (2): , Buenos Aires.

Soriano A., Leon R., Sala D. and Lemcoff J. (1984): La vegetacion de la depresion del Salado (Provincia de Buenos Aires) y su modificacion por influencia antropica, in *Hydrology on large flatlands* (ed. M.C. Fuschini Mejia), Proc. Olavarria Symp.**2**: 1011-1018, Unesco, Republica Argentina.

Starosolszky O. (1977): *Applied Surface Hydrology* (Water Resour. Publ., Littleton, Colorado).

Steiner J.T. (1980): The climate of the southwest Pacific region. A review for pilots, N.Z. Meteorol. Serv. Misc. Publ. No. 196.

Stezhenskaya I.N. (1966): Intraseasonal distribution of the runoff in years of different water levels in latitudinal zones of West Siberia, in *Geographical Information*, Issue 3, pp.87-92 (Inst. of Geog., USSR Academy of Sciences).

Sulam D.J. (1979): Analysis of changes in ground water levels in a sewered and unsewered area of Nassau County, Long Island, New York, *Ground Water* **17** (5): 446-455.

Sutcliffe J.V. (1974): A hydrological study of the Southern Sudd region of the upper Nile, *Hydrol. Sci. Bull.* **19** (2): 237-255.

Swedish Red Cross (1984): *Prevention better than cure,* Report on human and environmental disaster in the third world (Swedish Red Cross, Stockholm and Geneva).

Swiss Meteorological Institute (1978): Klimatologie der Schweiz, Beiheft zu den Annalen der SMA, Jahrg. 1977, v.18, part 2.

Szalay M. (1972): *Hydrometeorology and Water Balance* (Research Inst. for Water Resources Development, Budapest).

Szesztay K. (1979): Evapotranspiration studies for estimating man-influenced streamflow patterns under arid conditions, in *The hydrology of areas of low precipitation,* IAHS Publ. No. 128: 197-204.

Szollosi-Nagy A. (1976): Determination of expected losses due to sampling of hydrological records in time and space using Bayesian decision theory, WMO Operational Hydrology Report No.8 (WMO, Geneva).

Takasaki K.J. (1978): Summary appraisals of the nation's groundwater resources - Hawaii region, USGS Prof. Paper 813-M (Washington D.C.).

Takei A., Kobashi S. and Fukishima Y. (1975): The analysis of runoff from two small catchments in granitic hilly mountains, in *The Hydrological Characteristics of River Basins,* IAHS Publ. No. 117: 29-34.

Tamm C.O. (1958): The atmosphere, in *Encyclopaedia of plant physiology, v. 4. Mineral nutrition of plants* (ed. W. Ruhland), pp. 233-242 (Springer-Verlag, Berlin).

Tarar R.N. (1982): Water resources investigation in Pakistan with the help of Landsat imagery - snow surveys 1975-1978, IAHS Publ. No. 138: 177-190.

Taylor B.W. and Stewart G.A. (1958): Vegetational mapping in the Territories of Papua and New Guinea, conducted by CSIRO, Proc. Kandy Symp. on Tropical Vegetation, pp. 127-136 (Unesco, Paris).

Tejwani K.A., Gupta S.L. and Mathur H.N. (1975): *Soil and Water Conservation Research 1956-1971* (New Delhi).

Thornthwaite C.W. (1948): An approach towards a rational classification of climate, *Geogr. Rev.* **38**: 55-94.

Thornthwaite C.W. and Mather J.R. (1957): Instructions and tables for computing potential evapotranspiration and the water balance, *Publ. in Climatol.* **10** (3): 1-311.

Tietze K. and Geth M. (1980): The genesis of methane in Lake Kivu, *Geol. Rundschau* **69** (2): 452-472.

Tixeront J. (1979): Necessity of treatment of drought problems on integrated basis, Proc. Symp. Hydrological aspects of drought, New Delhi.

Tricart J.F. (1973): *Geomorfologia de la Pampa Deprimida* (INTA, Buenos Aires).

Troutman B.M. and Karlinger M.R. (1985): Unit hydrograph approximations assuming linear flow through topologically random channel networks, *Water Resour. Res.* **21**(5): 743-754.

Tschannerl G. (1979): Water management and environment in Latin America, UN Econ. Comm. for Latin America Report Vol.12.

UNEP (1985): International assessment of the role of carbon dioxide and of other greenhouse gases in climate variation and associated impacts, UNEP/WMO/ICSU conference statement, Villach, Austria.

Unesco (1964): Scientific problems of the humid tropical zone deltas and their implications, Proc. Dacca Symp. (Unesco, Paris).

Unesco (1969): Hydrology of deltas, Proc. Bucharest Symp., *Studies and Reports in Hydrol. 9* (Unesco, Paris).

Unesco (1971): Scientific framework of world water balance, *Tech.Papers in Hydrol.* 7 (Unesco, Paris).

Unesco (1978 a): *Atlas of world water balance* (Unesco, Paris)

Unesco (1978 b): World water balance and water resources of the earth, *Studies & Reports in Hydrol.* 25 (Unesco, Paris).

Unesco (1979): Map of the world distribution of arid regions, *MAB Technical Note 7* (Unesco, Paris).

Unesco (1987 a): *Manual on urban drainage design in urban areas*, 2 v. (Unesco, Paris).

Unesco (1987 b): Groundwater problems in coastal areas, *Studies & Reports in Hydrol.* 45 (Unesco, Paris).

Unesco/UNEP/FAO (1979): Water resources, in *Tropical grazing land ecosystems*, pp.87-92 (Unesco, Paris).

USGS (1984): National water summary 1983 - Hydrologic events and issues, USGS Water-Supply Paper 2250 (US Gov't Printing Office, Washington D.C.).

USGS (1985): National water summary 1984, USGS Water-Supply Paper 2275 (US Gov't Printing Office, Washington D.C.).

USGS (1986): National water summary 1985 - hydrologic events and surface-water resources, USGS Water-Supply Paper 2300 (US Gov't Printing Office, Washington D.C.).

USWRC (1978): *The nation's water resources 1975-2000, Vol.1: Summary* (US Water Resources Council, Washington D.C.).

Vachaud G., Vauclin M. and Colombani J. (1981): Bilan hydrique dans le sud tunisien. Caracterisation expérimentale des transferts dans la zone non saturée, *J. Hydrol.* **49**: 31-52.

Vacher H.L. and Ayers J.F. (1980): Hydrology of small oceanic islands - utility and estimate of recharge inferred from the chloride concentration of the freshwater lens, *J. Hydrol.* **45**: 21-37.

Vajner V. (1969): *The Geology of Mumbwa Areas* (Geol. Survey, Lusaka).

Valentin C. (1981): Organisations pelliculaires superficielles des quelques sols de région subdesertique, These, Université Paris VI.

Van Bavel C.H.M. and Hillel D.L. (1976): Calculating potential and actual evaporation from a bare soil surface by simulation of concurrent flows of water and heat, *Agric. Meteorol.* **17**: 453-476.

Van Breemen N. and Brinkman R. (1976): Chemical equilibrium and soil formation, in *Soil chemistry A. Basic elements* (ed. G.H. Bolt & M.G.M. Bruggenwert) (Elsevier, Amsterdam).

Van Burkalow A. (1982): Water resources and human health: the viewpoint of medical geography, *Water Resour. Bull.* **18** (5): 869-874.

Van de Griend A.A. and Engman, E.T. (1985): Partial area hydrology and remote sensing. *J. Hydrol.* **81**: 211-251.

Van der Leeden, F. (1975): *Water Resources of the World* (Water Information Center, New York).

Van den Wert P. and Kamerling G.E. (1974): Evapotranspiration of water hyacinth, *J. Hydrol.* **22**:201-212.

Volker A. (1966): Tentative classification and comparison with deltas of other climatic regions, Proc. Unesco Delta Symposium, Dacca, pp. 175-179.

Volker A. (1974): Hydrology and water resources development, in *Three Centuries of Scientific Hydrology* (Unesco/WMO/IAHS).

Volker A. (1979): Hydrology of various delta types, *Geologie en Mijnbouw* **58** (4): 387-396.

Volker A. (1983): River of south-east Asia: their regime, utilization and regulation, in *Hydrology of humid tropical regions* (ed. R. Keller), IAHS Publ. No. 140: 127-138.

Volker A. (1984): Hydrology of polders, in *Hydrology on large flatlands* (ed. M.C. Fuschini Mejia), Proc. Olavarria Symp.2: 641-680, Unesco, Republica Argentina.

Voropaev G.V. and Velikanov A.L. (1985): Partial southward diversion of northern and Siberian rivers, in *Large scale water transfers* (eds. G.N. Golubev and A.K. Biswas), pp. 67-83 (Tycooly, Oxford).

Vucic N. (1966): Influence of soil structure on infiltration and pF value of chernozem and chernozem-like dark meadow soil, Symposium on Water in the Unsaturated Zone, Wageningen **1**: 344-350, *Studies and Reports in Hydrology 2* (Unesco, Paris).

Vuichard D. and Zimmermann M. (1987): The 1985 catastrophic drainage of a moraine-dammed lake, Khumbu Himal, Nepal: cause and consequences, *Mountain Res. and Development* **7** (2): 91-110.

WAA (1987): *Water Facts 1987* (Water Authorities Association, London).

Wales-Smith B.G. and Arnott J.A. (1985): Evaporation calculation systems uses in the United Kingdom, in *Casebook on operational assessment of areal evaporation,* Report No.2, WMO No.635 (WMO, Geneva).

Walling D.E. (1979a): Hydrological processes, in *Man and Environmental Processes* (eds. K.J. Gregory and D.E. Walling), pp. 57-81 (Dawson, Folkestone).

Walling D.E. (1979b): The hydrological consequences of building activity: a study near Exeter, in *Man's Impact on the Hydrological Cycle in the United Kingdom* (ed.G.E.Hollis), pp. 135-151 (Geobooks, Norwich).

Walling D.E. and Kleo A.H.A. (1979): Sediment yields of rivers in areas of low precipitation: a global view, in *The Hydrology of Areas of Low Precipitation,* IAHS Publ. No.128: 479-493.

Walling D.E. and Webb B.W. (1981): Water quality,in *British Rivers* (ed. J.Lewin), pp. 126-129 (Allen and Unwin, London).

Walsh R. P. D. (1980): Runoff processes and models in the humid tropics, *Z.Geomorph.* **36**: 176-202.

Ward R.C. (1984): On the response to precipitation of headwater streams in humid areas, *J. Hydrol.* **74** : 171-189.

Warmerdam P. (1982): The effect of drainage improvement on the hydrological regime of small representative catchment areas in The Netherlands, in *Application of Results from Research and Experimental Basins* , Unesco Studies and Reports in Hydrol.32: 319-338 (Unesco, Paris).

Waterhouse B.C. (1984): Geological and geomorphological evolution of small islands, Pacific Region, Proc. Regional Workshop on Water Resour. of Small Islands, Suva, Fiji (Commonwealth Science Council Tech. Publ. No. 154), pp. 86-108.

Water Resources Planning Commission (1977): *Water Atlas of Taiwan* (Ministry of Economic Affairs, Taipei).

Watson K.K. and Whisler F.D. (1972): Numerical analysis of drainage of a heterogeneous porous medium, *Soil Sci. Soc. Amer. Proc.* **36**: 251-256.

WCED (1987): *Our common future*, World Commission for Environment and Development (Oxford Univ. Press).

Webb E.K. (1975): Evaporation from catchments, in *Prediction in Catchment Hydrology* (eds. T.G. Chapman & F.X. Dunin), pp. 203-236 (Aust. Acad. Sci., Canberra).

Weischet W. (1965): Der tropisch konvektive und aussertropisch advektive Typ der vertikalen Niederschlagsverteilung, *Erdkunde* **19**: 6-14.

Weischet W. (1969): Klimatologische Regeln zur Vertikalverteilung der Niederschlage in den Tropengebirgen, *Die Erde* **98**: 287-306.

Welch I.J. (1952): *Limnology* (McGraw-Hill, New York).

West H.L. and Arnold D.J. (1976): Assessment and proposals for the development of the water resources of Mali, a small tropical island, Proc. 2nd IWRA World Congr. on Water Resour., New Delhi, India, pp. 96-109.

Weyman D.R. (1974): Runoff processes, contributing area and streamflow in a small upland catchment, in *Fluvial Processes in Instrumented Catchments* (eds. K.J. Gregory and D.E. Walling), Inst. Brit. Geog. Special Publ. 6: 33-43.

Whisler F.D. and Watson K.K. (1972): The numerical analysis of flow in heterogeneous porous media. Proc. 2nd Symp. on Fundamentals of Transport Phenomena in Porous Media, Guelph, Canada, **1**: 245-256.

Widstrand C. (ed.)(1980): *Water conflicts and research priorities,* Water development, supply and management v.18 (Pergamon).

Wilson J.I. (1963): A possible origin of the Hawaiian Islands, *Canadian J. Phys.* **41**: 803-870.

Wilson M.F. and Henderson-Sellars A. (1983): Deforestation impact assessment: the problems involved, in *Hydrology of Humid Tropical Regions* (ed. R.Keller), IAHS Publ. No. 140: 273-283.

Wischmeier W.H. and Smith D.D. (1965): Predicting rainfall-erosion losses from cropland east of the Rocky Mountains, U.S. Dept. Agric. Handbook No. 282, Washington D.C..

WMO (1983): Operational hydrology in the humid tropical regions, in *Hydrology of Humid Tropical Regions* (ed. R. Keller), IAHS Publ. No. 140: 3-26.

WMO (1987): Tropical hydrology, WMO No. 655, World Meteorological Organization, Geneva.

World Bank (1978): Socio-cultural aspects of water supply and excreta disposal, Energy, Water and Telecommunications Dept., World Bank, Washington D.C..

Worthington E.B. (ed.)(1977): *Arid Land Irrigation in Developing Countries* (Pergamon, Oxford).

WRI/IIED (1986): *World Resources 1986*, World Resources Institute and International Institute for Environment and Development (Basic Books, New York).

Wyllie P.J. (1971): *The dynamic earth* (Wiley, New York).

Yoon Y.N. (1975): Correlation of the stream morphological characteristics of the Han River Basin with its mean daily and 7-day 10-year low flows, in *The Hydrological Characteristics of River Basins* , IAHS Publ. No. 117: 169-180.

Young C.P. (1986): Nitrates in groundwater and the effects of ploughing on release of nitrates, in *Effects of Land Use on Fresh Waters* (ed. J.F.G. Solbe), pp. 221-237 (Ellis Horwood, Chichester).

Young G.J. (1980): Monitoring glacier outburst floods,*Nordic Hydrol.* **11**: 285-300.

Zunker F. (1930): Behaviour of soils in connection with water (in German), in *Handbook of Soil Sciences,* v.6 (Springer, Berlin).

Index

476

Comparative hydrology